A Textbook of
Plant Ecology
Fifteenth Edition

With respects :

Dedicated to the memory of

my Father & Mother

(R.S. Ambasht)

my Grandfather & Grandmother

(N.K. Ambasht)

A Textbook of
Plant Ecology
Fifteenth Edition

PROF. R.S. AMBASHT

Ph.D., F.N.A.Sc., F.N.A.
Professor Emeritus and Ex-INSA Senior Scientist,
Banaras Hindu University, Varanasi

DR. N.K. AMBASHT

M.Sc., Ph.D.
Lecturer in Botany
Christ Church College, Kanpur

CBS

CBS Publishers & Distributors Pvt. Ltd.

New Delhi • Bengaluru • Chennai • Kochi • Kolkata • Mumbai
Hyderabad • Nagpur • Patna • Pune • Vijayawada

ISBN: 978-81-239-1600-2

First Edition: 1969
Thirteenth Edition: 1999
Fourteenth Edition: 2002
Reprint: 2004
Fifteenth Edition: 2008
Reprint: 2011, 2015, 2017, 2019

Published by **Satish Kumar Jain** and produced by **Varun Jain** for
CBS Publishers & Distributors Pvt. Ltd.,
4819/XI Prahlad Street, 24 Ansari Road, Daryaganj, New Delhi - 110002
delhi@cbspd.com, cbspubs@airtelmail.in • www.cbspd.com
Ph.: 23289259, 23266861, 23266867 • Fax: 011-23243014

Corporate Office: 204 FIE, Industrial Area, Patparganj, Delhi - 110 092
Ph: 49344934 • Fax: 011-49344935
E-mail: publishing@cbspd.com • publicity@cbspd.com

Branches:
• *Bengaluru:* 2975, 17th Cross, K.R. Road, Bansankari 2nd Stage,
  Bengaluru - 70 • Ph: +91-80-26771678/79 • Fax: +91-80-26771680
  E-mail: cbsbng@gmail.com, bangalore@cbspd.com
• *Chennai:* No. 7, Subbaraya Street, Shenoy Nagar, Chennai - 600030
  Ph: +91-44-26681266, 26680620 • Fax: +91-44-42032115
  E-mail: chennai@cbspd.com
• *Kochi:* Ashana House, 39/1904, A.M. Thomas Road, Valanjambalam,
  Ernakulum, Kochi • Ph: +91-484-4059061-65
  Fax: +91-484-4059065 • E-mail: cochin@cbspd.com
• *Kolkata:* 6-B, Ground Floor, Rameshwar Shaw Road, Kolkata - 700014
  Ph: +91-33-22891126/7/8 • E-mail: kolkata@cbspd.com
• *Mumbai:* 83-C, Dr. E. Moses Road, Worli, Mumbai - 400018
  Ph: +91-9833017933, 022-24902340/41 • E-mail: mumbai@cbspd.com

Representatives:

• Bhubaneswar	0-9911037372	• Hyderabad	0-9885175004	• Jharkhand	0-9811541605
• Nagpur	0-9021734563	• Patna	0-9334159340	• Pune	0-9623451994
• Uttarakhand	0-9716462459	• Dhaka (Bangladesh)	01912-003485		

Printed at:
J.S. Offset Printers, Delhi (India)

Foreword

I am pleased to observe that ecology is gaining ground in Indian Universities and that the ecologists are ever so eager to adapt newer approaches and findings to comprehend the discipline. The advances in this discipline indeed are so rapid that by the time a book comes out of the press the need of another book on the subject with a fresh orientation becomes apparent. The present work, therefore, is most welcome to students of ecology as it embodies the latest approach towards the functioning of the ecosystem without discarding much of the useful knowledge in ecology gained during the last 60 years or so.

I congratulate the author Dr. R.S. Ambasht who is my colleague and an accomplished ecologist for producing a wonderfully good work which evidently supersedes all other Indian books in the field. The book will be found suitable for college students both at the undergraduate and postgraduate levels.

Banaras Hindu University
June 28, 1969

R. Misra
M.Sc., Ph.D. (Leeds), F.N.A.Sc., F.W.A., F.N.A.
Prof. and Head, Department of Botany
and
Dean, Faculty of Science
Banaras Hindu University

About the Authors

R.S. Ambasht is Professor Emeritus and an Ex-INSA Senior Scientist at Banaras Hindu University. Formerly, he was Professor, Head, CAS Coordinator and CSIR Emeritus Scientist in Botany Dept., BHU, where he has taught ecology and environmental aspects for well over four decades and carried out researches. He has guided 29 Ph.D. students and published more than 200 research papers, reviews, book chapters, technical articles, books and edited internationally published books. He has made numerous academic visits to different parts of the world, presented papers, chaired and/or organised important sessions in Indonesia (1974 and 1977), Malaysia (1979, 1980 and 1992), Australia (1984 and 1996), France (1988), USA (1989 and 1992), Japan (1990, 1993 and 1996), Germany (1991), UK (1989 and 1994) and Canada (2000). He is the recipient of several honours and awards like chairing at several international congresses and conferences, Birbal Sahni Gold Medal of the Indian Bot. Soc., Platinum Jubilee Lecture Award of the Ind. Sci. Cong. and S. Saraswati National Prize of the UGC. He was the President of the National Institute of Ecology, Secretary and Treasurer of the International Society for Tropical Ecology and on the Board of the International Association, Ecology. He is the fellow of INSA (FNA) and National Acad. Sci. (FNA Sc.). He has visited famous Botanical Gardens of USA in Washington DC, Osnabruk and Bonn in Germany, Bogor in Indonesia, Kew in England and forests of islands of Sarawak (Malaysia) and Shikoku (Japan), Natural History Museum and Ecology Park in Japan and Smithsonian Natural History Museum in USA and joined an extensive tour of Queensland State Savanna (Australia). Prof. Ambasht is widely known for his researches in conservation, production and pollution ecology.

Navin Kumar Ambasht is a Lecturer in Botany at the Christ Church P.G. College at Kanpur. He obtained First Class B.Sc. Hons. (Gold Medal), First Class M.Sc. and Ph.D. degrees from the Banaras Hindu University. As the Senior Research Associate (Pool Scientist) of the CSIR, besides post-doctoral researches, he took B.Sc. and M.Sc. Classes at BHU. He also worked as a Visiting Scientist with the world famous environmental scientist Prof. Donat P. Hader at the F. Alexander University at Erlangen in Germany. He has published about 25 papers mostly in high impact world famous international journals and two edited books published by Kluwer-Plenum now Springer of New York, USA. He is the recipient of Banaras Hindu University Prize and M.M. Malviya Gold Medal and the Young Scientist Certificate of the Indian Science Congress (1996). He is well-known internationally for his researches on UV-B and tropospheric ozone enhancement impacts on crop plant.

Preface to the Fifteenth Edition

In the life of a University level science textbook to reach fifteenth edition is a landmark event. The book has served generations of students and teachers of ecology with distinction for nearly four decades. Dr. Navin Kumar Ambasht since joining as co-author in XI Ed. has added considerably to the updated contents. During the past few years, environmental pollution, deforestation, habitat degradation, desertification, climate change and biodiversity losses have drawn world attention of scientists and administrators alike. This updated edition includes important ideas, data and literature produced in the 21st century and they are referred at appropriate places in different chapters. These are in particular reference to Gaia hypothesis, light factor, cold and hot conditions, psychrophiles, antifreeze proteins, heat shocks, drought effects, oxygen deficiency conditions, halophytes, nitrogen level indicators, allelopathy, seed dispersals, litter decomposition, key-stone species, xenobiotics, biodiversity hotspots and ecological significance, UV-B and ozone stress in troposphere, etc.

We express our thanks to all those readers, colleagues and friends who have made useful suggestions. We are also thankful to Prof. Helmut Lieth (Germany) for providing his World Productivity map, Prof. Akira Miyawaki and (Late) Prof. Numata (Japan), Prof. Arthur McComb (Australia), Prof. Donat P. Haeder (Germany) for exchange of ideas during our academic visits abroad. Thanks are also due to Dr. Parvez E. Deen, Principal, Christ Church College, Kanpur for his keen interest, Dr. Vijay Kumar and Dr. Pravin K. Ambasht for discussion and suggestions, Dr. B.P. Ambasht, Dr. J.P. Ambasht, Dr. Akhilesh K. Ambasht, Mrs. Annpurna, Mrs. Sandhya, Mrs. Anupama, for their help and interest. Special thanks are due to Prakriti, Sukriti, Soumya and Master Aryan for providing a wonderful environment.

We express our thanks to Shri Satish Kumar Jain, Shri Vinod Kumar Jain, and Shri H.S. Poplai of CBS Publishers & Distributors for publishing these and earlier editions and to Shri Vivekanand Kuity of Students' Friends & Co. for publishing the first twelve editions. Suggestions are welcome.

R.S. Ambasht
N.K. Ambasht

Contents

CHAPTER 1

Introduction

Ecology, the science of environmental inter-relationships between organisms and environment, has now become much more important than ever before because of serious degradations in both the components of organisms and environment. There is a rapid loss of biodiversity i.e. accelerated extinction of a wide variety of taxa and a great many have become vulnerable, rare and threatened for extinction. Equally alarming is the rate of environmental pollution, habitat degradation and global climate changes. Additionally, the erosion of ozone shield and enhancing UV-B and ozone have all led to increased importance of ecology and from biology its ramification to various other disciplines of study and human activity. Today ecology has become a household word and a concern of common man.

The evolution of life on the earth has gone hand in hand with the evolution of environment. The environmental conditions of the earth before the advent of life or at about that time was very different than what it is today. The evolution of environment has always been influenced by the organisms. Man has evolved very late in the geological history of our planet earth when the environmental conditions with respect to temperature regimes, rainfall and atmospheric gases, etc. had become nearly similar to what is found today.

Man's activities in ancient to very ancient pasts have shown in undisputed terms, his understanding of environment and its impacts on his life. He learnt to modify it for a better adjustment. The art of producing fires, constructing houses for safety against sun, heat,

1

cold, rain and predators; clothing, agriculture had something to do with environment. The informal ecology has been there all through the human civilization. Even as a discipline of study, aspects which we now put in ecology were studied by geographers, explorers, taxonomists, naturalists and herbal doctors (Ayurvedic system).

As a science ecology is relatively a young branch of biology which deals with the interacting system of organisms and their environment. The term *Ecology* (old spelling *Oekologie*) is of recent coinage (E. Haeckel, 1869) and has been derived from Greek words *Oikos* meaning house and *Logos* meaning the study. Therefore, ecology is the study of organisms in their natural home or habitat. Ernst Haeckel regarded ecology as the knowledge concerning the economy of nature, the total relations of animal to both inorganic and organic environment. Ecology is generally *defined as the study of plants and animals in reciprocal relationship with their environment or external world*. It appears possible that Hanns Reiter had used the term 'Ecology' before Haeckel. Even before this term was coined some of the outstanding biologists of those days has described the ecological concepts such of food chains and population regulations (Leeuwenhoek in early eighteenth century) and on biological productivity. Hilaire gave the term 'ethology' a few years before the term 'ecology' was coined to denote the study of interrelationships between different organisms. In early literature some biologists laid greater emphasis on plant communities like Frederick Clements and some on environment like Karl Friederick. V.E. Shelford too in 1929 has defined ecology as the *science of communities*.

Eugene P. Odum (1971) has defined ecology as '*the study of structure and function of nature*' or simply '*environmental biology*'. F.E. Clements, known for his extensive contributions to plant ecology, has regarded ecology as *the science of community*. Famous British ecologist Macfadyen (1957) has emphasized that the purpose of ecology is discovery of the *principles which govern the relationships between 'plants or animals and their environment'*. Ecology is one of the basic divisions of biology like others—morphology, physiology, cytology, genetics, taxonomy, etc. But it differs from them essentially in two ways : (1) that it always comprehends along with organisms the non-living environment also and (2) that it deals with system of levels higher than the organisms in the biological spectrum of lev-

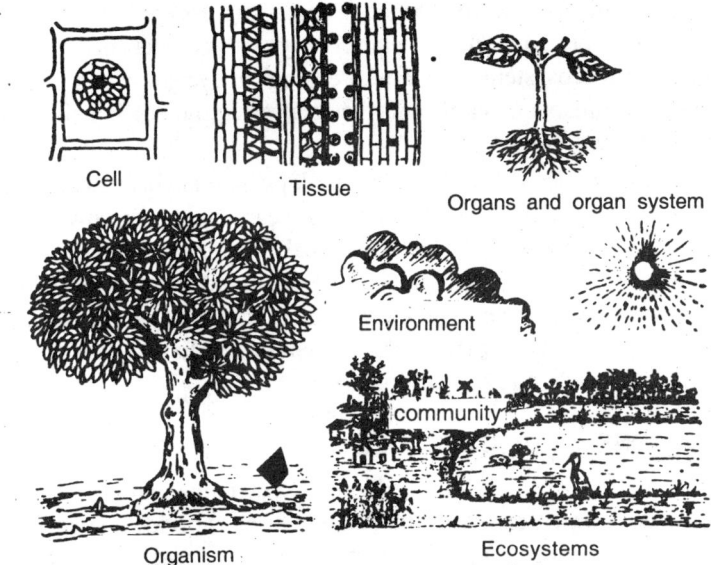

Cell   Tissue   Organs and organ system

Environment

community

Organism   Ecosystems

Fig. 1.1. Biological spectrum of levels of organisation. Ecology concerns mainly with the study of populations, communities, ecosystems and biosphere.

els of organisation of systems of *genes, cells, organs, organisms, population* and *communities* (Fig. 1.1).

By system we mean a *unified whole made of regularly interacting and interdependent components*. Organisms on the one hand and non-living parts like material and energy, etc. on the other form the major components of the system which regularly interact among themselves and remain interdependent. Thus we have genetic systems, cell systems, organ systems and organismic systems formed by the interactions of genes, cells, organs and organisms respectively with the matter and energy flowing into them. These systems are mainly studied in other disciplines of biology like genetics, cytology and physiology. *Population systems* are formed by assemblage of a large number of individuals of any one species (populations) in a habitat in interaction with their environment and the *ecosystems* are formed by interaction of assemblage of a variety of populations i.e. communities with their environment. These are the main structural and functional entities studied in ecology. Naturally populations and communities of all kinds of taxonomic entities like Algae, Fungi, Angiosperms, Arthropoda, Mammalia, etc., are studied particularly with respect to their functions in the ecosystem. The largest ecological system which includes all the organisms of earth and the total environment is called *biosphere* or *ecosphere*.

Like other divisions of biology, ecology has also been divided into plant and animal ecology. But in recent years there is a growing realization of the fact that in any biological organization the plants and animals are very much interdependent and plants and animals sharing the same habitats react with each other in many ways. Essential reference to animals as part of plant environment has, therefore, been frequently made throughout this book.

DIVISIONS OF PLANT ECOLOGY

Plant ecology is usually divided into (1) autecology and (2) synecology. *Autecology* deals with the ecology of an individual species and its population including the effect of other organisms and environmental conditions on every stage of its life cycle. *Synecology* deals with the ecology of plant communities. It involves the study of structure, nature, organisation and development of plant communities. Autecological studies of at least the more characteristic or dominant species of a community usually form a good basis for an understanding of synecology. The study of plant community structure is also called *plant sociology* or *phytosociology*. Other detailed and specialized aspects of ecology are termed variously to denote their specific nature such as *forest ecology, grassland ecology, freshwater ecology, marine ecology, desert ecology* etc. all of which are related to the different types of habitats. *Palaeoecology* deals with organisms and their environment in the geological past; *cyto-ecology* deals with cytological details in species in relation to populations in different environmental conditions; *conservation ecology* deals with application of ecological principles for a proper management of resources leading to high and sustained yield of useful biological materials for human welfare. Recently the term *resource ecology* is being increasingly used and this deals with plants, animals, water and mineral resources and their judicious management. Pollution ecology deals with the problems of environmental deterioration and ways and means of keeping the environments clean. Ecosystem ecology deals with both plant and animal communities along with their total environment. In this both structure and function of the ecological systems are studied in relation to space and time. Special emphasis is laid on the flow of energy from the sun to green plants to animals and decomposers etc. and the cycling of materials through the living systems. Ecological energetics and *production ecology* are the young branches of ecology. These deal with the mechanism and quantity of energy conversions and its flow through organisms, production processes, rate of increase in biomass of organisms in relation to space and time both by green plants (primary producers) and animals (secondary producers). This is measured in respect of *gross production* while the actual gain in production after deducting the loss due to respiration is called *net production.*

Various tools of other sciences, mathematics and statistics are increasingly applied in analysis of ecological systems. Application of radioactive materials, use of sophisticated instruments like photosynthesis systems, autoanalysers, gas-liquid chromatography, atomic absorption photometers, spectrophotometers, infrared gas analysers, flame photometers, elaborate field set up of electrical and electronic instruments to measure environmental factors, flow of energy, gas exchanges, use of computers in analysis of data, bomb calorimetry, culture of plants in environment controlled houses or phytotrons are some of the most modern means widely used in ecological research centres. Systems ecology including analysis and modelling are the youngest but most important fields of ecology which take the help of mathematics, cybernetics, statistics and computer based programmes, etc. to translate the biological systems into the language of mathematics. Thus as in mathematical equations, the nature and extent of change in biological systems can be predicted if this or that factor is added or deleted. This approach is holistic in which all the components are considered simultaneously instead of the factor-wise cause-effect conventional approach.

Thus we find that ecology is now a well defined science rooted in biology, but studied with the application of very exact sciences like physics, chemistry, mathematics and computers and it has ramifications spread to diverse disciplines of sociology, anthropology, environmental sciences, medicine, technology and many others.

DEVELOPMENT OF ECOLOGY

Global level

Before the agricultural civilization, man was essentially a hunter and food gatherer. He had mastered information on when and where to expect the presence of certain fruits, other food plants, animals that he hunted, the plants which healed the wounds, the affected bones, fever or other ailments, the plants with narcotic or drug properties, and many other things. There are excellent accounts about the natural vegetation and wild life in the Hindu epics like Ramayan and Mahabharat, preachings of Buddha and in writing of Aristotle and Theophrastus.

The formal ecology is only 150 years old. Burbour, Burk and Pitts (1980) have reviewed the progress of ecology in the western countries. F.H. Alexander von Humboldt (1769-1859), who had mastered knowledge in higher mathematics, chemistry, biology and geography, made extensive studies of the vegetation and environment, particularly through his long voyages from Europe to Peru, Mexico, Amazonia, Cuba and Venezuela. The first fourteen volumes of this book *Voyage aux Regions Equinoxialis* deal with the plants of these Latin American regions. J.F. Shouw (1789-1852) of the University of Copenhagen emphasised the importance of "temperature" factor in plant life and vegetation distribution. Kerner (1831-1898) in his book *Plant Life of Denube Basin* has given a wonderfully good account of vegetation, ideas about succession and about heritable and non-heritable variations in populations of species. A.A.P. De Candolle, Adolf Engler and Charles Darwin made great impact on the science of vegetation in the nineteenth century.

J. Warming (1841-1924), also at the University of Copenhagen for the first time synthesized morphology, physiology, taxonomy and biogeography into one holistic branch of science. He described major vegetation types, communities, dominants and subdominant life forms, the role of soil and climatic factors like temperature and water in delimiting communities. He gave the terms : *hydrophytes, xerophytes, halophytes* and *mesophytes*. A Schimper (1856-1901) of the Strassburg and then Bonn University explained about the causes of difference in regional floras and in his book on Plant Geography. He has very much stressed on soil and climatic factors. Among the early American ecologists, H.C. Cowles (1869-1939) and F.E. Clements (1874-1945) were most outstanding. Cowles taught ecology at the University of Chicago and published works on vegetation dynamics. He organised the Ecological Society of America in 1915. Clements at first worked at the University of Nebraska and published on the phytogeography of the region and the vegetation of North America. He gave a comprehensive account of plant successions. He also described the regional formation (large units of vegetation) associations and on seral communities. Clements in co-authorship with J.E. Weaver produced one of the most widely read book on plant ecology (1929 and 1939). Gleason in 1926 criticised the ideas of Clements on succession and association.

Christen Raunkiaer of the University of Copenhagen (1934) gave the concept of life forms in plants. Sir Arther Tansley, one of the founders and the first President of British Ecological Society (1914), gave the term **Ecosystem** (1935) and greatly influenced the growth of ecology. In France, J. Braun-Blanquet established the Zurich-Montpellier School of Phytosociology. Pearsall in UK worked on the lake ecology and among many ecologists that he trained, the two Indians *viz.* R. Misra and G.S. Puri on return to India founded the International Society for Tropical Ecology (1958) and Misra established the famous ecology centre at the Banaras Hindu University in 1937.

Eugene P. Odum of the Institute of Ecology, University of Georgia, USA has made tremendous impact on modern ecology mainly through his book *Fundamentals of Ecology* (3 editions), *'Ecology'* and *'Basic Ecology'*. He (1983) regards ecology as a 'hard' science in the sense that it involves tools of mathematics, chemistry and physics and a 'soft' science because of its root in natural sciences. Eugene, P. Odum, Howard, T. Odum, Ramon Margalef, Bernard C. Patten, George M. Van Dyne and Kenneth E.F. Watt have given quantitative perspectives to ecology during the sixties and seventies of 20th century. Gates, H.T. Odum, R. Lindman, Helmut Lieth, R.H. Whittaker, G.M. Woodwell and J. Phillipson have made notable contributions to ecological energetics, productivity and ecological models connected with them. F.H. Bormann, G.E. Likens, G.E. Hutchinson, J.S. Olson have worked out in great details the cycling of nutrients and water. Wiegert (1976) has edited the key papers on 'Ecological

Energetics' between 1942 to 1970. Bolin and Cook (1983) have edited a volume on "*The Major Biogeochemical Cycles and their Interactions*". Ambasht and Ambasht (1990-2005) have written the four editions of the book *Environment and Pollution* and Ambasht (1998) has edited *Modern Trends in Ecology and Environment*. Ambasht and Ambasht (2003) have edited two internationally multiauthored books, *Modern Trends in Applied Terrestrial Ecology and Modern Trends in Applied Aquatic Ecology.*

On populations there are a few important books such as the '*Population Ecology: A Unified Study of Animals and Plants*' by Begon and Mortimer (1981); '*Population Biology of Plants*' by Harper (1977). '*An Introduction to Population Ecology*' by Hutchinson (1978), and '*Introduction to Plant Populations*' by Silvertown (1987). Among some of the most important books on different aspects of ecology, a few are: *Basic Ecology* by Odum (1983); *Plants, Man and the Ecosystem* by Billings (1970); '*Plants and Environment*' and '*Plant Communities*' by Daubenmire (1974 and 1968); *Concepts of Applied Ecology* by De Santos (1978); *Concepts of Ecology* by Kormondy (1996); *Aims and Methods of Vegetation Ecology* by Mueller-Dombois and Ellenberg (1974); *Terrestrial Plant Ecology* by Barbour, Burk and Pitts (1988); *Communities and Ecosystem* by Whittaker (1975); *Introduction to Plant Geography* by Polunin (1960). Majumdar et al. (1998), *Ecology of Wetlands and Associated Systems*; Heywood and Watson (1995), *Global Biodiversity Assessment*; Margalef (1968), *Perspectives in Ecological Theory*; McKinney and Schoch (1998), *Environmental Sciences*; Perring et al. (1995), *Biodiversity Loss*; Schulze and Mooney (eds. 1994), *Biodiversity and Ecosystem Functions*. Schulze, Beck and Muler-Hohenstein (2004) have brought out an updated edition of Plant Ecology. Ambasht and Ambasht (2007) have produced the landmark fifteenth edition of this book.

A number of ecologists have started working on pollution effects on plant life particularly those caused by industrial fumes; liquid effluents discharged in agricultural lands, rivers and sea coasts and the work are appropriately cited in a separate chapter on this aspect.

During the International Biological Programme (1964-74) many ecologists have worked on the production (biomass and energy) ecology of terrestrial, freshwater and marine ecosystems with emphasis on systems analysis. Conservation studies of biological resources are also being actively carried out in most of the centres. A series of IBP hand books have been published. In the final phase of IBP, a synthesis of World data has been done and the Cambridge University Press has published a number of volumes on IBP results in the last few years. In two volumes on grasslands a number of Indian ecologists have synthesized data on tropical grasslands.

Ecological researchers are becoming more and more applied to human welfare due to such international efforts as those of IBP, MAB (Man and the Biosphere Programme of UNESCO), UN's Stockholm conference (June 1972) on 'Human Environment' and the Earth Summit at Rio de Janeiro in June 1992 and follow up action. IUCN (International Union for Conservation of Nature and Natural Resources), Switzerland, renamed now as World Conservation Union but the abbreviation IUCN is retained. Publication of a few very important journals as *Tropical Ecology, Indian Journal of Ecology, International Journal of Ecology and Environmental Sciences*, has helped in the propagation of new ecological knowledge developed in India and other tropical regions.

Roberts S., De Santo (1978) has given the vast array of career choices that could be available to an ecologist. Depending upon his special aptitude for field or laboratory studies, for plants or animals, or for theoretical or practical work he can select any one of them. But the essential feature in ecology is that whatever the field of detailed investigation an ecologist may select, he has to know some thing of other components of the ecosystem. De Santo has divided the ecological hierarchy into three—*zoological, microbial and botanical* and each could be studied at three levels—*cellular, organismic and community* (ecosystem). Further the habitat could be aquatic, aerial and terrestrial and in the climatic belts of *tropical, temperate* or *polar*. Further the system could be *natural* or *manipulated* and could be *clean* or *polluted* and the study could be *theoretical* or *applied*. Thousands of possible combinations of career within the above matrix could be available.

Indian level

In India ecological studies began with the descriptive accounts of forests by the officers engaged in forests services in the first two decades of the twentieth century. The first comprehensive ecological paper from Universities appeared in 1920 on the ecology of the Upper Gangetic Plains by Professor Dudgeon of Allahabad wherein he also discussed such ecological principles as the role of environment in succession of communities. With the return of Professor R. Misra to Varanasi from Leeds (UK) after his extensive researches on the ecology of English Lakes (1938), his ecological investigations on some aquatic plants (1944) ravines and eroded river banks (1944), methods of soil studies (1945) and low-lying areas (1946) and on the autecology of *Lindenbergia polyantha* (Misra and Siva Rao, 1948) were published. Phytosociological studies were initiated in the country by Bharucha and his students specially with regard to the biological spectrum of Mahabaleshwar (Bharucha and Ferreira, 1941). Puri (1950, 1951) made extensive forest ecological investigations.

On almost all aspects of ecology extensive ecological researches have, since then, been done by Professor R. Misra and his numerous students spread throughout the country. Misra and Puri (1956) have given a general account of literature on ecology in India. Further, Misra (1957) has reviewed the progress of ecological researches in the country and Misra and Singh (1971) have again reviewed the progress of ecology in India. Ambasht, Shardendu and Sikandar (1983) have reviewed the state of aquatic climate and productivity in India.

Autecological studies of grassland and wasteland herbaceous species like *Euphorbia hirta, Euphorbia thymifolia, Setaria glauca* (Ramakrishnan, 1959), *Cyperus rotundus* (Ambasht, 1964, Tripathi, 1965), *Xanthium strumarium* (Kaul, 1959), *Alhagi camelorum* (Ambasht, 1963), *Alysicarpus monilifer* (Maurya and Ambasht, 1973) have been done. Autecological studies on several medicinal plants have also been made at Varanasi. Quite a few interesting facts and phenomena have come to light from these works. Many cropland weeds have also been intensively studied for their autecology by Misra (1969). Marwah and Ambasht (1972) and Ambasht and Chakhaiyar (1979) have studied crop-weed competition and productive structure in wheat and mustard crop communities. Ambasht (1977) and Ambasht and Lal (1977) have given an account of adaptation in weeds on Ganga river bank and the role of temperature and water in regulating seed germination and distribution of *Chrozophora rottleri*—a weed of low lying floodable lands coming up during dry phase. Lal and Ambasht (1979 a and b) have studied seed output, germination, reproductive capacity, productivity, growth analysis and energy content in *Scoparia dulcis*. Ambasht and Lal (1979) have reviewed autecological findings on weeds.

Synecological works have been done in forests, grasslands deserts, freshwaters and other specialized habitats. The Himalayan forests have been studied by Puri (1950, 1951) and Mohan and Puri (1955, 1956). Misra and Joshi (1955) and Waheed Khan (1956) have given accounts of Madhya Pradesh forests. Puri (1960) has given a comprehensive account of Indian forests in his book *Indian Forest Ecology* which has been enlarged for a three volume series of which the first has been published (Puri, Meher-Homji, Gupta and Puri, 1983). Productivity studies in forests have been made specially in the Vindhyan forests by Misra (1969b) and Singh and Ambasht (1979, 1980). Pandeya et al. (1971) have described the forest ecosystems in the Narmada river catchment areas. In the Eastern Himalaya forest, Sharma and Ambasht (1984) have for the first time measured the rate of nitrogen fixation by different age group of root nodules bearing actinorhizal *Frankia* in the trees of *Alnus nepalensis*. They have also measured the rate of primary production in terms of biomass and energy in *Alnus* plantation forest. Ambasht, Singh and Misra (1982) have found that on the Vindhyan forest lands, solar energy fixation rate is better by grass dominated stands than shrub on tree stands particularly teak plantation stands. Sharma and Ambasht (1983) have developed a correction factor for computing the effective ground area in productivity studies for forests on sloping habitats. More than 90% of nodules of *Alnus* after death get decomposed within one year (Sharma and Ambasht, 1986).

Grassland communities have been studied in small patches of land for phytosociology, reproductive capacity, production in relation to a variety of ecological factors especially grazing (Ambasht and Maurya 1970 a, 1970 b, Singh and Ambasht, 1980). The productivity of grasslands protected against grazing is found to be quite high in *Dichanthium* and *Heteropogon* grasslands (Ambasht, Maurya and Singh, 1972). A number of productivity works in India have been published in the New Delhi symposium volume *Tropical Ecology with an Emphasis on Organic Production* (Golley and Golley, 1972) from grasslands, croplands, forests and freshwaters. Singh and Ambasht (1975 a, 1975 b) have worked out the interrelationships among community structure and productivity of grasslands, Ambasht and Singh (1979) have worked out interspecific associations among the grassland species. For Vindhyan Hills, Ambasht and Pandey (1981) have given detailed account of phytosociology and productive structures of *Aristida cyanantha* dominated grasslands. Misra (1983) has given a detailed account Indian savannas. Specialized habitats such as walls (Varshney, 1968) and eroded river banks (Ambasht, 1968, 1977) have also been studied ecologically for the vegetation they support. Ambasht, Singh and Sharma (1984) have evaluated the role of riparian herbs in the conservation of soils, nutrients and water. Desert ecology has been investigated by a number of people associated with the Central Arid Zone Research Institute in Rajasthan. Several ecologists have given accounts of freshwater vegetation (Misra, 1946, Ambasht, 1968, 1970). Varshney and Singh (1976) and Ambasht (1974, 1976) have dealt with the aquatic weeds and their control measures. Billore and Mall (1977) have made extensive studies on the biomass structure and nutrient dynamics in *Sehima* and three other grazing lands of Ujjain. Gupta and G.P. Mishra (1985) have worked out the energy budget of *Themeda* grassland at Jhansi. On a very broad regional basis for the Indian and South Asian savannas. Singh, Hanxi and Sajise (1985) have reviewed the distribution of five major Savanna types strongly governed by climate and latitude. These are : (1) *Sehima-Dichanthium*, (2) *Dichanthium-Cenchrus-Lasiurus*, (3) *Phragmites-Saccharum-Imperata*, (4) *Themeda-Arundinella* and (5) The temperate-alpine types. J.S. Singh (1973, 1976), Singh and Joshi (1979), Singh, Singh and Yadava (1979) and Singh and Yadava (1974) have made extensive contributions to Indian grassland ecology. M.K. Misra and B.N. Misra (1981, 1985) have analysed the grassland community structure in Berhampur, Orissa.

In later years there has been thrust mainly on pollution, conservation, rehabilitation and other applied aspects. Water and air pollution have been particularly studied such as the wetlands and rivers systems, and gaseous pollution due to NO_x, SO_2, HF, CH_4 production, enhanced UV-B irradiation impacts etc.

Primary productivity of wetlands in context of global climatic change has been described by Ambasht and Srivastava, N.K. (1991). Conservation and management as well as on rehabilitation on river corridors have been studied by Ambasht (1992), Ambasht and Ambasht

(1992), Ambasht and Shanker (1992), Ambasht and Srivastava (1994), and Ambasht, Kumar and Srivastava (1992, 1994). Soil erosion and movement of nutrients like nitrogen and phosphorus down the riparian slopes have been measured under the protection of vegetal covers by Kumar, Ambasht and Srivastava (1992a and b) and on water quality of rivers receiving factory effluents by Srivastava, Ambasht and Kumar (1991) and Srivastava, Ambasht, Kumar and Shardendu (1993). Recently John Mitsch (1994) has edited a volume 'Global Wetlands, Old World and New' which gives account of wetland ecology in different perspectives from all over the world and is based on papers presented at the International Wetland Conference in Columbus, USA (1992). Gupta (1992) has discussed the problems and reviewed the literature related to restoration of degraded watersheds. Singh (1992) has edited an excellent volume on '*Restoration of Degraded Land : Concepts and Strategies*', in which Ambasht and Shankar (1992) have described river corridor restoration.

Abrol, Wattal, Gnanam, Govindjee, Ort and Teramura (1991) have edited a volume containing 53 papers on different problems related to '*Global Climatic Changes on Photosynthesis and Plant Productivity*'. It includes material on UV-B effects, CO_2 enhancement effects and other environmental stress effects on agroecosystems and natural ecosystems.

Ambasht (1993), Ambasht and Srivastava (1992), Srivastava and Ambasht (1994a, 1994b, 1995) have studied nitrogen dynamics in tropical trees particularly in actinorhizal *Casuarina equisetifolia* plantations.

N.K. Ambasht and Agrawal (1994) and Ambasht (1998) have reviewed literature on the impact of UV-B irradiations on crop plants and Ambasht (1993, 1994(b), 1995, 1997 and 1998) has studied the impact of its enhanced dosages under field conditions on rice, maize and sorghum crops. He treated rice plants with UV-B radiations predicted at 20% ozone depletion and found a decreased photosynthesis, accompanied with decline in chlorophyll *a*, *b*, carotenoid and anthocyanin contents and rise in flavonoids. The enzymatic activity of catalase declined while of peroxidase increased. Ascorbic acid content was also reduced while phenolic contents increased (Ambasht and Ambasht (2005).

The latest thrust area in the field of ecology is the conservation of biodiversity. At the Earth summit or the United Nation's Conference on Environment and Development (UNCED) in June 1992, in Rio de Janeiro all aspects of biodiversity and problems of rapid extinction received much attention.

Reid (1992) has critically examined the issue 'Can the extinction crisis be stopped' in his paper on conserving life's diversity, nature's variety or gene pools is recognized now as a basic requirement for sustained growth and economic development. He has outlined the scope of biodiversity conservation through creation of knowledge, awareness, ethics and input of information for actions at farm, village, forest or laboratory in bioregional, national and international levels. However, it is to be clearly understood that the present day dimensions of biological impoverishments are being driven by strong forces and they cannot be changed by the current level of restorative actions. At most this may turn out to be only stopgap measures unless advances are made in dealing with overconsumption of resources, population growth, misguided resource management, policies, and social and economic inequities (Reid 1992). Ambasht, Srivastava and Ambasht (1994), Ambasht and Ambasht (1998a, 2002) have reviewed literature on biodiversity definitions, causes of their loss and extinction, ecological and economic importance and different methods of biodiversity conservation, both *in situ* and *ex-situ*. Ambasht and Ambasht (1998b and 2002) have reviewed the ecology of Indian wetlands, soil and nutrient conservation by vegetal cover and biodiversity with special reference to soil sub-system.

The Ecosystem

No organism or a species lives alone; always there are associates influencing each other and organising themselves into communities. The organisms of any community besides interacting among themselves, always have functional relationship with the external world or the environment. This structural and functional system of communities and their environment is called an *ecological system* or, in short the *ecosystem*. This is somewhat comparable to the non-technical word *nature*. The term ecosystem was first proposed by A.G. Tansley (1935) and since then a great deal has been written on the subject (Lindeman, 1942, Odum, 1963, 1971 and 1983, Billings, 1964, Misra, 1969, Macfadyen, 1954, Kormondy, 1996). Tansley meant the key property of ecosystem as interactions of plants and animals with their physico-chemical environment. Schulze et al. (2004) defined *"Ecosystems are thus networks of interactions between organisms and their environment in defined space"*. Of course, the size of an ecosystem is very flexible depending upon the nature of habitat and purpose of study. Although the term *ecosystem* is of recent coinage, yet the general ideas and concepts of a structural and functional entity of biotic communities and their environment were known much earlier. Almost in a similar meaning Karl Mobius (1877) had given the term *biocoenosis* and Forbs (1887) had called it a *microcosm*. The term biocoenosis was also used by early Russian ecologists like Dokuchaev and Morozov. Later, Sukachev (1944) used the term *geobiocoenosis* to make it more comprehensive to include the non-living earth segment and the living components forming a system. This idea of system or an interdependence of all components as envisaged by Tansley in ecosystem became more popular. Other equivalent terms of similar meaning are *biosystem* and *holocoen*.

The central theme of the ecosystem concept is that at any place where organisms live, there is a continuous interaction between plants, animals and their environment to produce and exchange materials. This means, that there are mechanisms for continuous absorptions of material by organisms for the purpose of production of organic materials and release and conversion of the organic material into inorganic form. The process is called *cycling of material*. The energy needed to do all this work comes from sun. No other organisms except green plants are capable of taking the solar energy and converting it into chemical energy. In this stored form other organisms take the energy and pass it on to still other organisms. During this process quite a good proportion of energy of the living system is lost. The whole process is called *flow of energy*. Thus, ecosystem may be comprehended to be of any dimension or size where the living and non-living systems are involved in continuous flow of energy and cycling of materials through the non-living and living components. The functional processes are related to the ecosystem structure and vice-versa. For instance the flow of energy is based on *trophic* or food and feeding structure of the ecosystem. It flows from the sun to green plants to herbivores and then to different levels of carnivores. The system is, therefore, a dynamic one where change is always taking place. In fact, the system grows in course of time. Howsoever may an ecosystem be balanced and self contained, it is necessarily an open system in as much as the primary source of energy (solar radiation) enters from outside; the basic raw materials of carbon, hydrogen, oxygen and nitrogen for the building of carbohydrates and proteins have global cycling patterns and these enter ecosystems through precipitation and atmosphere. These dissipate out and the materials partly get recycled and partly exported out of the system. This phenomenon shows that there are intricate feedbacks and interdependence at inter-ecosystem, biome and biosphere levels. All the time the two way movements, i.e., immigration to and emigration from all ecosystems keep on taking place.

Just as we study external and internal morphology and physiology to understand the structure and working of an organism and its processes; so also we can understand the ecosystem by studying the structure and the ecological functioning of all its components both living and non-living. The knowledge so gathered is integrated and reconstructed for a proper perspective of the ecosystem structure, function and dynamics as a whole.

There is a great bearing of ecosystem study on the economy and welfare of human society. The green plants utilize inorganic materials like carbon-dioxide, water, mineral, etc. and the solar energy and produce numerous organic substances. The plants grow and increase in weight as a result of organic production. Part of the material so built is lost by the green plants in the process of respiration. The left over organic matter is called *net primary production* as against the total or *gross primary production*. For running the machine of life all other organisms including man consume directly or indirectly the primary produce in the form of food. Thus the study of ecosystem is of greatest importance in all types and ranges of habitats. Realizing this fact, a global programme of extensive quantitative studies of diverse terrestrial, freshwater and marine ecosystems specially with respect to their organic productivity and conservation for the general welfare of mankind was launched. Under this 10 years (1964-74) International Biological Programme (IBP) ecologists in most of the countries of the world have studied various aspects of the ecosystem around them. Many more works on similar lines taking human interests and impacts are being studied under the Man and the Biosphere Programme (MAB) and the International Geosphere Biosphere Programme (IGBP).

Some common terrestrial ecosystems are named after the types of organism and habitat conditions such as *grassland ecosystem, crop ecosystem, forest ecosystem* and *desert ecosystem. Freshwater ecosystems* are usually named upon the size and nature of freshwater body such as *river ecosystem, pond ecosystem, lake ecosystem.* The largest and most uniform aquatic ecosystem is *marine ecosystem.*

There is a growing awareness on the part of both ecologists and governments towards the necessity of mass equation about the fundamentals of interrelationships between man and his environment. Man is a part of the ecosystem and not separate from it. His dominating influence more often than not, disturbs and degrades most ecosystems, making them less useful for future generations of mankind. Innumerable human practices have already rendered many ecosystems unproductive. A basic and fundamental education on national natural resources, the working of ecosystem, the *ecological interdependences,* the *ecological checks* and *balances,* the short and long range chain effect of an action on organisms, soil and climate etc. is therefore very essential if we want to check further damage to nature and to improve our resources and use them wisely.

Ecosystem is a functional system which in balanced condition is self-sufficient and self-regulated. But at the same time, several ecosystems are interrelated and often smaller systems combine to make a larger ecosystem. For example, a bird with large number of parasites on and within its body, carries a biotic community living in a very specialized environment. This mobile system is a part of tree ecosystem as the bird flies from tree to tree for food. The tree itself may be an ecosystem producing its primary food and providing materials to secondary producers or animals; living on it. The tree ecosystem in association with large variety of other trees, smaller plants and animals form the forest ecosystem. Forests are interrelated with other river and stream ecosystems or neighbouring town or village ecosystems where from visitors like man or his domestic cattle come and remove the forest products like timber, leaves, grazing grass, hunting game birds and animals.

Within a certain range of fluctuation in the environment or biotic components in any ecosystem there is always a natural tendency for counteraction in order to maintain the functional balance. There is a self regulatory mechanism in ecosystems of *checks* and *balances.* This is called *homeostasis.* For example in a balanced ecosystem, usually the balance between oxygen and carbon dioxide is maintained at certain level so that neither of the two becomes unsuitably high or low. The fluctuation in proportion of plant and animal populations tends to disturb this balance. Within a range, the ecosystem self-regulates the fluctuations. This oscillation of forces and counterforces acts only within a range, beyond which, the balance gets disturbed and a new structural and functional balance tends to develop or else the component organisms perish yielding place to new ones.

ECOSYSTEM COMPONENTS

There are a number of ways in which different components of an ecosystem should be classified. The commonest method is to divide it into two main segments, the living (*biotic*) and the nonliving (*abiotic*). The living components are usually separated in the space in predominantly two layers, the *green layer* represented by leaves above the ground level and in the upper photic zone of aquatic environment, known as *autotrophic stratum* and the second at the ground and just below or in lower layers of water bodies where the dead organic matter decomposes known as the *brown layer* or *heterotrophic stratum*. We can take up the (i) biotic and (ii) abiotic components separately.

1. Biotic components

Solar energy is primarily trapped by the chloroplast machinery of green plants to build organic material. Green plants take simple inorganic materials like salt, water, carbon dioxide, etc. and produce their own food. These organisms are, therefore, called autotrophs i.e. self food producing. All other forms of life which do not possess chlorophyll cannot produce their own food and depend upon others. Therefore, they are called *heterotrophs*. Fungi, most of the bacteria and animals derive their food directly or indirectly from the food produced by green plants and are examples of heterotrophs. Based on this fact of production and consumption, the green plants are called *producer* components. Among consumers some animals such as herbivores, like goat, cow, deer and rabbit eat the green plants and these animals are called *consumers of the first order*. Other carnivorous organisms eat the herbivores as a frog eats a grasshopper, a carnivore fish eats a herbivore fish. These are called *consumers of the second order*. There are in most ecosystems, some organisms that eat other carnivores like a snake eats a frog, or a bird eats all types of fish including carnivores. These are called *consumers of the third order*. Then there are the top carnivores like lion, vulture, etc. which eat other carnivores also but are not killed and eaten by other animals (Fig. 2.1).

The animals take plants or other organic material as food, break them down in their digestion process and produce new type of or-

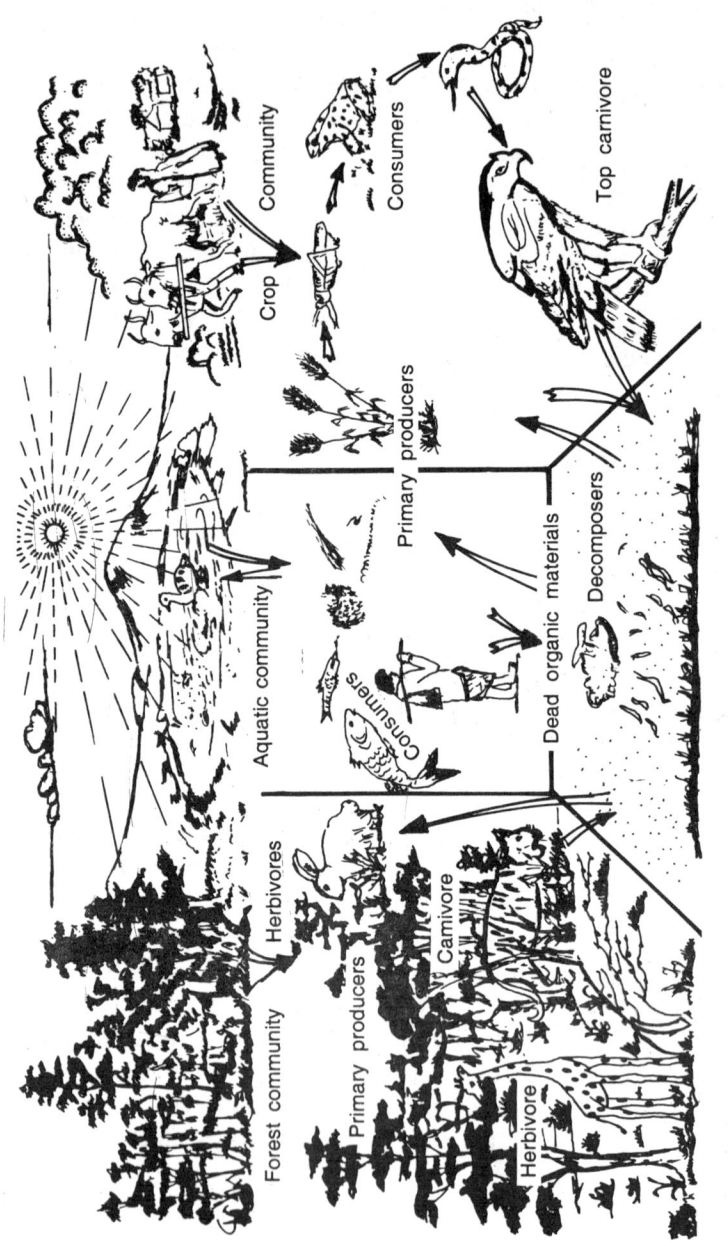

Fig. 2.1. Structure and function of forest, cropfield and freshwater ecosystems. The energy trapped by green plants is in each case passed on to herbivores and then to carnivores, etc. The dead organic matter is decomposed by decomposer organisms and brought back in usable form.

ganic materials in their body. Therefore, animals are also busy in the process of organic production. Based on this concept, the green plants are called *primary producers* and the heterotrophic organisms *secondary producers.*

Both the primary producers and secondary producers have their life cycles and new generation of populations develop while old ones die. What happens to the dead? If the materials so locked in the body of the organism are not returned to soil and atmosphere, the cycling of materials will stop and the earth will be full of dead organic matter. There is a continuous breakdown or decomposition of these organic materials taking place everywhere in all ecosystems. Some organisms with a specialized mode of nutrition and life, like fungi, bacteria, etc. constantly decompose dead organic materials into simple inorganic substances and during the process derive for them their food and energy. These are called *decomposers* or *reducers*. In fact decomposers are also a form of consumer who derive their food from dead parts of the primary and secondary producers. The role of decomposers is very special and important in the ecosystem as without their activity the entire cycle of materials is liable to get blocked. Decomposers are also called *scavengers*. Fig. 2.1 illustrates the ecosystem structure and function in forest cropland and pond ecosystems.

Some of the consumers live on or in other organisms (hosts) and derive their food from the host's body. These are called *parasites*. Some organisms prey upon other organisms like a frog preying upon a grasshopper and these are called *predators*.

There are certain other types of close biotic interrelationships among biotic components of the ecosystem and they are termed as *commensalism, mutualism* and *symbiosis*. Commensalism is mostly met within animals where two different types of organisms associate themselves together and one derives benefit from the other without adversely or favourably affecting the benefactor. This term in ecology includes relationships between different types of organisms for their requirements other than food also. Mutualism and symbiosis are the terms applied to very similar types of associations where one derives nutritions from another and in turn benefits the food supplier in some other way. For example, in lichens, the algae provides part of food for consumption by the fungal component. In turn the fungus affords protection to the algae against dessication, etc. Many of these relationships and terminologies of biotic components are based on food habits.

The quantity of biological materials in a given area or volume of space in an ecosystem is termed *standing crop*. If the quantity is referred in terms of weight it is called *biomass*. The functional aspects of species (of course with reference to the place of occurrence) is referred to as its *ecological niche*.

Odum (1983) has divided ecosystem into six components based on six functions. The biotic components are placed under three sections viz. (1) *producers* which include all types of autotrophic organisms; (2) *macroconsumers* or *phagotrophs* (phago = to eat) which include animals which eat or ingest other organisms; (3) *microconsumers* or *saprotrophs* (sapro = decompose) or *osmotrophs* (osmo = to pass through a membrane) which include heterotrophic microorganisms like bacteria and fungi, most of which decompose dead organic material. Wiegert and Owen (1971) have divided the heterotrophs into (1) *biophages* which feed on other living organisms and (2) *saprophages* which feed the dead organic matter. The disintegrating dead organic matter is also known as *organic detritus* (Latin word *deterene* means to wear away) and it may take just a few months or a year (in warm moist tropics) or several years (in peats and bogs) or still longer as in case of boreal and other cold regions. The disintegrating detritus results into *particulate organic matter (POM)* and *dissolved organic matter (DOM)* which play important role in the maintenance of the edaphic environment. Detritus is the medium on which microorganisms live.

Gaia hypothesis

In fact the role of organisms specially the microorganisms in controlling the geochemical environment of the earth over the long period of organic evolution has been the basis of *Gaia hypothesis*. *Gaia* in Greek stands for earth *goddess* and Lovelock and his associates in recent years have advocated that the present day environment with its high oxygen and low carbon dioxide, so necessary for life to flourish, have come to exist as a result of biological activity. Microbial populations in the soil maintain the soil environment such as aeration, pH, release of nutrients, breakdown of organic material, etc. Whether we accept Lovelock's Gaia hypothesis as such or not, no one can deny that organisms, especially the microbes have evolved with the passage of time in such a way as to occupy key position in keeping the earth's environment favourable for diverse life forms to grow and perpetuate well. Lovelock (2002) winner of 1997 Blue Planet 50 million Yen prize has given, *"The Gaia hypothesis views the biosphere as active adaptive control system able to maintain the Earth in homeostasis"*. In 1981, after considering several criticisms about global regulation by organism, he restated Gaia hypothesis as *"The evolution of organisms and their material environment proceeds as a single tight-coupled process from which self regulation of the environment at a habitable state*, appears as an emergent phenomenon". Way back in 1925, Alfred Lotka had written, "it is not so much the organism or the species evolves, but the entire system, species and environment. The two are inseparable.

2. Abiotic components

The abiotic components or the non-living environment are usually of two types—materials and energy. Materials are like water, mineral salts, atmospheric gases, etc. and energy is like light, heat, stored energy in chemical bonds, etc. The quantity of abiotic materials like minerals present at any given time in an ecosystem is termed the *standing state* similar to standing crop which refers to the quantity of biotic components. The materials continually keep on cycling i.e., entering into the living system and through death and decay returning to soil and atmosphere. This process is called *mineral circulation* or the *bio-geochemical cycle*. While the materials keep on cycling, fresh energy is continuously trapped from the sun by the green plants on the one hand and lost in space through respiration and loss of heat by all types of organisms on the other. Thus energy flows like one way traffic. The details of biogeochemical cycles of important materials are described elsewhere while dealing with the environment and details of energy flow are described in a separate chapter on ecological energetics.

Abiotic part is divided into three components viz. (1) *inorganic substances* like carbon, nitrogen, hydrogen, etc. which are involved in cycling, (2) *organic compounds* like proteins, carbohydrates, etc. which make the body of organisms and (3) the *climate* like temperature, light duration and intensity etc. (Odum, 1983).

TROPHIC RELATIONS

The structure and functions of terrestrial and freshwater ecosystems have been illustrated in Fig. 2.1. A landscape shows a crop field, forest and a small pond. The solar energy is trapped in all these three ecosystems by chlorophyll in green plants; carbon dioxide is taken from the atmosphere by terrestrial plants and from its dissolved condition in water by submerged part of aquatic plants and with the photosynthetic process green plants produce energy rich organic compounds. Necessary minerals for metabolism are absorbed in dissolved condition from soil solution or water. A large number of microscopic phytoplankton and other macrophytes do the primary job of energy fixation. They (the primary producers) are eaten by herbivores. In turn the herbivores are shown to be eaten by carnivores or predators in different ecosystems. The series of organisms fixing energy, eating and being eaten is called a *food chain*. Food chain can be traced in any ecosystem such as grass, grasshoppers, frogs, snakes and vultures in a grassland. Thus there are four to five steps or links in a food chain where food energy is transferred to the next trophic level. At each step a large portion of energy is lost as heat and only a small fraction goes to the eater. Therefore, the quantity of energy goes on decreasing at a rapid rate successively from producers to top carnivores. The ratio of energy intake and energy of the produced biomass i.e. of input and output is called *ecological efficiency*. This can be studied at any trophic level. The food chain starting from the green plants and through herbivores to the top carnivore is called as the *grazing food chain*. In another case the food chain may start from the dead organic matter being consumed by detritus feeding microorganisms which in turn are eaten by some other predators and this is called as the *detritus food chain*.

The food relations are not always so simple. In fact it is rather complicated and many organisms behave differently in their food habits. Trophic relationship between organisms of any ecosystem is not always in simple chain-like fashion but forms complicated network. This net-like trophic interrelationship is called a *food web*.

In a food web, one organism may be linked with several others in an interlocking of different food chains linked into a network. Man may occupy the position of a herbivore, a carnivore of the first, second, third or top carnivore levels. The shorter the food chain the greater is the ratio of energy of the primary producer reaching to the top consumer. Further, in all ecosystems not only the different grazing food chains get interlocked but the detritus food chain also gets interconnected. At each trophic level some energy is used for itself, some passes to the next trophic level organisms as food and some is routed to the detritus food chain by way of excretion and litter fall.

In some recent studies on food chains, some quite interesting phenomena have come to light. It is found that herbivores do not always cause a negative feedback on grasses, but on a community level often helped in their rapid regeneration as well. Dyer and Bokhari (1976) have found that the leaves of grasses eaten by grasshoppers regrow much faster than in grasses whose leaves are clipped by other means. These suggest that grazing animals through their excrete inputs on grazed grounds and the grasshoppers by their saliva help in positive way a quicker regeneration of the plants on which they feed.

Like six structural components, three of biotic and three of abiotic, Odum (1971) has divided ecosystem functions into six types. These are "(1) energy circuit, (2) food chains, (3) diversity pattern in time and space, (4) nutrient (biogeochemical cycles), (5) development and evolution and (6) control (cybernetics)". These aspects have been described in detail at appropriate places in different chapters.

Ecological pyramids

There is some sort of relationship respectively between the numbers, biomass and energy content of the primary producers, consumers of the first and second orders and so on to top carnivores in any ecosystem. These relationships may be represented in diagrammatic ways and are referred to as *ecological pyramids*. The ecological pyramids are of three categories : (1) of numbers, (2) of biomass, and (3) of energy or productivity. The shape of the first two may be upright pyramidal, or any other shape but the third is always upright pyramidal or triangle shaped.

1. Pyramid of numbers

This deals with the relationships between the numbers of primary producers and consumers of different order. At the base of such a figure is always the number of primary producers and the subsequent structures of this base are represented by the numbers of consumers at successive levels. The top represents the number of top carnivores in an ecosystem. In Fig. 2.2 two types of upright pyramids have been illustrated. A grassland ecosystem on the left and cropland ecosystem on the right are shown where the grass and weeds or the crop plants are in very large numbers. The number of rabbits in the grassland or grasshoppers in the crop field is usually less than the number of grasshoppers. The number of top carnivores is at least in the series of organisms forming a food chain. But there may be instances where the trend of decreasing number may not be true. Fig. 2.3 illustrates an instance where the number of primary producers (a tree) is less than that of herbivore birds feeling upon the tree fruits. The number of parasites like bugs and lice living and feeding upon the birds' body is still higher. Thus depending upon the size and biomass the pyramid of numbers may not be always pyramidal, it may even be completely inverted in shape. Odum (1983) has collected data from researches of a number of workers and given actual number of organisms at each trophic level in a grassland and a temperate forest. In a Southern Michigan grassland of USA, the pyramid of numbers is upright. The number of primary producers is 1,50,000 in 1000 sq. metre and the number of herbivores, carnivores and top carnivores decrease to 200,000, 90,000 and 1 respectively in the same area. But in an English Forest the number of herbivores is 150,000 for a mere 200 primary producers (tree) in 0.1 hectare, thus making the pyramid of an inverted shape (Fig. 2.7).

2. Pyramid of biomass

In order to explain the inverted nature of a pyramid of numbers, the idea of pyramid of biomass is given where the weight of primary producers forms the base. In Fig. 2.4, two types of ecosystems are shown where the pyramid of biomass is upright. The biomass of one tree is very high and even the biomass of a number of birds feeding upon the tree is far less than that of the tree. Similarly the biomass of even a very large number of parasites in and on the body of birds is far less. The pyramid of biomass, therefore, becomes upright in a case where it was inverted if numbers were considered. From an unpublished work of Golley and Child, Odum (1971) has stated that

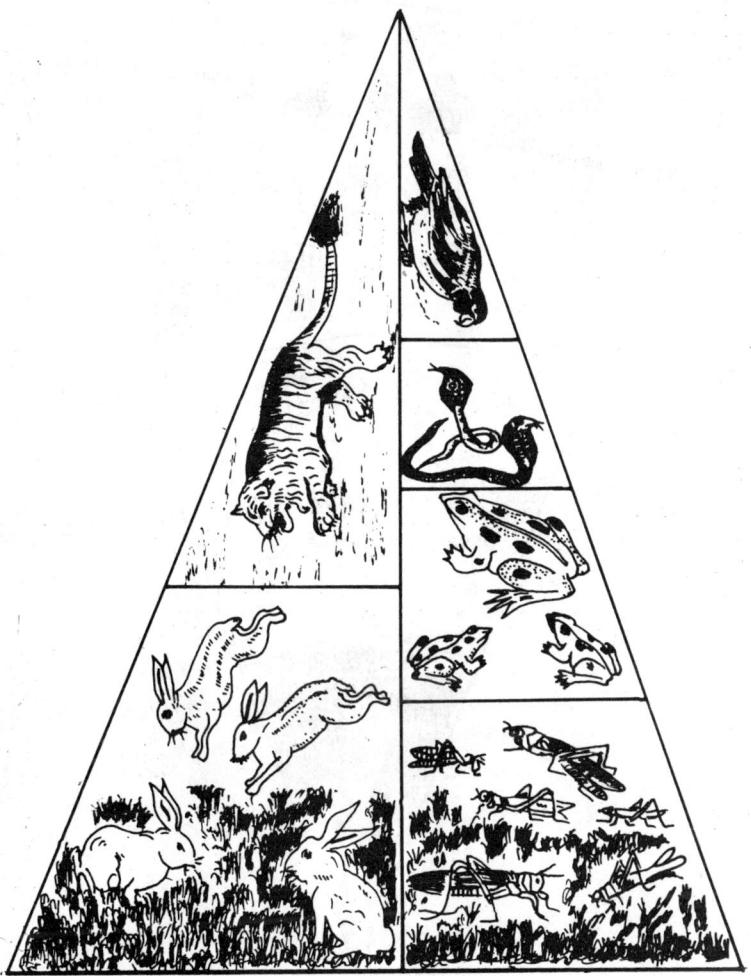

Fig. 2.2. Upright pyramid of numbers in a herbaceous ecosystem. The number of primary producers forming the base of the diagram is more than that of the consumer organisms of successive trophic levels.

in a Panama forest for 15 grams of heterotrophs in a sq. metre area 40,000 grams of primary producers are available (Fig. 2.7). But there can be instances where the pyramid of biomass also gets deformed or inverted as shown in Fig. 2.5. The biomsss of diatoms and other phytoplankton is quite negligible as compared to the small herbivore

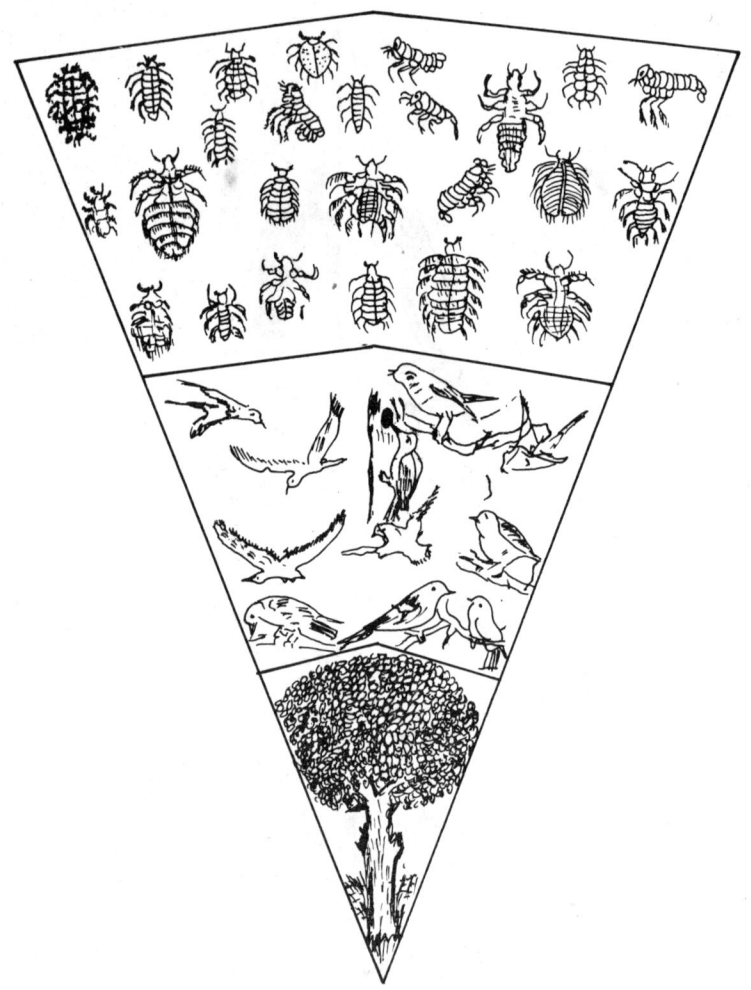

Fig. 2.3. Inverted pyramid of numbers in a tree-ecosystem.

fish that feed on them. The biomass of large carnivore fish feeding on small fishes is still higher. In English Channel, the biomass of primary producers is only 4 g/m^2 whereas that of the consumers is 21 g/m^2 (Harvey, 1950). In fact this is the case in most aquatic bodies. This can be explained if the time factor is also taken into account. The pyramid of energy explains this.

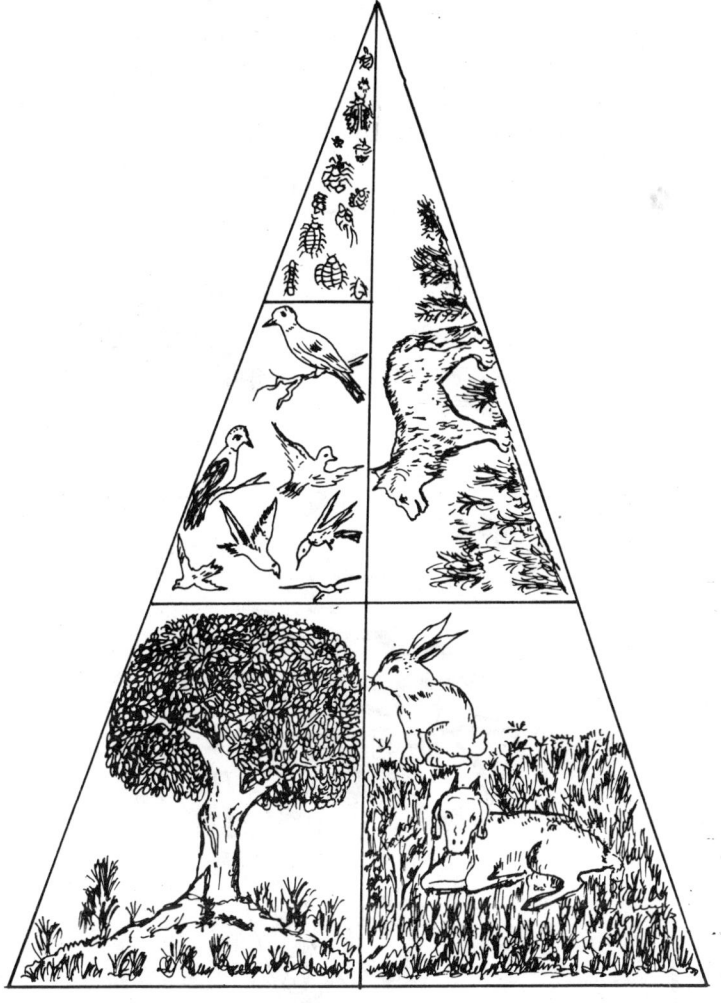

Fig. 2.4. Upright pyramids of biomass. The biomass of green plants are highest in forests and grassland and it decreases with successive trophic levels thus forming an upright pyramid

3. Pyramid of energy

As against the pyramids of numbers and biomass the shape of the pyramid of energy is always upright because in this the *time factor* is always taken into account. The pyramid of energy represents the total quantity of energy utilized by different trophic level organisms of an ecosystem per unit area over a set period of time (usually per square metre per day). In Fig. 2.6 organisms of two ecosystems—a terrestrial and an aquatic—are shown. The quantity of energy trapped

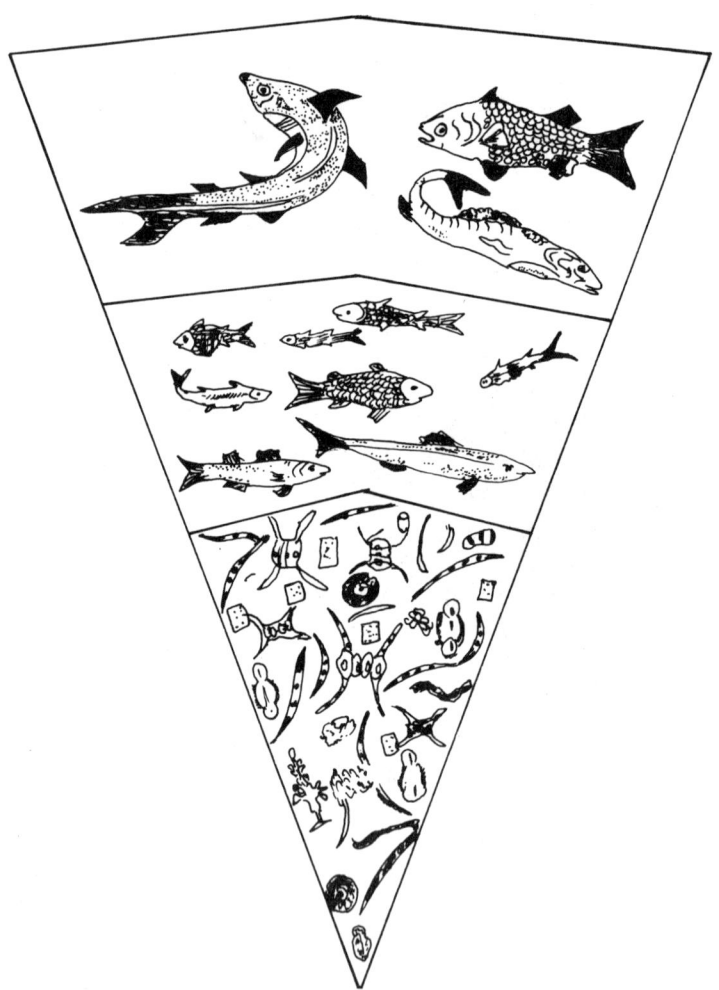

Fig. 2.5. Inverted pyramid of biomass in an aquatic ecosystem due to lower biomass of phytoplankton than of consumers in unit volume of water at any one time.

by green plants in an area over a period say one year is highest compared to that of organisms of other trophic levels and therefore, the base of pyramid is broad. In aquatic ecosystem also the populations of phytoplankton quickly complete their life cycle and sets of new generation of crops of phytoplankton are formed every few hours or days. The cumulative energy content that these generations of phytoplankton trap in course of a year is certainly much more than that of only a few generations of herbivore fishes in the corresponding time and space. The energy content of top carnivores (utilized in one year) is the least. Therefore, the pyramid of energy can never be of any other shape except upright pyramidal. In the event of destruction of primary producers or organism of any trophic level, the organism of next higher trophic level will automatically die for want of food (or source of energy) and ultimately the upright pyramidal shape is maintained. In Silver Springs in the USA the energy contents of different trophic level organisms as collected by H.T. Odum (1957) over a period of one year make an upright pyramid. The energy contents in $kcal/m^2/yr$ from 20,810 in primary producers decrease to 3368 to 383 to 21 in consumers of the first, second and third orders (Fig. 2.7). Thus a top carnivore for the organic production of 21 kcal needs a very broad base of green plants equivalent to 20,810 kcal energy.

PRODUCTIVITY CONCEPT

In the pyramid of energy a very important factor has been introduced and that is the *rate*. The rate of energy trapping by green plants governs the rate of production of organic material from simple inorganic substances in a given area over a given period of time. This rate of energy conversion or increase in organic biomass produced is called *primary productivity*. The total production as a result of photo or chemo-synthesis is called *gross primary productivity*. Some of the primary product is all the time lost by way of breakdown of organic matter in respiration. *Net primary productivity* is the rate of production when loss due to respiration is deducted from the gross production rate. Primary productivity refers to total increase in weight in all the parts—roots, stems, leaves, fruits, etc. as against agricultural productivity which refers to useful parts as grains or fodder parts. Ecosystem efficiency greatly depends upon the production rates of

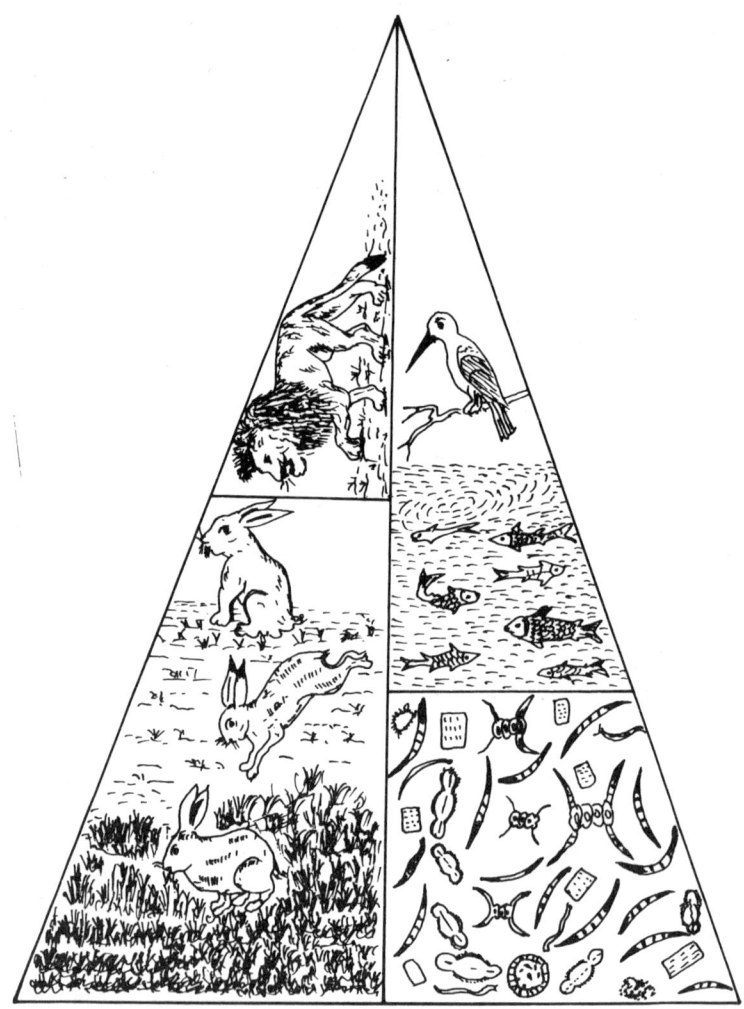

Fig. 2.6. Pyramids of energy in grassland and aquatic ecosystems. The cumulative energy contents utilized by primary producer is always higher as compared to energy utilization by successive trophic levels, over a period of time in a given area.

primary producers. Oceans form the largest ecosystem and its productivity differs in different climatic regions. On sea-shores the productivity per square metre per day may be 2 to 3.5 grams and in deep

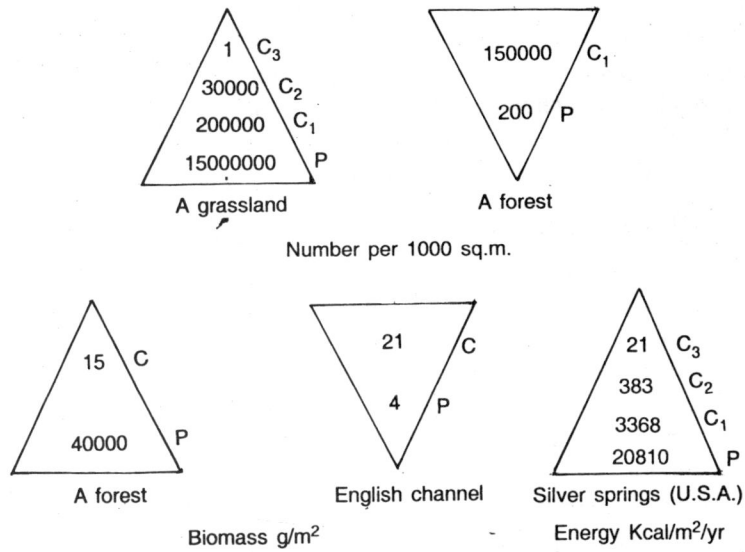

Number per 1000 sq.m.

Biomass g/m² - Energy Kcal/m²/yr

Fig. 2.7. Quantitative values of numbers, biomass and energy in grassland, forest and aquatic bodies at successive trophic levels (Data source, Odum, 1971).

seas only 0.5 grams. In highly productive lakes this value may be 4 to 10 grams/m²/day and even up to 50 grams for short periods in exceptionally favourable situations. The net productivity of crop plants (whole plant) is in the range of 0.3 kg to 1 kg or somewhat more for rice and wheat crops per square metre per year. Sugarcane is one of the very efficient converter of solar energy and its net productivity value ranges from 2 kg to 4 kg/m²/yr or even higher. All productivity values are expressed on dry weight basis of the organic matter or on the basis of energy content in unit time and space. Details on production is described later in Chapter 13.

SYSTEMS ANALYSIS AND ECOMODELLING

As already explained, a system is a unified whole, a structural and functional entity, made up of several interacting and interdependent components and the components of an ecosystem are different kinds of autotrophic plants, animals, decomposer organisms and the abiotic parts like soil, air, water, solar radiation, etc. By virtue of their interactions, there is a continuous flux or changes in the quantities of each of the components. The energy needed to bring about any change ultimately comes from sun but solar radiation by itself cannot be utilized in such interactions. The energy is first converted into chemical form in the bonds of a large variety of organic compounds. For the formation of such compounds many elements, inorganic salts and compounds are needed. It, at once, becomes clear that all ecosystems function through the process of flow of energy (starting with solar radiation) and cycling of materials. In ecosystem analysis, basically we therefore, need to know the biotic and abiotic components, their properties, the quantities of materials needed to build them, the rates at which the built up of produced organic matter is disappearing or getting used up in respiratory process, the quantity being passed on to next trophic level organisms, the driving forces of light, temperature, rainfall, soil structure, thickness of the profile, nutrient status etc. We also need to know the rate of litter fall, death of organisms, the rate of decomposition, mineralization and recycling of inorganic salts into the biotic components. From such worked out values, we learn in a very broad way about the working of any ecosystem. When we wish to learn in greater detail, we have to further analyse each category of broad components into subcomponents such as in the autotrophs, we record the names of different species, their density, frequency, biomass, etc. on a time scale, i.e., recording the pattern of changes and the driving forces in terms of measurable quantities. A very specialised role is being performed by each species and the dimensions or magnitude of such function differs under different set of ecological conditions even for the same species. This role is called as the ecological *niche* of the species. In view of the multiplicity of driving forces on each species, singly on individuals and collectively on their populations, it becomes very difficult to measure the niche dimensions. When we fail to make measurements, we simply describe ecosystems in qualitative terms. But when we are able to translate the complexities of interactions between the living and non-living components of ecosystems in the language of mathematics i.e., in terms of formulae, equations, graphs, etc. then we enter into the realm of system analysis. Representation of structure and function of ecosystems in different forms are known as *models*. Simple description of the system can be called as *verbal description model* and the graphic representation is called *graphic* or *picture* model. Models are simplified physical or abstract representation of real ecosystems which are very complex. Ecological modelling or *ecomodelling* is a very powerful tool aided by mathematics, cybernetics, data processing techniques and computers that help us in understanding and explaining the complexities of nature in a simplified mathematical framework. When expressed in the language of mathematics such as in equations and graphs, it becomes easier to predict changes in the end-result in any ecosystem on varying the driving forces. Instead of trying out different kinds of management techniques through trial and error, systems analysis and ecomodelling studies can, in future, pinpoint definite methods to achieve desired ends in the management of our resources and environments. Therefore, this branch of ecology is becoming more and more important as new problems about our environment are daily coming up. Through systems analysis, we can foresee or predict the adverse or good effects that would happen in future to our environment on account of our present activities. It would provide necessary insight to take corrective measures. But we must realise the limitation of this branch. Ecosystems are extremely complex bodies. Each component and sub-component functions and behaves differently to subtle changes in environment. Despite application of powerful computers, modelling is only a simplified and incomplete version of the actual picture of any ecosystem. Out of hundreds of interacting pathways, only a few physical and biological characteristics are possible to be translated into mathematical relationships or models.

In ecomodelling we must at first define the boundaries of the ecosystem to be studied. The components are then identified. Picture models are commonly made up of boxes and arrows. We then put the components and subcomponents, into boxes at appropriate places and connect them through arrows showing the direction and type of interaction between the boxes. Such a model is also called as compartment model. Interactions may be through transfer of matter and energy or other kinds of influences. In identifying the components and level of splitting into subsystems, we are guided by the nature of information we need. If we are to know the ecological efficiency, we have to measure input/output ratios of energy of the boxes, of primary producers, herbivores and carnivores. For nutrients we record the quantity and rate of transfer of the particular element in different boxes such as the boxes for soil subsystem, litter fall, litter disappearance, etc. For resource conservation such as of particular plant or animal species, or for water or soil we take into account that resource levels in relation to physical forces of environment on a time scale. These are more often represented in mathematical models in terms of equations instead of picture models. If the information on transfers between compartments are set in tubular arrangement we call it a *transfer matrix model*. The mathematical application of correlating the rates and kinds of changes in plant or animal growth performance, or population behaviour or other functions in relations to physical forces, is through the use of multiple regressions. With the use of appropriate regression equations an investigator can find out an approximate value of transfer functions between different compartments. But it must be borne in mind that even with the use of complex models, the values obtained are only near approximation of actual measured values of the real ecosystem.

Some ecomodels are prepared on lines of electrical circuits using different shapes of boxes and arrows representing energy sources, passive energy storage pools, heat sinks, etc. This is called *electrical analog circuit model*. Model based on not real data is called *simulation model*. The input of such forces into an ecosystem that affect the components (like sunlight or rainfall affecting plants), but not being themselves affected by the components are called *forcing functions*. If we take random variability in forcing functions, modelling becomes very complex and difficult and such a model is called *stochastic model*. But if chance variation in the forcing function is ignored, the model becomes mathematically less complicated and is called *deterministic model*.

STABILITY CONTROLS (HOMOEOSTATIS)

We have discussed two basic attributes of all ecosystems (1) cycling of materials described later in detail at appropriate place and (2) flow of energy described in a separate chapter. Another important attribute of ecosystems is the information content which enriches as the system matures. Based on information content there is a better physical and chemical communication and regulation of transfers and flows through the components, giving stability to the ecosystem. In another word the ecosystem is cybernetic in nature. *Cybernetics* is derived from the word *kibernetes* meaning *pilot* or *governor*. In mechanical gadgets or man-made cybernetic devices, the controls are external by men, but in the ecosystem, control is diffused almost at all levels within the system. The control provides (i) *resistance* against disturbance to maintain the steady state as well as (ii) *resilience* to recover to stable state quickly from the disturbed state. These two help to withstand disturbance and are usually not operative simultaneously in the same ecosystem. If a system is a resilient to some adverse factor, say against fire, then the component plants may not be fire resistant but would soon regenerate after the fire is over. On the other hand another ecosystem may be fire resistant, i.e. they do not burn fully and escape fire damages during usual fires. But if the fire is prolonged and the trees burn fully they would not recover after the fire is over, as these are expected to lack resilience against this factor. So the stability of some ecosystem may be on account of resistance and of certain others due to resilience. The third possibility may be due to both to a certain extent. The control mechanism is largely controlled by *feedback*. By feedback we mean that part of the output from any system is put back as an input. For instance if a money (principal) is deposited in bank it earns an annual interest. If part of the interest (output) is redeposited (input) in the bank the principal amount increases and in the next year it earns still more interest. Like this if part of the output of an ecosystem is used as feedback then the system grows (just as the principal amount grows) year after year. Such a feedback that allows growth is therefore called as *positive feedback*. The control mechanism prevents a system from overgrowth in a manner as a thermostat controls the temperature range beyond which it does not allow it to go, neither on the cooler nor on the warmer side. So in order to check overgrowth the mechanism of *negative feedback* is applied. For instance when the population of primary producers is increasing, it gets controlled at an optimum level by the rising population of herbivores. Next, the rising population of the herbivores cause a negative feedback effect on the plant population. Similarly the carnivores exert a negative feedback on the herbivore population. The positive and negative feedback phenomenon keeps on operating all the time among different trophic level populations in such a way to reach some kind of *steady-state*. The mechanism of achieving a steady state and continuing at that is due to homeostatic mechanism.

Homeostasis, therefore, refers to the *tendency* of the *ecosystems to resist change and remain in a state of equilibrium*. The science of *cybernetics* or of controls is largely dependent upon the feedback system and has its application in understanding homeostasis. Feedbacks are *positive* or *negative*. For instance in a balanced population a predator, like the snake eats up and reduces the prey population of frogs. It means that snake is interacting with frog in a *negative feedback* way. The frog population by virtue of its numbers is regulating the snake population. This is a *positive feedback* as the frog is responsible in a positive way. The positive and negative feedbacks strike a balance for maintaining an equilibrium. Negative feedback based on the experience or output of information generated earlier is responsible for self regulation or homeostasis. Whenever the population of any one kind of trophic organisms tend to increase beyond the carrying capacity of the ecosystem, negative feedback mechanism comes into force and brings down the population size to desired level. Man of course, is the greatest force of disturbing this mechanism of natural balance. Man needs air, water and food for his physiological needs. In quest of technological advancement for more and more of physical amenities, man is degrading the quality of air, water and such organisms that form the foundation of ecological pyramids of which he is a part. He is becoming rich in physical amenities but becoming poor in things of his physiological needs, leading to diseases and ecological backlashes. Homeostatic controls are achieved in nature after a long period of evolution of ecosystems and, therefore, man must exercise greatest restraint in not disturbing it or doing so to very little extent while undertaking technological projects. The development must be such that his physiological needs are fully protected i.e., the environment must remain clean with plenty of pure air, water and food.

In an electrical gadget different components are connected with network of fine wires and control knobs. In the ecosystems, too, we find that different components are linked by some invisible connections for an interdependent and interacting operations. in an ecosystem similar function is performed by a number of species, but each has a different efficiency and tolerance range to the environmental factors. So with the gradual evolution of ecosystem a number of alternative courses are developed to ensure success when the system faces different levels and kinds of perturbations, i.e. as the system matures, its homeostatic or control system becomes more efficient. Usually building up several alternative functional pathways, the diversity of species increases.

TYPES AND EXAMPLES OF ECOSYSTEMS

As it is clear from the foregoing descriptions, ecosystems are of so many different kinds and sizes and their classification may be only too general. One way is to classify them into natural ecosystems and manmade or man-modified ecosystems. Natural ecosystems can be distinguished into (i) terrestrial and (ii) aquatic. The terrestrial could be representative of forests, woodlands, savannas and grasslands from tundra region to temperate and tropical. In aquatic it can be freshwater and marine. In the marine, there could be oceanic, continental shelf, estuarine, deltaic, mangroves, coral reefs, etc. and in the freshwaters, ecosystems may be lotic like rivers and streams or lentic or lakes and other natural wetlands. Among the man-made or man-managed are, the agroecosystems, villages, towns and cities, industrial and mining areas, ponds, aquacultures, reservoirs, irrigation canals, etc. All these ecosystems have the basic structure and function of energy sources primary producers, consumers, decomposers with flow of energy and cycling of materials.

Light and Temperature Factors

The environment is a complex whole of so many interacting factors that influence every organism. All that surrounds and affects an organism is its environment. Plants are surrounded by soil beneath and air and many organisms above. The soil factor is also called the *edaphic factors*. Air or atmosphere surrounds the above ground parts and affects the plants in variety of ways. Factors like light, temperature, water and precipitation, etc. constitute the *climate* of a region. Besides *climatic* and *edaphic* conditions, the surrounding associated plants and animals, parasites etc. affect the growth of other plants. For instance the growth performance and grain yield of a crop plant like wheat are affected by the spacing between different individuals of the same species (intraspecific competition), by the quantity and type of weeds in the same field (interspecific competition), by the severity and kind of parasitic infections and by the type and frequency of grazing animals, etc. All these influences—both direct and indirect by other organisms are classed as *biotic factors*. In the modern ecosystem approach of ecology environment is divided into (1) *biotic* and (2) *abiotic* components.

The edaphic, climatic and biotic factors which constitute the environment affect plants, their population and community growth and dynamics in a holistic manner. But it is difficult to understand the mechanism of environmental influences unless we study the different components of environment separately.

The intensity and duration of light, temperature, the atmospheric conditions including the humidity and wind velocity, the quantity and pattern of precipitation, etc. constitute the climate of any place. There are a number of causative agents that affect each of these factors.

1. THE LIGHT FACTOR

Light is a form of radiant energy. It is an essential factor for the primary production of plant materials upon which all other living organisms depend directly or indirectly.

The solar energy that sustains all life on the earth is received in the form of electromagnetic waves. These are of different wavelengths between 2900 Å to 50000 Å or 290 mμ to 5000 mμ. Å = Angstrom = 10^{-10} metres. 1 mμ = 10 Å or 1 nm or 1/1000000 mm (or 10^{-9} metres).

The visible or luminous range i.e. light is between 380 to 720 nanometres of wavelengths and the chloroplast pigments absorb radiations from 330 and 740 nm.

Light is a form of energy and hence it can be converted into other forms of energy such as heat. The quantity of light is usually expressed in terms of gram calories received in a unit area over a period of time such as g cal/m^2 year.

The intensity of brightness of light is measured in terms of candle power. The light intensity from a standard candle at one foot distance is called foot candle (F.C.) and at one metre distance a metre candle or lux. One F.C. is equal to 10.76 lux because as the light travels it spreads and its intensity decreases inversely to the square of distance travelled.

Effect of light on plants

Light affects and regulates plant life in a very large variety of ways. The most important role of light is in photosynthesis where the chlorophyllous tissues use light energy to build energy rich complex organic compounds from simple low energy inorganic substances. Light is abundantly received on the surface of earth and on an average approximately only 2 to 3 per cent of this solar energy is used in primary productivity. However, in deep shade under trees or under water, light becomes limiting below which phytosynthesis is not sufficient or effective for growth. *Compensation point* is that intensity of light at which the rate of phytosynthesis is just sufficient to meet the requirement of respiration. At this point the dry weight of plant does not increase. This point differs from species to species and in the same species in different individuals of different age. Based on their relative preference for natural growth in bright or diffused light the plants are broadly classified into *heliophytes* and *sciophytes*. Heliophytes grow in open sun and sciophytes in shade. But most of the plants are not very rigid in their requirement of light intensity. Such heliophytes as can also grow in shade though not so well are called *facultative sciophytes* and those which fail to grow in shade are called *obligate heliophytes*. In the same way there are *facultative heliophytes* (sciophytes which may also grow in light) and *obligate sciophytes* (sciophytes that fail to grow in sun). In heliophytes, leaves are usually small and thick with pronounced mesophyll and several layered palisade, often vertically oriented with bifacial (upper and lower sides) occurrence of palisade cells. In sciophytes, mesophyll is not pronounced and leaf size is big. In heliophytes chloroplasts have smaller thylakoid grana than in sciophytes. High light intensity has a destructive effect and presence of protective pigments called *Chymochromes* occur such as in anthocyanin rich young reddish leaves. High light intensity for prolonged duration the photosynthesis is reduced by reversible *photoinhibition* and non-reversable *photodestruction* processes. Details of physiological and biochemical aspects are described at length by Schulze et al. (2004). Overexcitation produce chemical radicals which react with oxygen and produce membrane and protein destroying reactive oxygen species.

Light factor regulates to a great extent the canopy and leaf adjustment in terrestrial communities and species stratification in aquatic habitats. In surface floating leaf plants the petioles are flexible and long which helps in fully covering the water surface area by leaves. This is beneficial for the species but harmful to submerged plants as the light is drastically cut down at the surface itself. The dominant species occupy maximum space at canopy level and utilize maximum light.

In tropical rain forests the composition, stratification, life form, etc. are very much adjusted to the light intensity received at different layers. The tall trees receive full light, smaller trees and shrubs get less light and ground vegetation is adjusted to rather low light intensity. The climbers reach the top layer and get full sunlight. Not only the quantity of light is lower in lower regions in multistoreyed forest but its quality may also become quite different due to selective absorption or transmission of certain wavelengths through the canopy of leaves. The reduced light intensity and variation in temperature and wind velocity appreciably affect metabolic activity. Leaves *reflect*, *absorb* and *transmit* through them different wavelengths in different proportions. A large part of infra-red is reflected (about 70%) whereas ultraviolet (U.V.) range is least reflected by leaves (about 3%). Leaves covered with hairs reflect more light. U.V. radiations are retained in epidermal regions, whereas the photosynthetically active radiations pass through and reach mesophyll and get absorbed in chloroplast pigments. About 40% of the light may pass through thin leaves but in thick ones only about 10% of the light filters down. Red and green wavelengths pass down on floor of dense forests in greater proportions.

The arrangement of leaves in all kinds of communities are in such a way as to utilize the radiations with maximum efficiency. They are more on the peripheral part of the canopy receiving full sunlight. The inner core region of canopy where only 1 to 25% of the light reaches have much less leaves. In most cases, photosynthesis is effective up to as low as 2% of the full sunlight and many thallophytes survive upto even 0.5% to 0.1%. The relative quantity of light reaching different strata of vegetation depends upon the stature and density of dominant species. In closed forests about 70 to 80% is held up by trees, 10 to 15% is reflected, and the remainder 10 to 12% is available for ground flora and 2 to 3% reach the soil surface. In grasslands, savanna and croplands, reflection is a little more (about 20 to 25%) and the rest is almost equally shared by upper and lower leaves. Broad leaved dicot crops, however, absorb more of light on top layer or canopy casting dense shade on ground.

The light intensity stratification is more marked and sharp in aquatic bodies. The intensity decreases rapidly with depth. Thus the plant distribution in ponds and lakes is distinctly governed by light factor. The occurrence of reed swamps on marsh, submerged species in shallow regions, rooted species with floating leaves in somewhat deeper zones and free floating forms in deep regions and manifestations of ecological adaptations of life form chiefly in response to variation in light intensity. Phytoplankton move up and down in water and adjust their position during the 24 hr. cycle in relation to diurnal fluctuation in light intensity.

In industrial areas or wherever smoke is heavy the light intensity is remarkably cut down. Plant leaves are doubly affected, due to atmospheric smoke and due to the dust particles that cover the leaf surface and reduce light entering the leaf.

Light also affects the opening and closing of stomata and thereby regulates CO_2 and O_2 exchange between plants and atmosphere. All the primary productivity is thus largely governed by light directly as a source of energy and indirectly through CO_2 exchange regulation. The rate of transpiration is also regulated by the stomatal opening and closing.

As already indicated, light intensity under natural condition especially in tropical regions for most parts of the day is much more than the optimum requirement of plants. Thus even lower leaves of trees in shade are able to carry production process with high efficiency. But intense light in injurious for many species. In certain plants like maize and sugarcane which follow Hatch and Slack pathway of photosynthesis and which are mostly tropical in nature, the net photosynthesis rate goes on increasing even up to the summer brightness level of light intensity.

Under bright sun the arrangement of chloroplasts within the cells of palisade tissue in leaves is along the vertical walls. Thus chloroplasts escape high intensity as light falls on them from an angle and on one side only. Phytoplankton or microscopic water plants move down during bright part of the day to escape excess light. Excess light also results in overheating of plants which is again injurious. Under natural condition hardly two parents of light energy is used because other factors become limiting. Plants growing in open sun have as a result developed certain common features in their morphology, anatomy and physiology. For example the stems become more compact and hard with more branching and shorter internodes. The leaves are thicker and short with closely arranged small stomata. Often the leaves are hairy and palisade tissue is well developed but chloroplasts are fewer in number.

Besides morphological responses, various plant functions such as germination of seed, leaf fall, initiation of flowering etc. are governed by the light factor. It regulates germination of seeds in many species in different ways. Some seeds require darkness for germination, some require light for the purpose and some are indifferent to this factor. The mechanism of effect of light on physiological activation of seeds for germination is not well understood in many species.

Photoperiodism

The length of day and night periods in the twenty four hour cycle also regulates the leaf fall, flowering and other phenological processes in plants. This phenomenon is called *photoperiodism*. From the view point of flowering, those plants which show photoperiodic response are classified into : (1) *short day* plants and (2) *long day* plants. Those species in which phenological events are indifferent to length of light or dark periods in daily cycle are called (3) *day neutral* plants. The photoperiod or day length varies with seasons and the extent of variation is dependent upon the latitude. Equatorial region has around twelve hours of day and twelve hours of night. From equator to the North or to the South the summer days become longer and night shorter than twelve hours and the reverse is true in winter. The greater the distance from equator, the greater will be the departure in day and night length from twelve hours. It is now more conclusively found that the photoperiodic regulations of flowering is governed actually by the length of continuous night (or dark) period and not the day length. Even then the original term 'photoperiodism' is being retained in view of its popularity. Different species have different critical photoperiods which usually range between 11 to 14 hours of day or 10 to 13 hours of dark phase. The *critical* photoperiod is key point in photoperiodism. A species which flowers only so long as the photoperiod is less than the critical period or the continuous dark phase is more than the difference between 24 hours and the critical photoperiod, is called a *short day* plant. Such plants fail to develop flowers when the dark period falls below the required length or the continuity of the dark phase is broken. Similarly the species which flower only when the light period or day length exceeds their critical photoperiod or the night length becomes less than the difference between 24 hours and their critical photoperiod are called as *long day* plant. One fact must be remembered that the classification of plants into *short day* and *long day* is based on the fact that a species flowers only if the day length is less than or more than the *critical photoperiod* and not the *twelve hours period.*

Plants distributed in different geographical regions, in course of their long evolutionary history, have become ecologically adjusted to the light period of the place. Even within a species there are examples of photoperiodic ecological races as in *Xanthium strumarium* (Kaul, 1959).

The ecological understanding of responses of plants to different conditions of light is of great practical significance. For instance, the crop of betel leaves (or *paan*) in India is grown in artificial shading. This results in greater expansion of leaf with less development of stiff or hard tissue and less of chlorophyll. The betal leaves that are soft, palatable and yellowish white in colour fetch a higher price.

The knowledge of photoperiodism is of great practical utility in selection of species and seasons of their cultivation depending upon the plant parts which are economically important. For example, the beautiful flowering herbaceous plants of higher latitudes are widely cultivated in Indian gardens during the winter season when night period is long enough to induce flowering. In the sugarcanes (short day) vegetative growth is more important for sugar and hence if the continuity of long dark phase is broken by a few minutes of illumination, the flowering is checked. This results in greater cane growth and higher sugar yield. Similarly the Maryland mammoth variety of tobacco, because of photoperiodism, does not flower at Maryland (USA) and as such the leaf is larger than at latitudes where photoperiod is favourable for flowering.

Many phenological events are regulated by the rhythms of light and dark hour lengths in the 24 hour diurnal cycles. Experiments have shown that some of these rhythmic responses may be in response to not exactly 24 hours cycles but about this period. A term *circadian rhythm* is applied now a days for diurnal rhythms that may not be of exactly 24 hours.

Leaf is the principal organ that receives light stimulus. Even a few leaves, if artificially subjected to desired level of photoperiod, cause flowering responses. For instance when a leaf of long day species of *Nicotinia sylvestris* is grafted on short day plant of *Nicotiana tabacum* and the plant is exposed to long days, it begins to flower even though the plant is a short day one. Similarly, if just a few leaves, of the short day plants of *Kalanchoe blossfeldiana* in a long day situation, is kept under short light period by covering them with bag for part of the day, the plant begins to flower. Thus, whatever flowering hormone that is produced at the desired day length, even in minutest quantity synthesised by one or two leaves, is sufficient to bring about flowering in all branches of the plant.

Some of the common examples of short day plants are *Nicotiana tabacum, Dahlia variabilis, Cannabis·sativa, Chrysanthemum indicum, Kalanchoe blossfeldiana, Xanthium strumarium and Helianthus tuberosus*; and of long day plants are : *Allium cepa, Beta vulgaris, Daucus carota, Nicotiana sylvestris, Papaver somniferum, Vicia faba* and *Avena sativa.*

Among the main adaptational responses shown by plants to high-light intensity are the shorter internodes with stout stem and with well developed vascular tissues, more of branching, smaller and thicker leaves, small stomata, high stomatal index (number of stomata per unit area) and more number of hairs on pubescent (hairy) leaves. Leaf blades orient themselves at an inclined angle to the source of sunlight and have chloroplast arranged vertically in the mesophyll. Root growth, on the other hand is longer and well branched and the root: shoot ratio is higher in sunny plants. Such plants have higher respiration rate, lower water content (on dry weight basis) and higher salt content as compared to plants growing in shade or dull light. Sunny plants flower more profusely and seeds have usually higher energy content per gram dry weight. These plants are more resistant to climatic and biotic hazards like drought, heat and infections. The above characteristics are true in most cases but there are exceptions too.

Measurement of light intensity

The intensity or brightness of light is usually measured by an instrument called *photometer* or *luxmeter*. This consists of two parts, a photoelectric cell and an amperemeter. The light energy falling on a photocell generates electric current in proportion of the light intensity and it is read in the amperemeter. The meter is so calibrated that the needle directly gives the light intensity in *lux* unit. The luxmeter used to measure light intensity under water has a microamperemeter connected through a long electric cord to a photocell. The photocell is sealed in a transparent waterproof case. This helps in taking the light intensity readings outside at the shore or in a boat while the photocell is lowered to different known depths of water in such a way that the photoreceptor disc remains in a horizontal position.

Precaution : The sensitive surface of the photocell should not be exposed to direct sun without adequate filters of known values. Direct sun rays damage or solarize the instrument.

Radiometers are used to measure the quantity of light. There are jet black and bright white discs which completely absorb and reflect the light respectively. The black disc assumes higher temperature than the white disc due to absorption of energy. Various devices to increase the sensitivity of the instrument are being made these days.

Secchi disc is a black and white painted circular disc of about 20 cm diameter used to measure the depth of water upto which the light intensity is sufficient to distinguish the black and white segments of the disc. This gives an indication about the turbidity of water.

2. THE TEMPERATURE FACTOR

Temperature is a measurement of the degree of hotness. We are very familiar with this factor of environment. Summer and winter seasons are expressions of rhythmic change in temperature. Heat like light, is a form of energy. The radiant energy received from the sun is converted to heat energy. Heat is measured in calories. One calorie is equal to the heat required to raise the temperature of one gram of water from 15.5 to 16.5°C. The degree of hotness or *temperature* at the point water freezes into ice is 0°C or 32°F and at the point water boils into steam is 100°C or 312°F. Heat travels in space in the form of *radiation*, in *liquids* through *convection* and in *solids* through *conduction*. After the radiant energy hits the surface of earth it is converted into heat energy. The re-radiated waves have a heating effect and as such the atmosphere is warm near the earth's surface and as we go up it is progressively cooler. Besides this altitudinal gradient of temperature, there is a latitudinal gradient of temperature from hot tropics to progressively cooler belts of temperate and polar regions.

Plant growth and metabolism are profoundly affected by temperature. Unlike in warm blooded animals the temperature of plant parts are regulated by the temperature of the surroundings. Some morphological structures and physiological process like transpiration lead to some difference between plant and environment temperatures. Temperatures also affects the rate of transpiration and other physiological processes. Increase in temperature increases the capacity of air to hold more moisture in vapour form. This result in higher difference between vapour pressure deficits and consequently the rate of transpiration increases.

Temperature also regulates the activity of enzymes which in turn regulate physiological processes. The temperature at which physiological processes are at their maximum efficiency is called *optimum* temperature. The minimum, maximum and optimum temperatures are called *cardinal* temperatures. The cardinal temperatures vary from species to species and within the same species or individuals from part to part and in the same part at different age. Like chemical reactions, the rate of biochemical reactions in plants also increases the temperature to a certain extent. The ratio of rate of reaction at difference of 10°C is called the *temperature coefficient* or Q_{10}. Various periodic phenomena of plants like seed germination, vegetative growth, reproduction, etc. are also temperature regulated.

Most of the land plants thrive in quite a wide range of temperatures from almost freezing to about 55°C and these are called *eurythermic* plants. But their optimum range is not so broad and show best growth between 20°C to 30°C. Aquatic environments usually show less fluctuation in temperature and hence many aquatic plants thrive in a narrow temperature range and these are called *stenothermic* plants. Some algae like *Chlamydomonas nivalis* and diatoms live on snow. The range of temperature beyond which on either lower or upper side the species dies is its *lethal limits* and the range of temperature in which its metabolism is active and favourable is called as *activity limit*. Extremes of temperature, both on cold and hot sides damage the membranes, destroy the proteins and denature thermolabile enzymes thus disrupting nucleic acid metabolism.

Vernalization

In a certain variety of wheat, it has been noticed that when grown in spring it does not flower unless it has passed in vegetative phase a winter season. It was thought that low temperature effect is essential for the production of some stimulus needed for flower initiation. The seed presoaked and subjected to about two weeks of almost freezing temperature when sown in spring produced seeds in the same season. This type of effect of low temperature on seeds that hasten flowering is called *vernalization*. Many annuals and biennials and a few trees of cold climates need a certain period of low temperature so as to flower in the following spring. If the winter is not cold, flowering is inhibited or delayed.

In order to create vernalization effect, seeds are soaked and allowed to germinate and then cold treatment is given. Presence of oxygen is also necessary. Devernalization (or removing the cold treatment effect) can be achieved by keeping vernalized seeds or seedlings at high temperature or exposed to low oxygen or high nitrogen and carbon dioxide atmosphere. Some people regard that cold treatment results into the synthesis of *vernalin* that may change into florigen or into gibberellins thus inducing an early flowering.

Thermoperiodism

The temperature of any place changes constantly from hour to hour in a diurnal cycle. Many plants, have in course of time got adjusted to the rhythmic diurnal cycle in temperature changes for several phenological events. This regulation of phenology to diurnal thermal ranges is called *thermoperiodism*. For example seed germination in many species is found to be better under natural alternating high and low temperatures of day and night than in constant temperature chambers. In some tomato varieties alternating temperature of 26°C and 18°C promotes higher fruit production and better vegetative growth. Ambasht and Lal (1977) have shown that in alternating temperature sets the time required for seeds of *Chrozophora rottleri* to germinate is reduced considerably as against in sets kept at favourable but constant temperature.

Many plants of temperate regions show best growth when the day temperature is about 10 to 15°C higher than the night temperature. In rare cases as reported in the African plant *Saintpaulia ionantha* the growth is best when night temperature is higher than day temperature. In most of the tropical plants such thermoperiodic responses are not well developed and growth remains best within a narrow difference of day and night temperature.

Effect of temperature on vegetation pattern and composition

The type of vegetation both in structure and composition is widely different in hot equatorial belt, warm tropics, relatively cool temperate and the permanently cold polar regions. The most striking difference in these geographical belts is the difference in their temperature besides light intensity, photoperiod and other ecological factors. The hot and humid equatorial and tropical regions are full of evergreen forests. These are extremely rich in variety of vegetation and different life forms of plants occupy the aerial space in several layers. In fact the tropical rain forests are the richest ones in diversity and density. A large variety of dicot trees, shrubs of all stature and size grow intermixed and adjusted compactly in limited space. The temperate belt has low temperature and the region is rich in forests of coniferous trees like pine and deodar and dicots like oaks, birches and chestnuts. Polar region has, on accouts of extreme cold condition, less free water available due to snow formation. The plants are dwarf and grow sparsely.

This effect of variation in temperature on vegetation with the same photoperiod is clearly evident in the Himalayas. The lower altitudes in the warm and humid Himalayas have rich growth of sal (*Shorea robusta*) trees with patches of *Dalbergia sissoo* (*Shishum*), *Eugenia* sp., etc. As the altitude increases temperature falls roughly at the rate of about 7°C per thousand metres. With the fall in temperature at altitudes above 1500 meters the temperate species like *Pinus roxburghii* and species of *Quercus incana* begin to appear. Further up, around 3000 metres *Pinus wallichiana, Cedrus deodara* and *Abies pindrow* become dominant. The tree height gradually decreases in temperate belts. Above 4000 metres the growth is almost negligible. Shrubs like *Rhododendrons*, grasses and several dicot herbs like *Saxifraga, Primula* and *Anemone* replace trees. At about 5600 metres altitude the temperature falls below freezing point all the year round. Under such a snow covered habitat only some highly specialized algae, some lichens and mosses grow for a short period. Higher altitudes are devoid of vegetation.

Different species are adjusted to different temperatures. A temperature not suited to one may be most suited to another. In not so cold climates, even a few hour of frost damages crops of potatoes, tomatoes, etc. The low temperature kills the growing buds. However, plants adjusted to extremes of cold are not killed even when their sap is frozen solid. Especially the perennating parts like bulbs and turions remain frozen in temperate lakes. They become active with the rise in temperature in spring.

Injury due to extremes of temperature

Both high temperature and very low temperature in air and soil cause widespread damage in many plants and some have developed adaptations to overcome such unfavourable situations.

In very cold temperatures usually below freezing point many species become dormant and biological activity is reduced to an extremely low level. Generally in most of the species prolonged freezing results in precipitation of proteins, dehydration and death of protoplasm. Most of the tropical plants succumb to cold temperatures even above freezing point, but plants of cold temperate region endure low temperatures much below freezing. Old parts resist cold injury whereas young buds quickly suffer due to frost. In most crops which are likely to suffer in case of frost, it is advisable to irrigate frequently to reduce adverse effects. Prolonged winter renders a high viscosity in protoplasm causing slow water diffusion through root tissues. Evaporation is also reduced and despite abundant water below the upper leaves suffer in case of frost, it is advisable to irrigate frequently to reduce adverse effects. Prolonged winter renders a high viscosity in protoplasm causing slow water diffusion through root tissues. Evaporation is also reduced and despite abundant water below the upper leaves suffer from drought and often wilt, a phenomenon called *parch blight*. Raunkiaer's famous classification of plant lifeforms, described in a later chapter in some detail, is also based principally on the adaptive ability of plants to perennate against cold injury. In many plants of cold climates the salt content of cell sap is quite high which lowers the freezing temperature considerably. Such plants escape damage on account of freezing to as low as $-10^{\circ}C$ or even more. Cold or chilling leads to ice formation in the cell and in intercellular spaces. Frost damages the membranes. Tropical plants are not adapted to harden their tissues and suffer most. Photosynthesis is disturbed and reactive oxygen species (ROS) are produced. Many plants have developed mechanisms to counteract and detoxify ROS effects. Xanthophylls, cytosols, ascorbates and flavonoids have important roles to play in detoxification. Formation of *antifreeze* proteins are reported mostly in animals and some plants and they check formation of ice crystals. Frozen soil and roots fail to supply water lost by leaves during transpiration.

Similarly when temperature increases above the optimum level the enzymatic activity is retarded followed by chlorosis and any further increases in temperature results in death of younger tissues. In certain tropical trees the surface of the stem is whitish which reflects most of the radiation and below the thin surface layer green tissues are present which photosynthesize and meet food requirement when leaves have fallen. In summer, in most deciduous trees of the North India, old leaves, are completely replaced by young, small and thin leaves. Thin young leaves transpire rapidly and thus reduce the leaf temperature to a certain extent. In many trees the stem also gets covered with thick insulating cork which protects living tissues inside against high temperature. In tropical forests and savannas, fire is an important factor which kills most of the ground vegetations, but trees escape due to insulating bark that prevents the inner living cells from high temperature. Even the ground vegetation reappears after the fire, since perennating buds located on underground parts escape damage and become active in favourable season. Organisms adapted to cold are called *psychrophiles*, to hot conditions are *thermophiles* and intermediate conditions as *mesophiles* (Schulze et al., 2004). Seeds are adapted to survive wide range of prolonged extremes of cold or hot conditions. Some are adapted to survive fires also. Frosts cause freezing of water present in plant tissues. Tissues are hardened in cold region and prolonged frost situations to withstand the damaging effects of freezing. Warm conditions enhance and cold conditions, retard the biochemical reactions and biological processes. Temperature extremes also affect biomembranes and lipids. *Hyperthermy* or high temperature above the optimal range for a brief a few minutes (or even long period) within tolerable limit, known as *heat shock*, brings about metabolic changes and production of heat shock proteins (HSP). After about 6 to 8 hours of the shock, HSP synthesis stops and normal metabolism is re-achieved i.e. the 'switched off' '*house keeping genes*' are gradually 'switched on'. Shulze et al. (2004) have described and discussed the switch 'off' and 'on' mechanisms of HSP synthesis and return to normal conditions.

Temperature is also an important factor in regulating parasitic infection and its spread.

Temperature measurements

Measurement of temperature is usually done by thermometers. For special requirements various modifications of ordinary thermometers have been made.

Maximum-minimum thermometer is a special type of U-shaped thermometer in which one arm records the maximum and the other the minimum temperatures. In each thermometer an iron pin is left at the maximum mercury level which indicates the extremes of temperature during the period the instrument is kept for one set of observation. A piece of magnet is used to bring down the pins to existing mercury level before setting the instrument for the next record of diurnal temperature fluctuation.

Whereas the above instrument records only the two extremes, maximum and minimum temperatures, another instrument called a thermograph continuously records temperature on a graph sheet usually for one week. The recorded temperature on the graph is called a thermogram. In this instrument there is an incomplete metallic ring which expands and contracts with the change in temperature. This is adjusted to a pen which moves up and down with the contraction or expansion of the ring. The pen is filled with a special ink which does not dry on the nib. A recording drum makes one rotation in one week. This is covered with a graph sheet duly marked for temperature on one side and days and hours on the other; thus the moving pen continuously records the temperature of any place where the instrument is kept. Often the same instrument has a device for recording relative humidity, the instrument is then called a *thermohygrograph* (Fig. 3.1).

For measuring temperature in water *thermocouples* and *thermistors* are used. This is based on the principle of thermocouples. In the thermocouples electrical potential generated due to a sharp function of two dissimilar metals is measured. This potential difference is di-

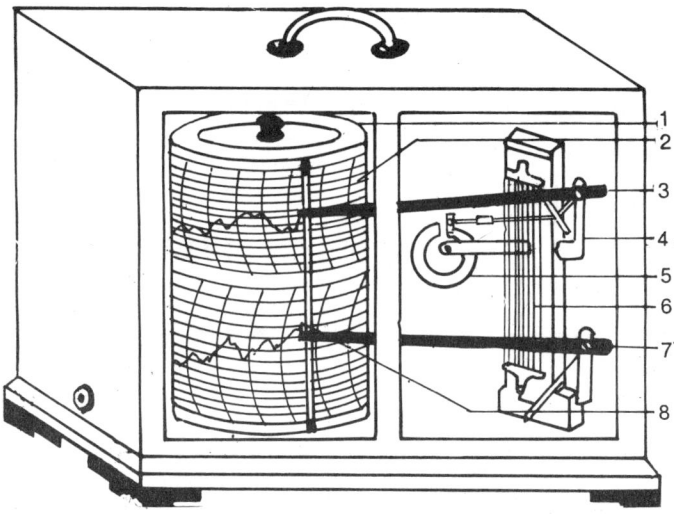

Fig. 3.1. Thermohygrograph, the pen continuously records temperature and relative humidity on graph paper wrapped on a slowly revolving drum, 1. revolving drum, 2. graph, 3. temperature recording pen, 4. lever, 5. temperature sensitive metal ring, 6. hair strings, 7. humidity recording pen, 8. nib.

rectly proportional to the temperature of the surrounding in which the metals are coupled. Thermistor or the *thermally sensitive resistor* on the other hand records the resistance offered to an electric current passing through sintered metallic oxide mixtures. The sealed glass globule is lowered in water to the depth at which the temperature is to be recorded. The potentiometer needle directly gives the temperature. This is a very sensitive instrument and its electrode should be handled with care and preferably kept covered with a soft rubber cap when not in use.

4

Water and Hydrological Cycle

Water is by far the most important substance necessary for life. All the physiological processes take place in the medium of water. Water is the universal solvent and nutrients enter into the plant body in dissolved condition. Thus water helps in nutrient absorption. As an essential constituent of photosynthesis water itself is needed in the manufacture of carbohydrates. Protoplasm, the very basis of life is made up mostly of water.

Water is a very important ecological factor. It regulates the plant environment a great deal. Both the rates of intake and transpiration of water are greatly affected by other environmental factors.

Water is present everywhere. The earth's surface is covered for more than 70 percent of its area by water in form of sea, lakes and rivers. Water is present below the land surface in the form of permanent water table which is used by the deep roots of big trees and by crop plants through artificial means of irrigation from wells and tubewells.

Water is also present in the form of ice on large scale on South and North Poles, on the tops of high mountains and extensive areas of temperate parts of the earth. Plants use this water during the brief summer when ice melts and the plants in such habitats suddenly burst in life.

Based on data from Hutchinson (1957), Odum (1971) has prepared a diagram giving quantities of stored and moving water in the form of hydrologic cycles in terms of geograms (= 10^{20} grams). The oceans have 13,800 geograms, on which 3.8 geograms evaporate annually. 3.4 geograms of water returns in the form of rain and the rest is replenished by run off from land to sea. Ice caps on mountain tops and at the South and North Poles have as much as 167 geograms of water in form of ice. About 0.13 goegram moves in the form of vapour and clouds.

There is rapid cycling and movement of water throughout the globe (Fig. 4.1). The free water from sea, rivers or lakes or other water bodies constantly keeps on evaporating into the atmosphere. Plants absorbs a great deal of water and transpire most of it into the atmosphere. The tremendous quantity of atmospheric water evaporated from the sea moves according to the wind direction and falls in the form of rain. On land this gets distributed some to fresh water bodies, some infiltrates to the permanent water table and some remains in top soil to be partly used by plants. The extra water runs off in the form of streams which converge and join to form rivulets and

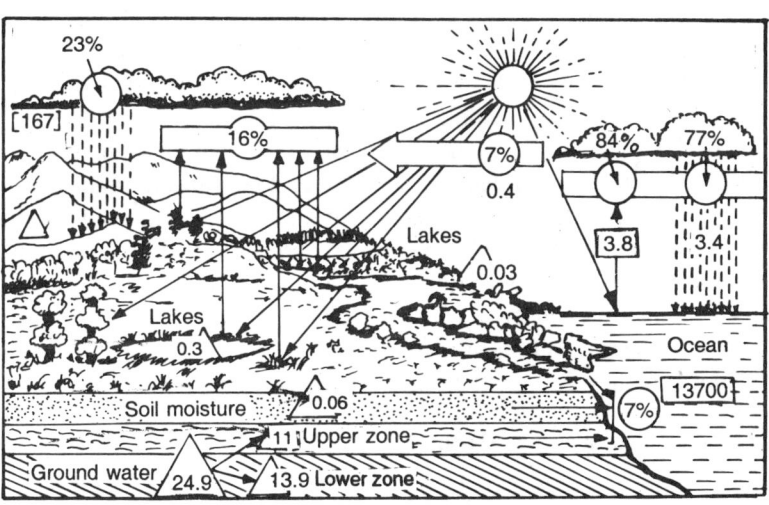

☐ Storage geogram (IO^{20} q)
O Precipitation evaporation and flow percentage
Δ % of fresh water

Fig. 4.1. Hydrological cycle showing water storages in terms of geogram and percentage of freshwater and flow rate for evaporation, precipitation and runoff.

rivers. Water is eventually returned to the oceans. In order to understand the ecological importance of water, therefore, each phase of water in a hydrologic cycle should be studied. About 0.005 percent of the total water is all the time moving in the cycle. The mean annual rainfall on a global basis is 85.7 cm. It is roughly estimated that the land receives 23% and oceans 77% of the total rainfall. Through evaporation and transpiration oceans supply 84% and the land 16% of water vapour to the atmosphere. The remaining 7% difference of land is returned to oceans by surface runoff such as by rivers (Simmons, 1974).

The hydrologic cycle can be well understood by starting from the ocean which has about 87% of the total water supply on the surface of the earth (about 817 million cubic miles or one quadrillion acre-feet. This is fantastic quantity. Solar radiation warms the surface and about 83,700 cubic miles of water in fresh form is evaporated from the saline stock on the ocean every year. An equivalent quantity, of course returns to the sea and the cycle continues endlessly in a balanced way. About 93% of the evaporated ocean water cools and falls back in the form of rain on the sea, and about 900 cubic miles of water moves along with wind currents over land surface and joins the cloud stock of over double this quantity of water obtained through evapotranspiration from land. All these rain down on land surface where the 900 cubic miles of water returns to oceans through rivers and underground flows and the rest about 2100 cubic miles of water recharges the soil only to be evaporated and transpired back into the atmosphere in course of another year. During the annual cycle oceans receive highest quantity of water in October when its level rises by 1 to 2 cm due to addition of about 7.5×10^{18} cm^3 water whereas on land it is highest in March-April (Simmons, 1974).

Atmospheric water and rainfall

In the atmosphere moisture is found in two forms : (a) invisible vapour form and (b) visible cloud or fog form.

The invisible or vapour form of atmospheric moisture is quite an important factor and is usually reported in weather reports in terms of *relative humidity*. Simple *humidity* or *absolute* humidity is the moisture content of the atmosphere. The capacity of the atmosphere to hold moisture in vapour or invisible form is dependent upon the temperature of the atmosphere. It increases with the increase in temperature. The relative humidity is the quantity of moisture as a percentage of the maximum quantity that the air can hold at the prevailing temperature. The relative humidity is dependent upon two factors—the moisture content and the temperature of the atmosphere. Thus, the relative humidity may change without any addition or depletion of moisture if the temperature changes. If the atmosphere becomes warmer the relative humidity decreases and if it cools, the relative humidity increases. In winter we commonly observe, in early morning, the formation of dew. This is due to the cooling of the atmosphere. The relative humidity gradually increases not because of any addition of moisture but because the capacity of the atmosphere to hold moisture decreases with the fall in temperature and reaches 100% or saturation point. Further cooling results in condensation of vapour into water and this temperature/moisture point is called the *dew* point.

Due to cooling of the atmosphere by different physical methods the clouds and fogs form if sufficient quantity of moisture is present in the atmosphere.

When cooling takes place in moist air due to its upward movement in colder belts of atmosphere, the relative humidity increases and clouds form due to condensation of vapour into water or ice droplets. It rains when, due to further cooling or some other physical means, droplets aggregate into drops large enough to fall.

If the cooling of air is near the land or water surface the hazy atmosphere due to condensation of fine droplets is called *fog*. After sunrise the temperature of the atmosphere gradually increases. This increases the vapour holding capacity of air and the fog again vanishes in the air into invisible form. This is a very common experience in the Northern India on winter mornings. The atmospheric moisture, specially in form of cloud, intercepts a good deal of light and heat radiation and thus plants are affected indirectly also by the clouds etc. The atmospheric moisture saturation deficit, that is the degree of wetness or dryness, influences plant life by regulating the rate of water loss in the form of evaporation and transpiration. Parasitic fungi become abundant in moist weather and reduce primary production. Dew is an important source of moisture to plants in the winter season. Often in forests, the floor receives rain from trees in the form of drips from tree leaves due to accumulation of condensed water on winter mornings. Certain plants, specially epiphytes and lichens absorb moisture directly from the atmosphere.

Rainfall or precipitation of atmospheric moisture is the ultimate source of water for plants. Different regions of the earth receive different quantities of rainfall depending upon geographical features and the availability of moisture laden wind. The quantity, duration and intensity of rainfall profoundly regulate the plant life of any place. Even in areas cleared of natural vegetation, agriculture is to a large extent dependent upon the rainfall. Irrigation engineering is only a management of water resource primarily obtained in form of rainfall. The chief cause of rainfall is the cooling of warm air on rising to heights which causes condensation. It may be the *cyclonic type* where a large area of atmosphere moves on in whirling fashion and rises up vertically. In other types rising of air may be due to physical barriers like mountain ranges. In India the rainfall is mostly of a special pattern of moist air movement called *monsoon*. During the air masses moving from the Arabian sea to the west coasts of India and from the Bay of Bengal to the Eastern parts of the country they become extremely moist. The air movement is also governed by the pattern of landmass and hill ranges. Maximum rain falls in Assam, Bengal and the Western coast. The quantity of rainfall decreases from Bengal to Bihar to Uttar Pradesh (Fig. 4.2). Vegetation pattern is closely associated with the quantity of rain. In Assam dense forests are of mixed type with tall evergreen trees.

In temperate and boreal (polar) climatic region melting snow and ice are also important sources of moisture.

Only a part of the rain water is used by plants. The rest is lost in many ways like evaporation and runoff. Thus there is a difference between the actual rainfall and the effective rainfall. Evaporation is governed by the moisture content and temperature of the atmosphere and hence in *effective rainfall* the total rainfall in relation to temperature is taken into consideration.

Surface water

The quantity of water that a soil holds or that infiltrates depends upon the the properties of soil and the type and density of vegetation covering it. For example in bare area the rain drops beat the surface

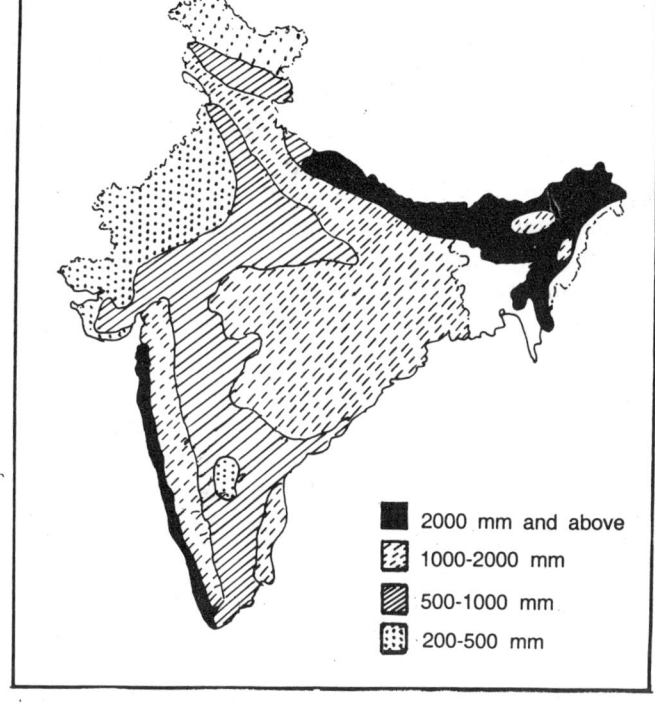

2000 mm and above
1000-2000 mm
500-1000 mm
200-500 mm

Fig. 4.2. Map showing range of rainfall in different parts of India.

lifting soil particles into the air. Lighter or smaller particles are raised highest in each beating and as such the finer particles are selectively deposited on the surface. The clay particles closely fit in the pore space and check water percolation. This results in horizontal movement of water in form of run off resulting in loss of the effective rainfall. The vegetation intercepts the beating effect of rainfall and thus water is gradually soaked in soil where from plants use it over a long period.

The degree of slope is another factor for water loss. On hill slopes the soil is artificially cut and prepared into small flat plots one over the other like steps of stairs for the purpose of agriculture. This retards flow rate of water and provides opportunity for the water to percolate down. Soil erosion is also checked. This type of practice is called *terrace* cropping.

Soil moisture

Plants take their supply of water from soil except in aquatic habitats like ponds and lakes. Soil particles of varying sizes ranging from coarse sand to clay (2 mm to less than 0.002 mm diameter size) are packed together in various proportions. The dead organic matter on decomposition becomes fine powder of organic matter or humus. The size and nature of packing of soil particles in different proportion determine the pore percentage space and its size. These regulate the quantity of rain water that can be soaked, infiltered and held in the soil. The larger pores allow a free percolation of water due to the force of gravity. This is called *gravitational water*. A huge quantity of fresh water is stored in form of ground water. The water saturated layer of ground water is called the *water table*. The moisture held in soil pores of very small size due to capillary action is called *capillary water* and it remains readily available to roots up to a certain soil moisture tension. There is a force of adhesion between soil particles and thin film of moisture and this checks some moisture from evaporation even in dry conditions. This is *hygroscopic moisture*. The pore space not occupied by water contains soil air. The soil air also, like external atmosphere, contains water vapour.

The absorbed water, besides participating in a variety of physiological reactions also keeps the leaves turgid. Atmospheric dryness causes a rapid transpiration and as a result water content of leaf tissues falls and the leaves wilt temporarily. The leaves again become turgid with absorption of moisture. However, when the soil moisture falls below a certain level depending upon the type of soil and plant species, the leaves wilt permanently and any saturation of atmospheric moisture does not make the leaves turgid again unless the soil is quickly irrigated. In plant the actual water content is the difference between the quantity of water absorbed and the quantity transpired and used in metabolism from aerial parts. The ratio between these values i.e. absorption and transpiration is called the '*water balance*'. This water balance is widely different in different species, or in same plant in different seasons.

Like other ecological factors water requirements of different species differ. But in most plants both excessive wetness and excessive dryness are harmful. In soil a balance of water allows a suitable level of soil air. Both moisture and air are necessary.

Depending upon their water requirements or more specifically upon the quantity of water available in their habitats the plants are divided ecologically into three categories : (1) *hydrophytes*, (2) *mesophytes* and (3) *xerophytes*. Hydrophytes are those plants which live partly or fully submerged in water at least for some weeks. Mesophytes grow on land with a moderate supply of soil moisture. Xerophytes are those plants that live in dry or arid land.

Classification of fresh water environment

There are large varieties of habitats whose environment is predominantly characterized by excess of water. They are chiefly divided into two categories, the *marine* and the *fresh water*.

The common freshwater bodies are pools, lakes, streams, rivulets, rivers and marsh-lands. Although a lake or a river is separated from other such aquatic bodies yet on account of uniformity in environmental conditions the plants are also very much the same. In fact hydrophytes of even different continents are nearly the same everywhere.

The aquatic environment has been classified in a number of ecological ways (Fig. 4.3). The standing water bodies are called *lentic* (like lakes and ponds) and running water bodies are called *lotic* (like rivers and streams). The stationary and fast running conditions influence plant growth profoundly. In the latter condition only such forms survive which develop a very effective attaching mechanism, extremely flexible tissues that escape injury from physical force of water current and have streamlined external morphology to offer least resistance. Depending upon the temperature factor the upper warmer region of lakes is called *epilimnion* which is rich in plants, the deep cold almost plantless zone is called *hypolimnion* and the intermediate zone is called *thermocline*. On the basis of productivity and fertility the lakes may be *eutrophic* or highly productive. These are shallow, rich in nutrients and rich in plants. The other type is *oligotrophic* in which the lakes are deep, poor in productivity and low in nutrient concentration. Another type is *dystrophic* characterized by bogs with low calcium carbonate, low nutrients and high humus contents. The zones of aquatic bodies are also classified on the basis of availability of light. The marginal shallow region where abundant light is available

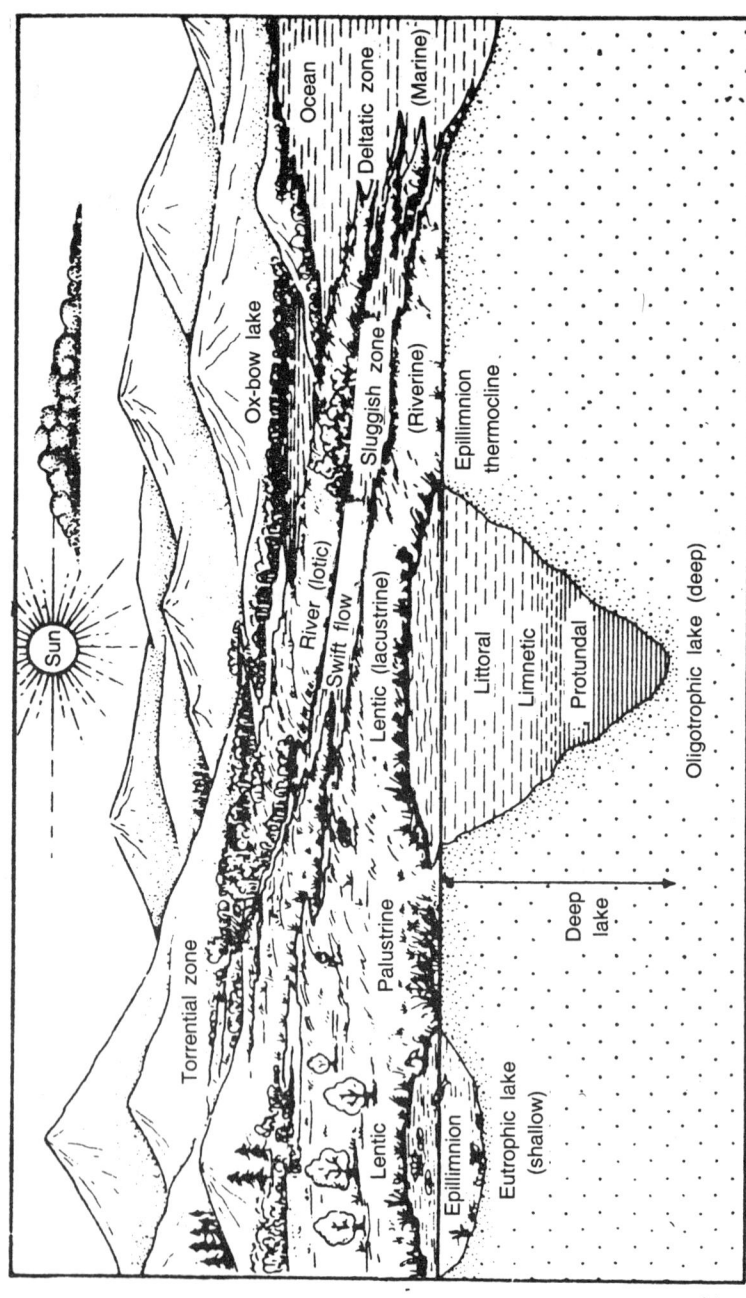

Fig. 4.3. Various kinds of lentic and lotic water bodies with their zonation.

for plant growth is called *littoral* zone, the deeper layer where light condition is just sufficient for plant growth or say at compensation point is called *limnetic* zone. Deeper than this where plant life is not possible due to paucity of light is the *profundal* zone. In shallow ponds and lakes the entire area is of littoral type (Fig. 4.3). The upper illuminated zone is also called as *trophogenic*, the lower dark zone as *tropholytic* and intermediate zone as *compensation depth*.

Temperature of water changes on surface and deep layers with change in season, wind and wave action, convection currents, change in the density of water (temperature regulated). Temperate lakes have warmer surface layers and cold deep layers in summer because the warmed up surface water does not go down. But in winter with progressive cooling the dense cool water travels down until 4°C. When it cools further the downward movement stops since density of water decreases and gradually surface water freezes whereas lower layers remain comparatively warmer in liquid state. Thus in temperate lakes and rivers two inversions in temperature stratification take place in a year and they are called *dimictic*. Lakes in warmer region never cool so much as to cause complete circulation and only one overturn of water is undergone in one year and hence are called *monomictic*. In some warm humid areas there is irregular circulation and this is called *polymictic*.

In marine environment the classification is somewhat different. The shallow shore region is called *neritic zone*. This region is visited by tides that periodically wash the shore and therefore, the *neritic zone* is further divided into upper or *supratidal*, the intermediate region where high tides reach or *intertidal*, and the lower *subtidal* regions. The portion of sea away from land is called the *oceanic zone*. The deeper slopes in continuation of the shore upto a depth of two thousand metres are called the *bathyal zone* still deeper zones in oceanic belt are called *abyssal* zone.

1. HYDROPHYTES

Plants of predominantly wet areas have been classified by Iverson (1936) into (i) *limnophytes* which grow fully in water and (ii) *amphiphytes* which grow in shallow partly inundated lake margins with aerial emergent plant parts. All such plants which usually germinate in water and spend all or part of their life cycle in water are called *hydrophytes. Hejney* (1960) has distinguished three kinds of hydrophytes. (i) *Euhydatophytes* where the vegetative parts are inside water and the floral parts may remain inside or outside the water, (ii) *Hydato-aerophytes* where vegetative parts are partly submerged and partly aerial such as floating leaves and the inflorescence is aerial and (iii) *tenagophytes* for plants that grow in amphibious conditions of wet and dry phases. Luther (1946) has taken the character of plant attachment as the important factor of distinguishing hydrophytes into (i) *Haptophytes* which remain attached to submerged rock surfaces like algae and bryophytes, (ii) *Rhizophytes* which remain attached to substrate and send roots deep inside the mud and (iii) *Planophytes* which are not attached but remain free floating. Microscopic planophytes are called *planktophyes* and larger ones as *pleustophytes*.

Classification of hydrophytes

On the basis of life forms hydrophytes may be broadly classified into *phytoplankton* and the *macrophytes*.

Macrophytes are predominantly vascular plants. They are further divided usually on the basis of their habit and location in ponds or lakes. They may be (1) the marginal emergent plants, (2) the rooted submerged hydrophytes, (3) the rooted hydrophytes with floating leaves and (4) the free surface floating hydrophytes. Some of the common macrophytes are shown in Fig. 4.4 to 4.14.

(i) The *marginal emergent hydrophytes* are found around ponds and lakes where the level of water has gone down considerably and only 2 to about 15 cm of basal part of plant body is immersed and the rest lies straight up emerging in the air. The area of distribution of such plants extends down as the pond water recedes on drying. The common Indian species of this category are (1) *Eleocharis plantaginea,* (2) *E. pallustris, Isoestes coromandelina* and other *Isoetes* spp. (Fig. 4.6), (3) *Typha* sp., (4) *Cyperus* spp., (5) *Fimbristylis*, (6) *Polygonum* sp. (Fig. 4.4), (7) some wild *Oryza*, (8) *Zizania*, etc.

(ii) *The submerged rooted hydrophytes* : From the periphery to the marsh zone, depending upon the size and depth of the pond, different life forms and communities get distributed in different regions. The

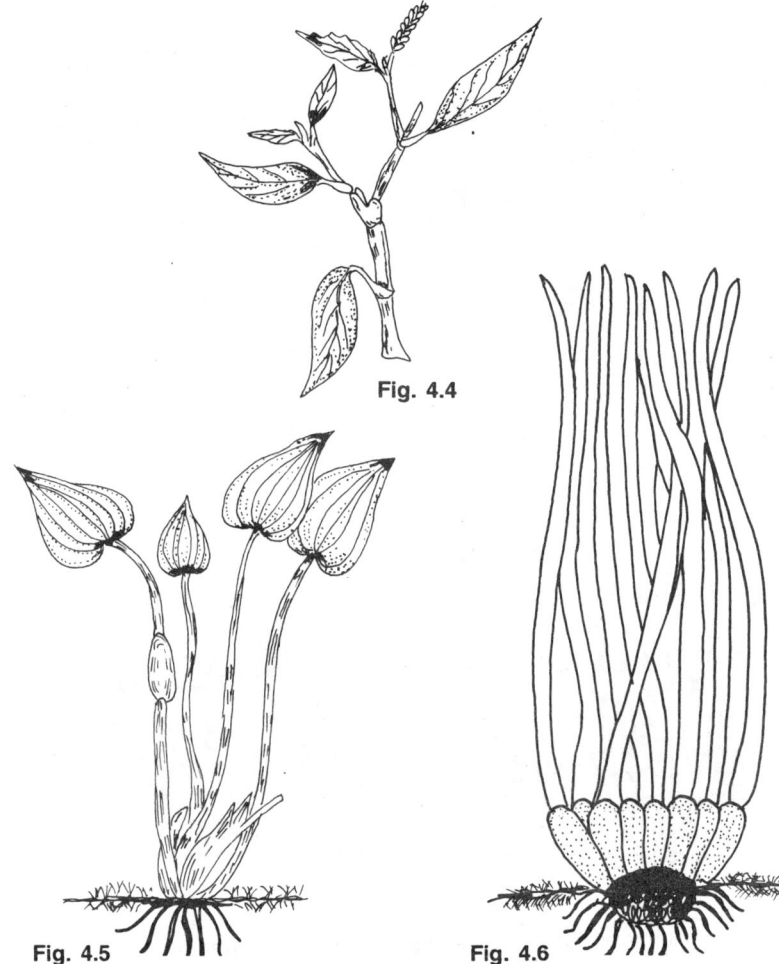

Fig. 4.4

Fig. 4.5

Fig. 4.6

Fig. 4.4 *Polygonum amphibium* and **Fig. 4.5.** *Monochoria vaginalis.* These grow on marshy margins of lakes and have hollow or lacunate tissue in stems and roots. **Fig. 4.6.** *Isoetes* sp. a pteridophyte growing in very shallow pond margins.

rooted hydrophytes which remain submerged are restricted to shallow region where light is abundantly available right upto the bottom. The submerged hydrophytes are divided into :

    (a) Plants with long stem covered with leaves and with roots

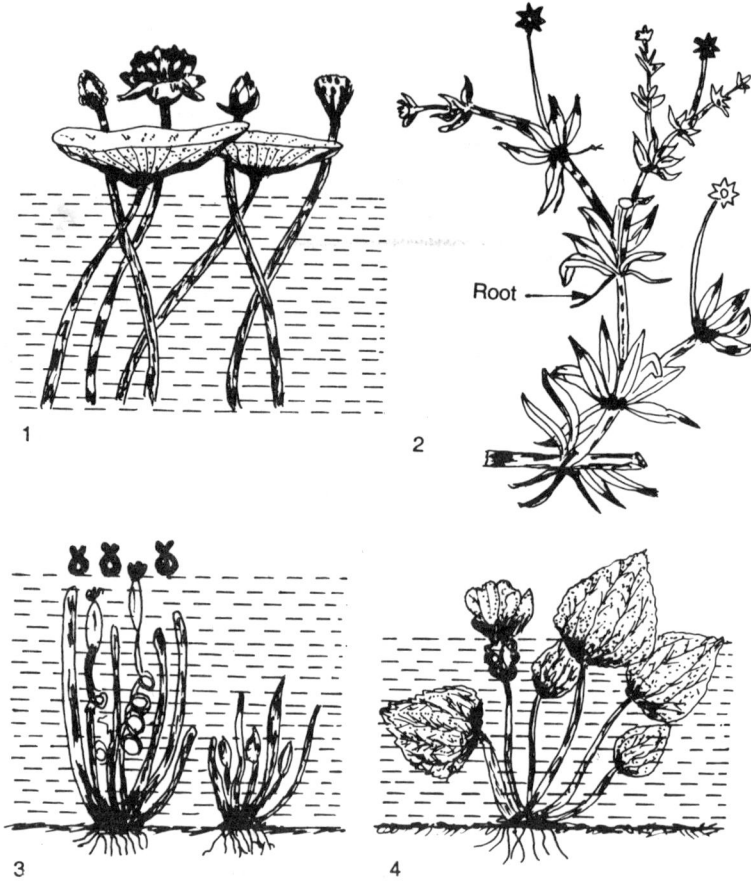

Fig. 4.7. Hydrophytes of shallow freshwaters: 1. *Nelumbo nucifera* with long and flexible petiole and peltate circular leaves and emergent flowers; 2. Submerged plant of *Hydrilla verticillata*; 3. *Vallisneria spiralis* with coiled flower stalks which uncoil and increase in height on increase in water level and 4. Submerged plants of *Ottelia alismoides*.

arising from nodes. Examples are *Hydrilla* (Fig. 4.7), *Lagarosiphon*, *Pontamogeton pectinatus* and *Najas*.

(b) Tuberous stem with cauline leaves e.g. *Vallisneria* (Fig. 4.7), some species of submerge *Aponogeton* etc. In these plants the leaves are ribbon shaped, thin and filmy. In general the submerged leaves do not bear any stomata.

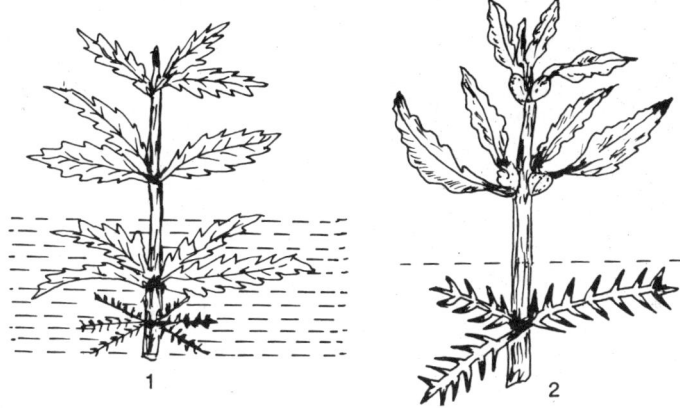

Fig. 4.8. Partly submerged plants of (1) *Limnophila aquatica* and (2) *Myriophyllum* sp. showing heterophylly. Submerged leaves are thin and hairy whereas aerial leaves are broad.

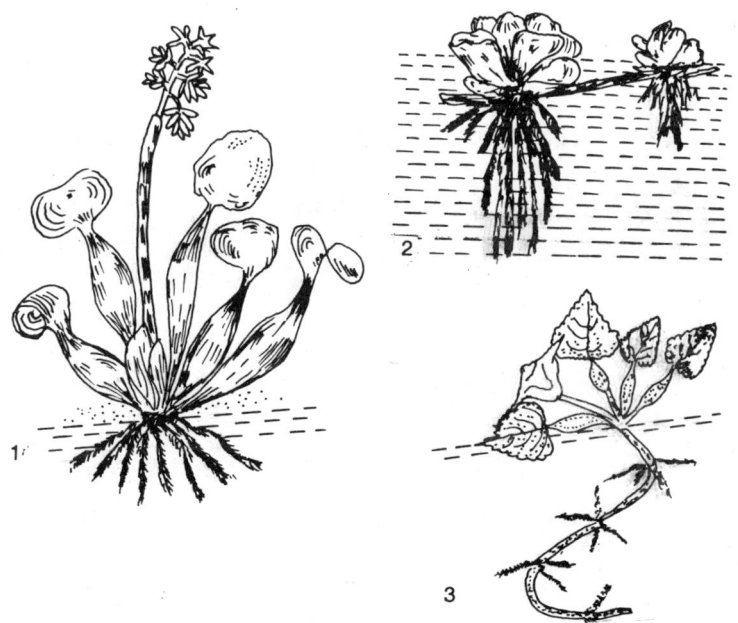

Fig. 4.9. Free floating (1) *Eichhornia crassipes*, (2) *Pistia stratioites* and (3) rooted *Trapa bispinosa* showing swollen and spongy petiole and prominent roots.

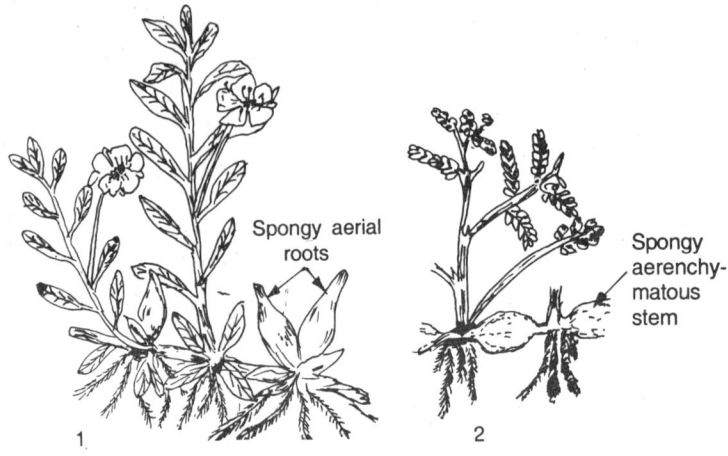

Fig. 4.10. (1) Stems running over the surface of water, producing spongy roots in *Jussiaea repens* and (2) spongy stems in *Neptunia* sp.

Fig. 4.11. Fresh water community dominated by surface floating leaves of *Nymphaea*. Flexibility of petiole beneath the water surface helps in adjustment of leaf lamina in the available space on the water surface.

(iii) *The rooted hydrophytes with floating leaves* : This type is also restricted to more or less shallow region of pond or lakes say up to three metres depth. The leaves emerge above the water surface and float. This group of plants may be

Fig. 4.12. *Victoria* sp. growing in a pond of Botanical Garden at Bogor (Indonesia). The big circular leaf found in this plant, supposed to be the largest among the plant kingdom, floats on water surface due to combined effect of peltate shape; strong mechanical tissues radiating out the petiole, upturned margins and air spaces (photo: R.S. Ambasht).

fixed in mud by rhizomatous structures with floating leaves on long flexible petioles e.g. *Nymhaea stellata* (Fig. 4.11), *Nelumbo nucifera*, (= *Nelumbium speciosum*) (Fig. 4.7). *Aponogeton* sp. etc. or (ii) by a trailing type of stoloniferous structure where leaves are fixed on short petioles e.g. *Nymphoides* and *Potamogeton natans*. In *Victoria amazonica* a giant waterlily plant (Fig. 4.12)· the leaves are largest in plant kingdom. They are supported on water surface on a strong petiole from which mechanical tissues intermixed with lacunate tissues radiate out in the circular lamina. The upturned margins are like the rim of a plate.

(iv) *The free floating hydrophytes* : These occur scattered everywhere in ponds and lakes and keep on changing their position due to water or wind currents. Usually in small ponds such species are aggregated more in the centre where other categories of hydrophytes are few. In form and structure free floating hydrophytes are (i) large with rosette of big leaves floating above the water in the air as in *Trapa bispinosa* (water chestnut) and *Eichhornia crassipes* (water hyacinth) (Fig. 4.9). The roots are also abundant and long, with prominent root pockets. (ii) Slightly smaller forms are *Pistia* and *Salvinia* with spongy floating leaves. (iii) The other variety is of very small surface floating plants with the body reduced to thallus like structures e.g. *Lemna, Wolffia, Spirodela* and *Azolla*. Some species may remain just below the surface water or come up periodically and sink down for perennation e.g. *Utricularia*.

Ecological adaptation in hydrophytes

All features of specialization that help hydrophytes to adjust to the direct and indirect effects of excessive water all around the plant body and to other concommitant adversities in environment are referred to as ecological adaptation. The adaptation may be structural, in the external (morphological) and internal (anatomical) body organisation or may also be functional (physiological). It is more convenient to describe the adaptations in hydrophytes for different aquatic habitats.

1. Adaptation in emergent species

In the emergent species growing on marginal areas on shore the under-mud parts of stem and root suffer from paucity of oxygen. Plants growing in water logged soil, also experience these conditions. Respiration is an essential function of all living cells and tissues and oxygen is needed for aerobic respiration. In marsh plants like *Oryza* sp. (rice) the roots have also developed the physiological adaptation of respiring anaerobically (without the use of external oxygen). The aerial, under the water or mud covered plant portions have aerechymatous tissues in gaseous continuations as in *Cyperus rotundus* (Ambasht, 1964) and *Cladium mariscus* (Conway, 1937) through which all the parts receive a supply of oxygen absorbed from the air.

In many plants which normally grow in mesophytic condition but get partly submerged for few weeks in the rainy season, new roots showing hydrophytic characters may develop. In *Saccharum benghalense* the old roots with abundant aerenchyma are formed when plants remain under water or in muddy conditions.

In many plants which normally grow in mesophytic condition but get partly submerged for few weeks in the rainy season, new roots showing hydrophyic characters may develop. In *Saccharum benghalense* the old roots with abundant aerenchyma are formed when plants remain under water or in muddy conditions.

In some species like *Xanthium strumarium* growing in partially submerged condition the roots show dimorphism. The roots are rather strong and tough and spread in mud extensively. The adventitious roots that develop on submergence float freely in water and sometimes get covered with epiphytic growth of green alga like *Oedogonium*. The photosynthetic oxygen bubbles of the algae settle on the *Xanthium* roots and may thus help in oxygenating the system (Ambasht, 1962).

2. Adaptation in free floating species

The free floating species usually grow in water are fairly rich in nutrients. The necessity of root for absorption and anchorage does not arise in many floating species since the entire plant is in contact with water. An erect system like a stem is also unnecessary for such a life form. In most free floating plants like *Wolffia, Lemna, Azolla*, etc., the root and stem are very poorly developed. Leaves are the most characteristic and well developed parts which do most of the function of a full plant like absorption, maintenance of high buoyancy due to abundant aerenchyma, efficient photosynthesis and vegetative reproduction. However, some of the very fast growing free floating species like *Eichhornia crassipes* have extensive root growth. These roots too have abundant lacunate tissue full of air spaces (Fig. 4.13).

In many species where some leaves are submerged and some aerial, then show a different shape and size to suit the situation. Polymorphism or different shapes in the leaf of the same individual is called heterophylly (Fig. 4.8).

3. Adaptation in other hydrophytes

Besides emergent and free floating species many other types of aquatic plants occupy different positions and these also have a wide variety

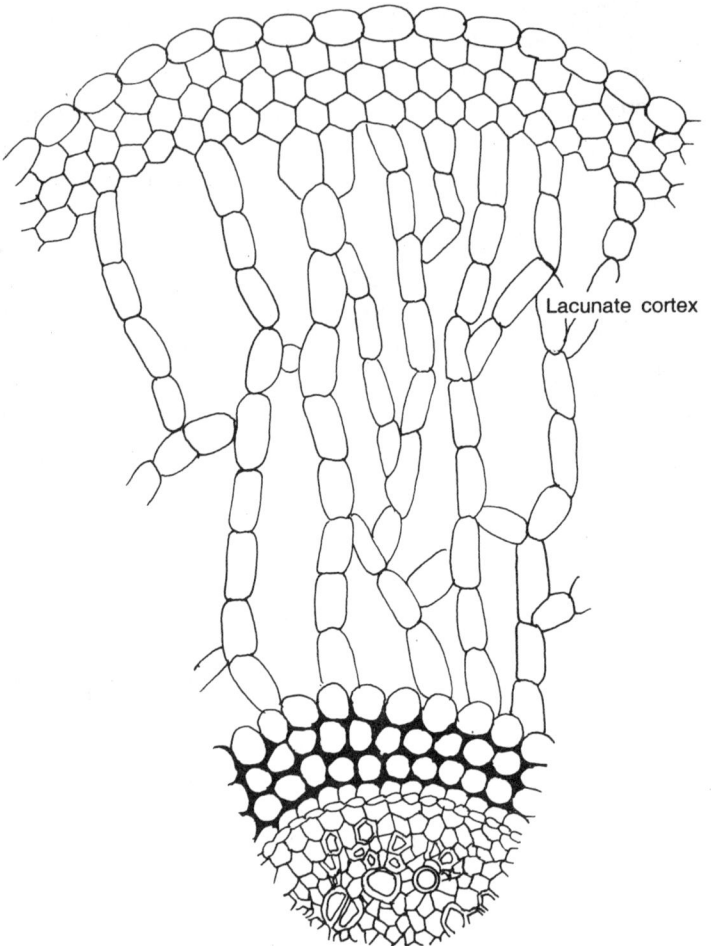

Fig. 4.13. T.S. of the root of *Eichhornia crassipes* showing extensive development of air filled lacunate cortex.

of adaptational features. In some plants their basal parts are firmly fixed in mud and some parts are above the bottom mud occupying submerged or above water surface position.

Adaptational features of under-mud parts

The chief adaptational features of under-mud parts such as roots, rhizomes or bases of shoots are against an unstable substrate and against an upward pull by the above mud parts on account of their buoyancy. Therefore, the under-mud parts are usually well developed and heavy. The total weight of these parts often exceed that of the rest of the plant body and in some species of shallow region like *Equisetum flaviatile* and *Phragmites communis* this weight may be about five times that of the above ground parts.

Adaptation in submerged organs

In submerged plants the adaptations are chiefly in response to high density of the medium (i.e. water) and low light intensity. The leaf and other shoot organs are provided with cavities or hollow spaces called lacuna. The lacunate tissue is filled with air which provides buoyancy to the plant. With the aid of buoyancy the shoot tends to pull up and thus the plant maintains its expanded shape by the aid of air filled tissues. In this way the necessity of mechanical tissues such as sclerenchyma or hard secondary wood, etc. to keep plant erect and in shape does not arise. Therefore, the presence of lacunate tissue filled with air and absence of mechanical tissues are two chief features of adaptations (Fig. 4.14 and 4.15). Leaves are generally thin and filmy. Chloroplast is abundant in equidermal cells because light intensity is low. The stomata are usually absent or if present they are functionless. The function of stomata in land plants is to facilitate gaseous exchange for photosynthesis and for water loss by way of transpiration. In submerged plants the question of transpiration does not arise. Carbon dioxide is also available in dissolved condition or additionally in form of bicarbonate ions. These forms of carbon can pass in the plant body directly from epidermal cells. The chief adaptational features in these respects are absence of cuticle, abundance of chloroplasts in surface cells and absence of stomata. Another physical problem in submerged condition is the mechanical force of water current. Many species show dissection of leaves into thin segments, or undulating margins or *fenestration* which means formation of a number of holes or open rings in the leaf through which water current flows and the leaf escapes damage. Dissection of leaves reduce the chance of physical damage from water current and increases surface area for a more efficient absorption of materials and increased photosynthetic activity.

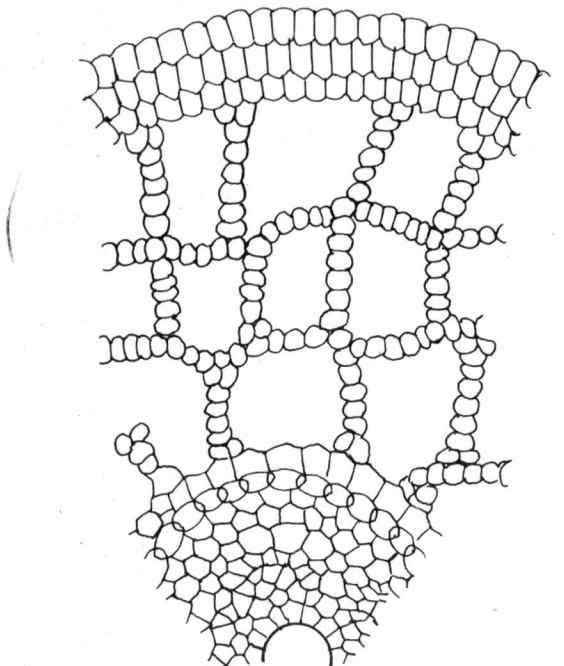

Fig. 4.14. T.S. of stem of *Hydrilla*. Absence of mechanical tissues, reduction of vascular tissues and abundance of lacunate tissues are chief features of adaptation.

Adaptational features of floating leaves

The floating leaves in most species are quite similar in shape and features of adaptation. The leaves are peltate, usually circular in shape and have a strong leathery texture. The upper surface of the leaf is such that water glides off and does not adhere. The leaf margin is strong enough to resist tearing due to pressure of wind or water current. The petiole is quite flexible so that the leaves may move a little this way or that and adjust the leaves in the available space on the water surface. the lower surface in touch with water is usually deficient of stomata. The upper leaf surface has stomata for easy exchange of gases. Anatomically, the lower side of the leaves is usually lacunate or airfilled. This helps in keeping the leaves afloat. The upper side is full of chlorenchymatous tissue. Despite abundance of air tissues the leaves maintain their shape due to presence of supporting

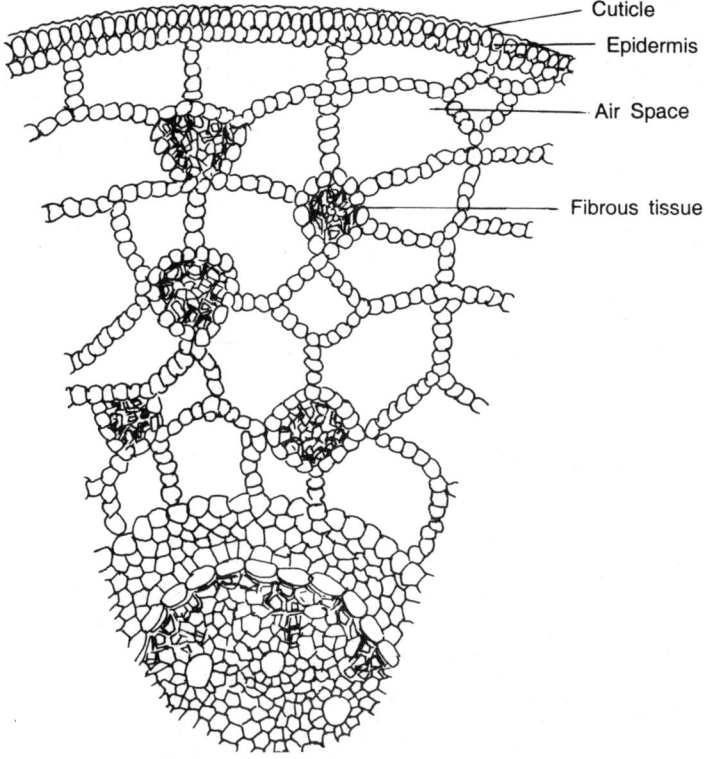

Cuticle

Epidermis

Air Space

Fibrous tissue

Fig. 4.15. T.S. of the stem of *Potamogeton natans* showing profuse lacunate tissue with supporting fibrous bundles.

mechanical tissues or chemical crystals like *sclereids* (Fig. 4.16). Against fluctuating water level, the petioles show a remarkably fast rate of elongation so that surface floating leaves do not suffer from submergence for a long period when the water level rises.

(2) MESOPHYTES

Mesophytes are plants that normally grow in habitats where water is neither scarce nor abundant. In such habitats the pore space in soil is occupied almost equally by water and soil atmosphere. Such a balanced condition of water and gases is very suitable for plant growth and hence in mesophytic condition, we find the best growth of for-

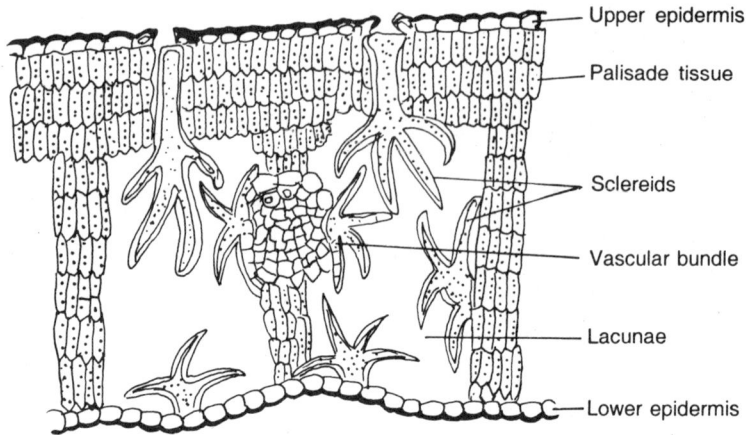

Upper epidermis
Palisade tissue
Sclereids
Vascular bundle
Lacunae
Lower epidermis

Fig. 4.16. V.S. of *Nymphaea* leaf. The leaf remains on the water surcace. The lower lacunate tissue affords buoyancy, the sclereids provide mechanical support to upper heavier layers of chlorenchymatous tissues that perform the bulk of photosynthesis being in contact with atmosphere and direct sunlight.

ests or crop plants. Both excessive water and excessive dryness create a large variety of situation adverse to plant growth and only a few species, specially adapted to such environment can grow there. However, in mesophytes, no such adaptation is necessary unless the habitat is specialized in some other way.

Mesophytes are very extensive on the surface of land and most crops like wheat, maize, barley, pea, gram or sugarcane or species in grasslands, meadows, tropical and temperate forests are all mesophytes. The roots in mesophytic species are quite well developed and young roots are abundantly covered with hair. There is a balanced proportion of mechanical tissues in roots as against their absence in hydrophytes and excessive predominance in xerophytes. The same is true for stems. The leaves are well developed, usually broad and only moderately thick. Stomata are well developed and present usually on the lower surface. The chloroplasts are also abundant and usually dark green in colour.

(3) XEROPHYTES

Xerophytes are plants that usually grow in extremely dry or arid conditions. There is no hard and fast line between the water status as habitats for meso- and xerophytes and the term is only a relative one. There is a considerable degree of controversy and difference of opinion on the definition and classifications of xerophytes. Some authorities regard only those plants which actually suffer from deficiency of growth water as xerophytes. Others maintain that all species that grow in normally arid or dry habitats whether or not they suffer from deficiency of growth water in their substrate should be broadly placed under xerophytes. Deficiency of water affects adversely, both directly and indirectly the normal function and structure of plants. Xerophytes or plants of normally dry habitats may be of a large variety from the standpoint of their life-form and structural and functional adaptation. Unlike hydrophytes, which grow only in free water, the xerophytes do not always show preference for dry habitats for a good growth. Xerophytes are only adapted to grow in dry habitats and assume characteristic dominance in desert, etc. due to absence of other types of plants. But when better moisture conditions become available, the production rate and total biomass increase in xerophytes. The dry habitats for the purpose of water availability to plants may be of two types—(i) the actually dry habitats and the (ii) wet habitat where from water is not available to plants on account of high osmotic concentration of salts or freezing temperature, etc. From the standpoint of habitat features the xerophytes may be of the following types :

  (i) *Psammophytes* are the plants that grow on sand or small pebbles usually in dry habitats like *Acacia senegal, Alhagi camelorum*, etc.

 (ii) *Lithophytes* are the plants that grow on rocks where even in good rain, water does not stay. Species of lichens, *Linaria ramosissima, Selaginella* sp. and certain Euphorbias are common examples.

(iii) *Psychrophytes* are plants that grow in soil which become very cold and the availability of water is little on account of frequent freezing. In India, such plants are restricted to the Himalaya.

(iv) *Halophytes* are the plants that grow on highly saline soils where water is poorly available to plants on account of high osmotic concentration in such habitats. Mangrove plants in the Sundarbans of India show various types of adaptations. Common examples are *Brugeièra polyrrhiza, Avicinia* sp., *Rhizophora* sp.

All xerophytes have to endure or escape the periodic cycle of actual or physiological dry phase (or drought) and this is done in a variety of ways.

1. Plants that escape dryness in external and internal environment

In several types of habitats with a prolonged dry season and short wet season there grow a variety of ephemeral annuals that germinate, grow, flower and produce seeds with a brief wet period of a few weeks. With of the onset of the dry season the seeds become dry and dormant and remain buried in soil for the rest of the period. The seeds again become activated with the first showers of next wet season. Thus these plants although, growing profusely in relatively dry lands have their biological clock so adjusted as to complete the life cycle within the short wet season. They do not actually suffer from dryness during their growing phase. The dry phase is passed as seed or other propagule (diaspore) stage. Therefore, these species are grouped as drought escaping. Drought escaping ephemeral annuals are found in the warm sandy deserts with a short rainy season, or on higher altitudes of the Himalaya, or part of cold temperate regions where snow melts for a short period in the summer season only. The main ecological adaptation of such xerophytes is the ability to complete their life cycle within a short period of six to eight weeks or so. Further these annuals can withstand aerial dryness to a good extent. The root system of these plants is usually more extensive than the shoots which remain very small. Although native of arid lands these xerophytes do not suffer from paucity of water either in their external or internal environment.

2. Plants that suffer from dryness in external environment only

Another type of xerophytes commonly found in deserts is of succulent plants or simply the *succulents* like several cacti and euphorbias (Fig. 4.17 and 4.18). These display an important type of adaptation shown by several diverse groups of plants growing in deserts through-

Fig. 4.17. Succulent xerophyte—*Opuntia*, the stem is flattened and the leaves are reduced to spines. The succulent tissues have abundant mucilage.

out the world. Succulence results in a high capacity of tissues to absorb and retain water despite dryness in the external environment. The parenchymatous cells which are thin walled enlarge considerably and their chemical make up is such that they quickly take in a large quantity of water in the wet season. This results in swelling up of the tissues. Water is held tenaciously and loss of water is very slow. Thus succulence results in a quick storage of water during brief rains and slow utilization and little loss of this important material during the rest of the season. This type of adaptation overcomes actual water deficiency in the plant tissues and, therefore, many ecologists are inclined to regard the succulents as not true xerophytes. However, the fact that succulents grow characteristically in deserts and grow there all the year round is sufficient to put them in this group. Succulence in stem is common in Cactaceae and Euphorbiaceae and in leaves of Agavaceae. In roots, it is found in *Ceiba parviflora*.

Fig. 4.18. *Euphorbia* sp. a cactus like xerophyte (cactoid) with succulent stem. Latex is abundant inside the stem.

The major features of adaptation in succulents towards xeric conditions are as follows :

The root system is characteristically shallow. It branches uniformly on all sides and secondary roots are laterally very extensive but extend only a few centimetres below the soil surface (Fig. 4.19). This type of extensive shallow root spread is of great significance in maximising the efficiency of water absorption. In deserts the permanent water table is normally very deep and soil remains dry up to a few metres. The sporadic and light showers saturate the top few centimetres only and thus extensive surface spreading roots exploit to the maximum these moist zones of the soil. Some plants also show succulence in roots.

The stem in most of these plants becomes swollen and succulent. Many of these belong to the families Cactaceae, Euphorbiaceae and Asclepiadaceae. The Cactaceae plants as a whole are called *cacti* while

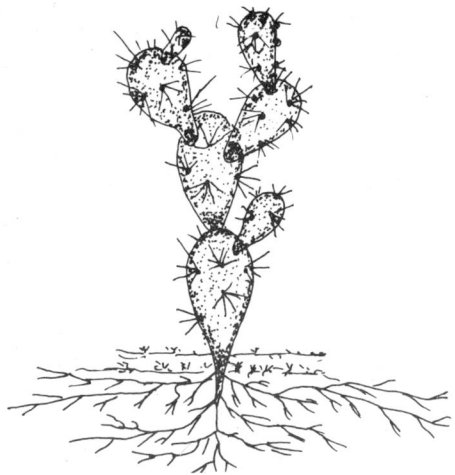

Fig. 4.19. Diagram showing shallow but laterally extensively spreading roots of *Opuntia* in desert condition.

other succulent cactus-like plants not belonging to this family are called *cactoids*. The succulent stems are usually green and do most of the photosynthesis. In some like *Opuntia* sp. the stem is flattened and all the function of the leaf is performed by the stem. Leaves in *Opuntia* are reduced to spines. The green tissues have stomata upon their surface. A very striking adaptational feature is that stomata in many of these plants close in day time when otherwise water loss through stomatal transpiration is likely to be very high. The carbon dioxide requirement for photosynthesis during the day when stomata are closed is supposed to be met from internal carbon dioxide produced in anaerobic respiration of organic acids. The stem epidermal surface is covered with waxy material which prevents water loss by way of cuticular transpiration.

In many desert plants like *Agave* and *Aloe*, the leaves become thick leathery or abundantly succulent. In these the stem is reduced. Here too the rate of growth is very slow. In many plants like certain euphorbias leaves fall with the onset of the dry phase or in many cacti, the leaves are reduced to spines. In deserts or dry lands, paucity of water results in sparse growth of plants and this results in greater grazing pressures on these plants by herbivorous animals.

Absence of broad leaves and abundance of spines further protect the xerophytes from the ravages of grazing.

Succulent plant parts retain the capacity to regenerate and grow in a favourable season even after being detached from the parent plant for several months or years. Succulents are in fact among the best adapted plants of deserts to resist drought conditions by their ability to build sufficient buffer stocks of water in their body and by reducing the water loss to bare minimum.

These features result in a better ability for them to compete with other plant species. In fact *Opuntia* when introduced in India from the New World became so aggressive as to infest cultivated fields. Its quick vegetative regeneration and continued growth even in summer made it impossible to eradicate physically the noxious weed. Ultimately an insect *Cactoblastis cactorum* had to be imported which fed on this plant. This biological control is now successful in checking its spread.

Succulent xerophytes suffer from deficiency of water in extermal environment only.

3. Plants that suffer from dryness both in internal and external environment

Plants of this type are generally referred by many ecologists as true xerophytes. These are non-succulent perennial plants that actually suffer from water scarcity both in their external as well as internal environment. Herbs like *Alhagi camelorum, Solanum xanthocarpum, Saccharum benghalense*, shrubs like *Capparis sepiaria, Capparis aphylla, Zizyphus numularia, Calotropis* sp., *Salvadora* sp. and small trees like *Acacia senegal, Acacia arabica, Prosopis spicigera, Prosopis juliflora* are some of the common Indian xerophytes of this category.

Special features of adaptation in all these plants and plant parts are for a better water economy. These features can be conveniently described taking roots, stems, leaves, etc. separately.

Adaptations in roots : The root system in the majority of such perennial xerophytes are very deep. The main tap root grows straight down to several feet and there is a big disproportion between root length and shoot height. The aerial part of *Alhagi camelorum* may be only a few centimetres in height but its roots penetrate several metres to the moist zone (Fig. 4.20). Similarly shrubs and small trees like *Prosopis cineraria* (= *P. spicigera*) may have roots reaching to more than 15 metres. In fact actual xerophytic roots need detailed investigation with regard to their spread and depth. Another feature of root adaptation is that in many species the roots have the capacity of perennate and regenerate even when the shoot is dead or cut away by grazing animals. This feature helps in continued growth of the species in dry situations because regeneration through seed is possible only in the wet season. The third feature is that roots in many species are covered with a thick layer of cork-like tissue which protects the internal living cells against hard, dry and hot soils. Rate of tap root elongation is one of the fastest in these plants which greatly facilitate the young tip region to remain in moist soil in deeper layers. Many plants show high osmotic concentration also in their root tissues.

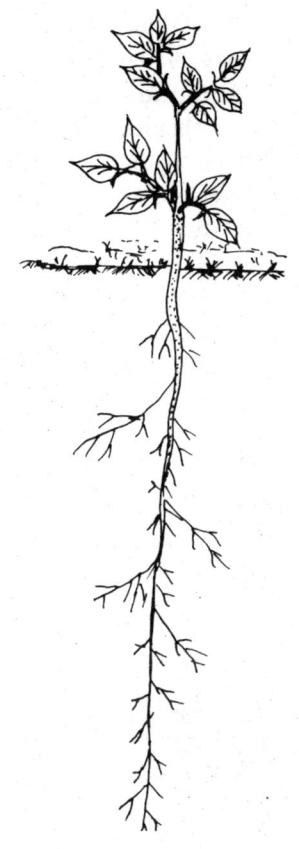

Fig. 4.20. *Alhagi camelorum*, a common weed of sandy soil. The roots go down to a few metres.

Adaptational feature in stems of non-succulent perennial xerophytes are not of many diverse types. These are essentially slow growing on account of nonelongation of cell and their compact arrangement. This reduces the total surface area. The stem is usually covered with corky layers which reduce water loss. Presence of gums and resins in many of these plants as in species of *Acacia*, also helps in formation of layers which prevent loss. Many bushy xerophytes of this category are covered with spines which incidently provide protection against grazing animals. The plants are widely spaced and the rate of growth is slow due to paucity of water. It has been clearly demonstrated that many deserts if provided with adequate water supply become highly productive. The general shape of stunted main stem with a several thin bushy branches is very much the same in taxonomically widely different genera.

Most of the stems of this category of plants are hardy, thin and full of mechanical tissues. In *Casuarina* tree modified stems are photosynthetic with the stomata situated in sunken position in the furrows (Fig. 4.21). There are abundant hairs in the furrows which reduce water loss by way of transpiration.

The leaves are usually small and fall during the prolonged dry season. Most of the features of adaptations shown by leaves are towards reduction in the rate of transpiration and this is specially true when available moisture is scarce. As against this the succulents usually transpire at slower rate even when moisture is abundant. The perennial non-succulent leaves withstand for a longer time at permanent wilting level and recover even after a long interval of drought, moisture is again available.

Many plants that do not shed their leaves have many structural adaptations. These include waxy coating on the epidermis, vertical orientation of palisade tissues, development of sclerenchyma (these

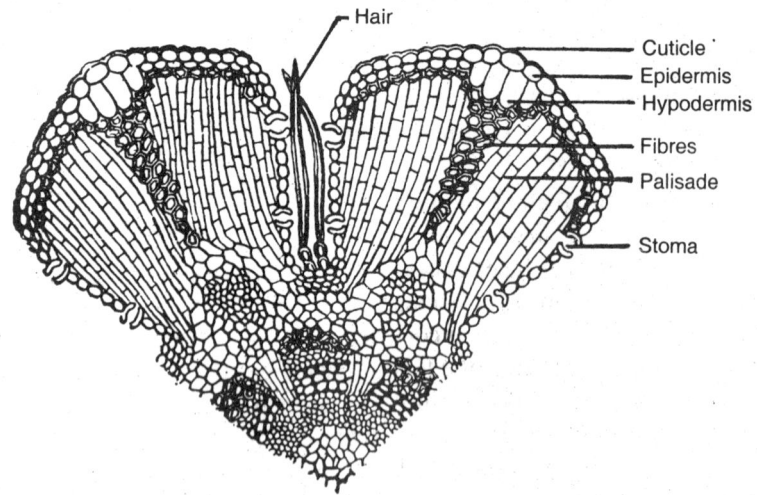

Fig. 4.21. T.S. of the stem of *Casuarina equisetifolia* showing strong xerophytic characters such as prominent growth of mechanical tissues, sunken position of stomata and prominent hairs.

plants are called sclerophyllous), presence of hairs on the surface and presence of cavities or pits on the lower surface and presence of stomata sunk within these cavities, etc. The waxy or cutinized surface forms a waterproof layer and checks cuticular transpiration. Further, it protects the leaf against physical injury and drooping at the time of permanent wilting. By the vertical orientation of chlorophyllous palisade tissue the chloroplasts receive sunlight at an angle and this is advantageous specially on bright hot sunny days. Sclerenchyma development maintains the rigidity of shape in the absence of adequate water in the tissues.

The role of hairs covering the leaf surface has been attributed to their forming an insulating layer between moist leaf epidermis and dry atmosphere. The atmosphere in the pits on the lower side becomes saturated due to some transpiration through the stoma located in them. This air in the saturated pit remains protected against the current of dry wind because of the coat of epidermal hairs and the sunken position of the stomata. An atmosphere of saturated or moist air in the cavity outside the stomatal pores reduces the rate of transpiration much as a wet cloth does not dry rapidly in moist air in the rainy season. Such sunken cavities are also found in *Nerium* leaves (Fig. 4.22).

In some grasses leaves roll up like a ribbon in a dry season such as in *Ammophila* (Fig. 4.23). Some species of *Selaginella* also show this habit. An open expanded leaf is effective in photosynthesis and dry matter production in a favourable season whereas rolling in dry season reduces the surface area of the transpiring part and protects the greater part of the leaf from direct light. In desert plant species the leaves die in the dry season but do not fall and instead cover the plant and protect it from direct bright light by providing shade. The cover of dead leaves, further acts as clothing or insulating material to the plant against dry wind.

Microphylly that is reduction in the size of leaf blade has been regarded by many as an effective way of conserving moisture because of the reduction of transpiring surface area. Experimental studies show the opposite result as in microphyllous plants the number of leaves increases and the total surface area per unit volume of space is rather high. However, another advantage that has been attributed to

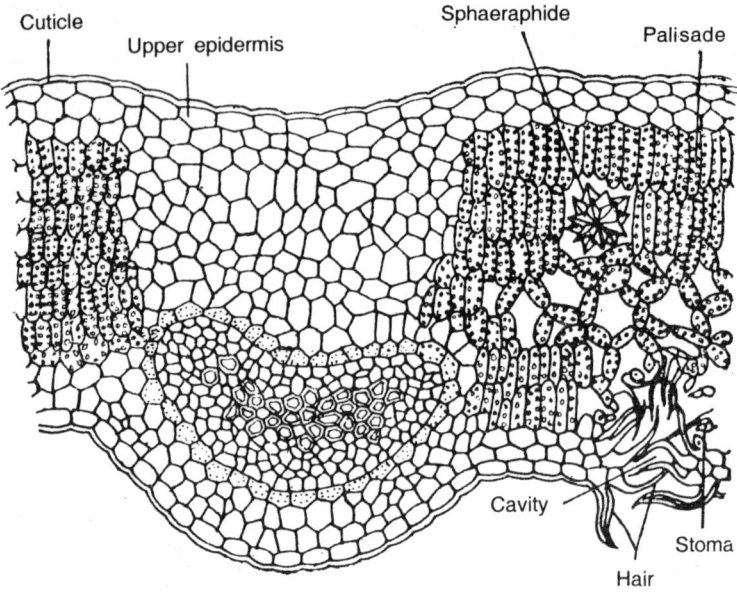

Fig. 4.22. V.S. of *Nerium* leaf showing xerophytic characters such as sphaeraphides; cavities on lower surface, sunken position of stomata and hairs.

microphylly is that smaller leaves do not get overheated in bright sun as do broad leaves.

Old ecological literature is full of descriptions of structural adaptation of numerous 'ypes in xerophytes directed towards conservation of moisture. Almost every morphological modification has been associated to some physiological feature towards water economy. But recent experimental approach in quantitative

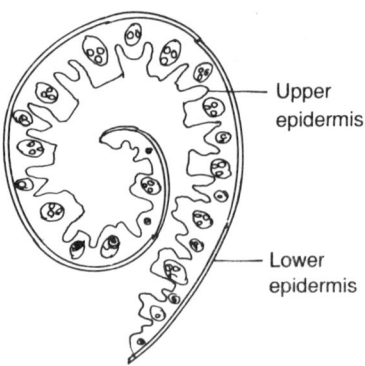

Fig. 4.23. Rolling of leaf in *Ammophila arenaria*. This prevents transpirational losses of water.

estimation of water loss, etc. reveals that a number of speculative explanations were incorrect. The perennial nonsucculent xerophytes transpire quite rapidly if ground water is available and such plants are there in deserts or arid areas because of their ability to remain inactive in long periods of drought.

Dehydration and drought lead to shrinkage of protoplasts, change in the concentration of solution in the cell and in extreme conditions damage the biomembranes and denature the proteins. Plants adapted to drought show more extensive roots to exploit better the scarce soil moisture and reduction in the water loss through transpiration, hairy growth on leaves and leaf shedding during dry phase. At biochemical level Abscicic acid (ABA) plays important regulatory functions including in the stomatal movement. In the CAM plants adapted to desert climate the stomata close during day time and it reduces or stops transpiration while during night the stomata are open. For photosynthesis CAM plants get CO_2 from internal sources. Malic acid stored in vacuoles is released and on reaching chloroplast is decarboxylated by malic enzyme and then the CO_2 is fixed by Rubisco during day time. Thus the CAM plants have a much better water use efficiency than the C_3 and C_4 plants.

HALOPHYTES

A very specialised category of plants which suffer from water deficiency mainly on account of high content of salt in the soil, mud or free water which check water absorption by plants is called halophytes. Therefore, this category of plants is in physically wet but physiologically dry (i.e. water not easily absorbable) condition. In the delta and estuaries of rivers and at many intertidal belts on sea coasts a specialised kind of halophytes grow. These are called as mangrove plants. Among the dry and desert salt halophytes are *Tamarix dioica, Suaeda fruticosa* and *Cleome brachycarpa*. Among the mangrove plants are *Rhizophora mucronata, R. conjugata, Avicennia officinalis, Acanthus, ilicifolius, Ceriops candolleana, Bruguiera gymnorrhiza, Excoecaria agallocha, Sonneratia, acida*, etc. In India mangrove vegetation is mainly confined to coastal areas of Kerala, Maharashtra, Tamil Nadu, Andhra Pradesh and Bengal. These plants show a number of adaptational features, both structurally and physiologically. Succulence is the more common feature accompanied with very high osmotic concentration of the cell sap that enables the plants to absorb and retain necessary quantity of water. In the under-mud layers highly anaerobic condition prevails and in some cases aerial respiratory roots come out of the marsh. These are called *pneumatophores*. High salt content, tidal waves and poor oxygen are harmful for seed germination and young seedings.

Halophytes are plants that grow in high salt content in soil or overlying water, particularly the common salt - sodium chloride while other non-salty habitat plants which cannot tolerate high salt are called *glycophytes*. Sodium and chloride ions are aggressive osmolytes. Halophytes have developed mechanisms of excreting or eliminating salt from the cytosol to the surface of leaf or in specially formed large vacuoles, or bladder hairs. Drought stress tolerance is achieved by certain genes which produce the desired protective proteins.

Some mangrove seeds germinate while still attached with the parent plants in the aerial environment and fall down in such a way as to get readily fixed in mud after attaining a certain size. The process is called *vivipary*.

Various aspects of mangrove and estuarine vegetation in South-East Asia have been described in an edited publication by Srivastava, Ahmad, Dhanarajah and Hamzah (1979).

Measurement of rainfall

The quantity of rainfall is usually expressed in terms of millimetres or centimetres per year. A rain gauge is a simple instrument used to measure rainfall. It consists of a drum, a cap attached to a funnel and a measuring cylinder (Fig. 4.24). It is permanently fixed in the open where every morning usually at 8.00 the cap is removed and quantity of rain water in the graduated cylinder which has entered in the previous 24 hours is noted. Effective rainfall is regarded as that part of the precipitation which becomes available to plants for absorption. It is measured by deducting the losses through evaporation from soil and plant surfaces, the runoff and gravitational flows from the total rainfall.

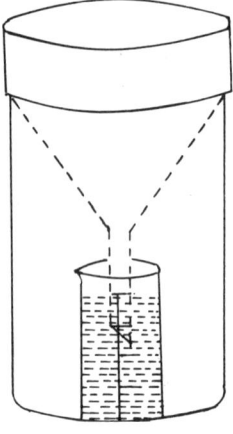

Fig. 4.24. Rain gauge.

Measurement of relative humidity

Relative humidity is measured by a number of instruments. The commonest are as follows :

Psychrometer : This consists of a pair of thermometers, the bulb of one of which is wrapped with a piece of cotton wick for the purpose of wetting. Therefore, the instrument is also called a *wet and dry bulb thermometer*. From the wet bulb evaporation of water takes place on free contact with the atmosphere. Evaporation results in cooling of the bulb. In dry atmosphere of low relative humidity there will be greater evaporation and consequently greater cooling or fall in temperature. If there is no depression in the temperature it means that the R.H. is 100%. The difference between the dry bulb thermometer and the wet bulb thermometer readings gives the fall in temperature. A reference to a conversion table (Table 4.1) about the fall in temperatures at any given atmospheric temperature gives the relative humidity value. In order to bring a rapid and free contact of moving air to the wet and dry bulb, the instrument is either whirled or an inbuilt fan is provided. Many modern makes provide a conversion chart upon the instrument itself to do away with the consultation of a separate table.

Hygrographs and hygrometers : These instruments are meant for directly reading the relative humidity of the atmosphere. The hygrograph is usually meant for continuous and automatic recording of relative humidity values over a period of time say for one week on each setting. This instrument is based upon the property of hairs especially human hair to contract or expand on contact with humid or dry atmosphere. Several strands of human hairs are tightly fixed to the terminals of a frame. The frame is attached to a pen by means of a lever. As the hair contracts and expands, the pen records lines on a very slow moving piece of graph paper wrapped on a drum that revolves one round in one week. Generally the drum is placed in a glass covered chamber while the sensitive hair frame is in contact with the atmosphere. Usually the hygrographs are combined with temperature recording devices and this double purpose instrument is called a *Thermohygrograph* (Fig. 3.1) and the graph sheet showing temperature and relative humidity readings is called a *thermohygrogram*.

Table 4.1. Relative humidity table

Dry bulb deg.	0.5	1.0	1.5	2.0	2.5	3.0	3.5	4.0	4.5	5	6	7	8	9	10	11	12	13	14	15	16	18	20	22
								Difference between readings of wet and dry bulbs in degrees centigrade																
2	92	83	75	67	59	52	43	36	27	20														
4	93	85	77	70	63	56	48	41	34	28	15													
6	94	87	80	73	66	60	54	47	41	35	23	11												
8	94	87	81	74	68	62	56	50	45	39	28	17												
10	94	88	82	76	71	65	60	54	49	44	34	23												
12	94	89	84	78	73	68	63	58	53	48	38	30	14	12	4									
14	94	89	84	79	74	69	65	60	55	51	41	33	21	16	10									
16	95	90	85	81	76	71	67	62	58	54	45	37	24	21	14	7								
18	95	90	86	82	78	73	69	65	61	57	49	42	29	27	20	13	6							
20	95	90	87	82	78	74	70	66	62	58	51	44	35	30	23	17	11							
22	96	91	87	83	79	75	72	68	64	60	53	46	36	34	27	21	16	11						
24	96	92	88	85	81	77	74	70	66	63	56	49	40	37	31	26	21	14	10					
26	96	92	89	85	81	77	74	71	67	64	57	51	43	39	34	28	23	18	13					
28	96	92	89	85	82	78	75	72	68	65	59	53	45	42	37	31	24	21	17	13				
30	96	92	89	86	82	79	76	73	70	67	61	55	47	44	39	35	30	24	20	16	12			
32	96	93	90	86	83	80	77	74	71	68	62	56	50	46	41	36	32	27	23	19	13			
34	97	93	90	87	84	81	77	74	71	69	63	58	51	48	43	38	34	30	26	22	18	10		
36	97	93	90	87	84	81	78	75	72	70	64	59	53	50	45	41	36	32	28	24	21	13		
38	97	93	90	87	85	82	79	76	73	70	65	60	54	51	46	42	38	34	30	26	23	16	10	
40	97	94	91	88	85	82	79	76	73	71	66	61	56	52	48	44	40	36	32	29	25	19	13	
42	97	94	91	88	85	82	79	77	74	72	67	62	57	53	49	45	41	38	34	31	27	21	15	
44	97	94	91	88	86	83	80	77	74	73	68	63	58	54	50	47	43	39	36	32	29	23	17	12
46	97	94	91	89	86	83	80	78	75	73	68	64	59	55	52	48	44	41	37	34	31	25	19	14
48	97	94	92	89	86	84	81	78	76	74	69	65	61	56	53	49	45	42	39	35	33	27	21	16
50	97	94	92	89	87	84	82	79	77	75	70	65	62	57	54	50	47	43	40	37	34	28	23	18

5

Atmosphere and Gaseous Cycle

The thick gaseous mass enveloping the earth, the atmosphere is essential for all living beings. Carbon dioxide and oxygen are directly exchanged between organisms and the atmosphere. Other constituent gases also influence life directly and indirectly in several ways. The principal gaseous constituents of the atmosphere are Nitrogen about 79%, Oxygen about 21% and Carbon dioxide 0.03%. Water, in varying quantities spatially and temporally, is present both in an invisible form as vapour and visible form as cloud and fog. Very small particles of dust are also constantly floating in the atmosphere. These also influence the organisms both directly and indirectly by their influence on almost every aspect of climatic factors. Microscopic forms like bacteria, viruses, fungal spores and pollen grains are the biological constituents of the atmosphere. Atmospheric pollution in the form of smoke from industries brings a variety of gases to the atmosphere.

Green plants give out oxygen during the day time and this gas is used in respiration by all organisms all the time. The process of photosynthesis thus regulates the oxygen and carbon dioxide balance in nature. Nitrogen which is abundantly present in the atmosphere is not used directly by plants or animals except by some bacteria and blue green algae. Nitrogen is however, an indispensable material for all living things. It is a basic material of all proteins and nucleoproteins and it is obtained by green plants primarily from soil in a combined form. The atmospheric gases and other suspended contents affect plant metabolism. Air currents and wind also have a number of physical and physiological influences both directly on the plant body and indirectly through their effect on other climatic factors. The ecological aspects of the atmosphere, can be described under suitable headings as follows.

CARBON

Carbon dioxide concentration in the atmosphere is only 0.032% which is less than that needed by most of the plants for their maximum efficiency in photosynthesis. Experiments show that with the same moisture content and sunlight intensity as is available in an open field on sunny days the rate of photosynthesis increases several times if the carbon dioxide concentration is artificially increased, say in a glass house. The CO_2 is the principal source of carbon needed in the building up of all organic world through photosynthesis of green plants and goes back into the atmosphere by the breakdown of these organic compounds through respiration by plants or organisms that eat living or dead plants. There are various stages through which carbon cycles in the biosphere from abiotic to biotic and back to abiotic (Fig. 5.1). The quantity of carbon fixed by green plants globally is in the range of 9×10^{13} kg per year. In ocean water, carbon dioxide remains dissolved in huge quantities estimated to be over 50 times that of the atmosphere. This regulates the atmospheric carbon dioxide between atmosphere and sea on the one hand and between atmosphere and organisms on the other.

In the atmosphere the CO_2 varies from hour to hour during the day. In the morning CO_2 concentration is on the high side because of accumulated respiration of the community in absence of the opposite process that is photosynthesis in darkness. The O_2 content gradually increases by noon and CO_2 content correspondingly decreases. In a manured field, due to the richness of organic fertilizer the microbial population increases considerably in the ground. The CO_2 evolved in microbial respiration diffuses out of the surface soil and this also readily increases the production rate (photosynthesis) of crops. Therefore, organic manuring is doubly advantageous, firstly, it increases the nutrient status of soil and secondly it increases the CO_2 content in the atmosphere immediately above the ground.

The CO_2 content of the soil atmosphere has a very significant influence on the life of soil organism and on the roots of higher plants. CO_2 released through life in soil does not readily diffuse out and this

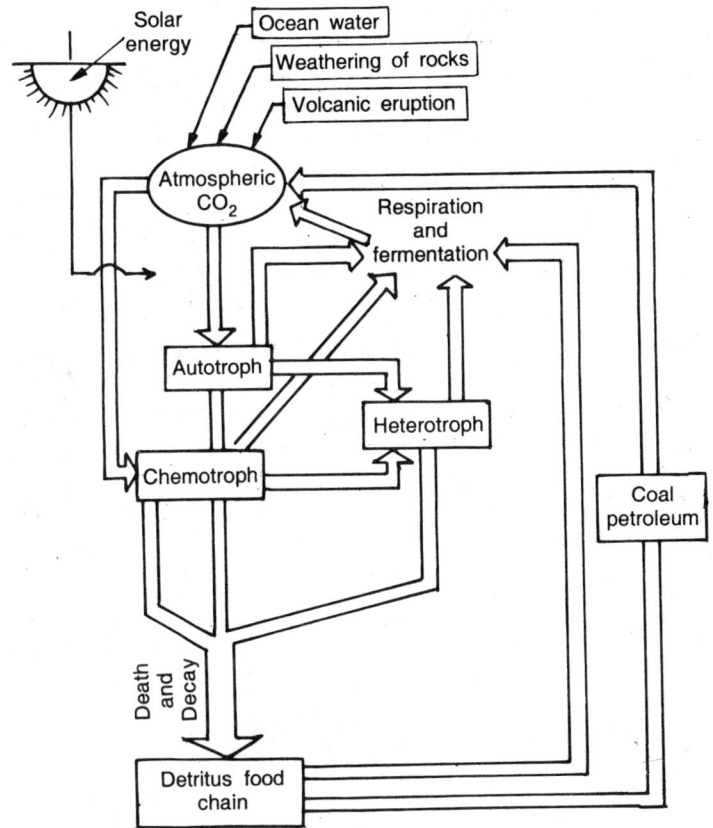

Fig. 5.1. Carbon cycle. Diagramatic representation of movement of carbon between atmosphere, organism and soil.

considerably increases the CO_2 content at the expense of the other contents like O_2. Therefore, slow diffusion of CO_2 results in poor aeration and a retarded metabolic activity. Periodic irrigation flushes out the excess CO_2 and fresh atmosphere diffuses in soil on the drying up of the soil. For a rapid movement of water and gases the soil should have adequate pore space of about 50% with non-capillary and capillary sizes in good proportions. Through the self-regulating feedback mechanisms in the different pathways of the carbon cycle the proper balance is maintained. The rapid pace of exploitation of carbon fixed over long geological periods in the form of coal and petroleum is responsible for the return of huge quantities of carbon dioxide to the atmosphere. How long the natural system will be able to maintain the carbon balance or equilibrium is an important question facing scientists. It is feared that atmospheric carbon dioxide may increase by 25% of the existing content soon.

The trend of CO_2 increase based on accurate measurements at Mauna Loa observatory in Hawaii clearly shows that it has risen from 0.0315% in 1958 to 0.0335 by 1985 (Fig. 5.2). This gas was as high as 90% before the advent of life on this earth as N was only 2% then and oxygen was only in traces. But with the origin of life, and then of chlorophyll, CO_2 kept on decreasing. It is estimated by isotopic carbon ratios, the CO_2 content in atmosphere was 0.027% or 270 ppm in 1850 and since then it is steadily increasing, more rapidly in recent decades. Wavy line of CO_2 increase in Fig. 5.2 is due to seasonal rise and fall every year.

Scientists have succeeded in estimating the fraction of carbon

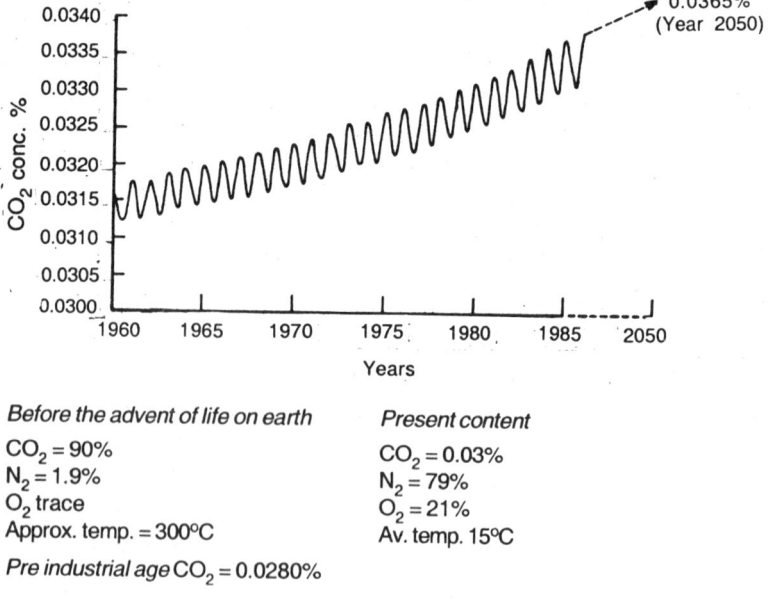

Before the advent of life on earth

CO_2 = 90%
N_2 = 1.9%
O_2 trace
Approx. temp. = 300°C

Pre industrial age CO_2 = 0.0280%

Present content

CO_2 = 0.03%
N_2 = 79%
O_2 = 21%
Av. temp. 15°C

Fig. 5.2. The pattern of rise in CO_2 on global basis.

dioxide in the atmospheric stock that has been added by the burning of fossil fuel. The fossil carbon is radioactively dead but atmospheric carbon has a certain proportion on radioactive C^{14} formed due to cosmic bombardments of the earth (Revelle et al., 1965) and very accurate measurements of CO_2 content by infrared gas analysers of pollution free places like mountain tops in Hawaii islands and in Antarctica have revealed a very definite increase in carbon from fossil fuel sources. This rise in atmospheric carbon dioxide is taking place despite a good deal of regulatory effect of oceans. Rise in carbon dioxide increases the heat absorbing capacity of atmosphere much in the same way as glass houses do, that is they permit the radiations to pass through and strike the earth, but once converted into heat and reflected upward the heat waves are absorbed by CO_2 rich atmosphere and cause rise in temperature. Many scientists are alarmed that such rise in temperature may cause melting of polar snow, sea level rise and an overall flooding of coastal low lands. But others have argued that during fossil fuel combustion, fine droplets of liquids also gets dispersed into atmosphere and these aerosols reflect sufficient radiations as albedo of earth to cause cooling and compensate green house heating effect of increased CO_2 content. But there is a definite trend of temperature rise in recent years.

On quantitative basis the carbon cycling is shown in Fig. 5.3. Oceans in totality account for an estimated 38980×10^9 tons of which 580×10^9 tons are in the upper zone of ocean water. Its annual exchange rate with the atmosphere is balanced for about 90×10^9 t yr^{-1}. It is estimated that total carbon in fossil fuels and in shales is 12000×10^9 of which recoverable is 7500×10^9 t. The terrestrial biosphere have a stock of 1760×10^9 t of which an annual exchange between plants + animals with the atmosphere is 56×19^9 t yr^{-1}. However, an entirely one way addition from land to atmosphere by way of fossil fuels in the order of 5×10^9 t yr^{-1}, and this is primarily responsible for increase in overall CO_2 content and global warming.

OXYGEN

Oxygen which is in abundance (20.94%, Machta and Hughes, 1970) in the atmosphere is another indispensable material for life. This is needed in respiration, a process through which the stored energy is released for use of various biological processes in the body of organ-

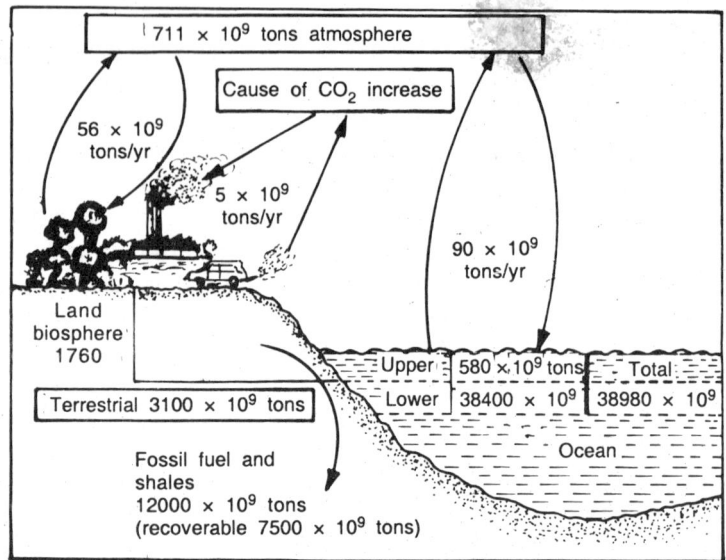

Fig. 5.3. The global carbon storages and major exchanges.

isms. Oxygen is constantly taken in by animals and plants from the atmosphere. Aquatic plants receive their oxygen requirement from a very limited concentration of this material dissolved in water. In ponds, lakes and marshy habitats the oxygen content may fall to become limiting. Oxygen deficiency causes a number of effects on plant life such as stunted growth, death of normal roots and formation of new adventitious roots with a greater proportion of air spaces, thin cell walls and poor root branching. Anaerobic respiration becomes more pronounced. Excessive carbon dioxide concentration which usually occurs in oxygen deficient soils also has detrimental effects on roots. The detailed description of plant responses to soil atmosphere is describer later in the section dealing with soil features.

Oxygen gas is present in atmosphere in sufficiently large quantities and there is no possibility of oxygen deficiency on global scale even if we burn all the organic matter including fossil fuel. The CO_2 content marginally rises, but the reduction in oxygen content will not affect the physiological needs of man and other organisms. Details of water pollution is described in the chapter on Environmental Pollution.

Soil atmosphere in the pore space supply oxygen for root and microbial respiration. There is a continuous gas exchange between the atmosphere and soil pores and the rate becomes faster when soil respiration is fast. In marshy condition or water covered conditions atmospheric recharge of oxygen is slowed leading to oxygen deficiency. Plants like lotus and other floating leaved ones actively transfer air through the lacunate (aerenchyma) tissues from aerial leaves to roots in the water and undermud region. Otherwise oxygen diffusion through the water column or in water logged soil is extremely slow. If oxygen is not deficient for metabolic purpose, it is called *normoxia*, if deficient to considerably reduce the activity, it is *hypoxia* and if oxygen becomes almost absent, the condition is called *anoxia*. Oxygen deficiency in soil leads to negative redox potential or highly reducing state. In India with monsoon climate, for certain periods most of the soil gets flooded and reach a condition of hypoxia. Rice seeds adapted to primary hypoxia germinate in poor oxygen but wheat seeds fail to do so. The control is at enzymatic level. Aquatic plants have the ability to elongate the internodes at a very fast speed upon the rise of water level. It can be easily demonstrated in practical classes. An individual plant of rice or other floating or aerial leaf water plant can be placed in a tall measuring cylinder or glass pot and filled with an overlying column of 20-25 cm of water. The plants internodes elongate, by 15-20 cm in the next 24 hours and reach above the water surface. Elongation is regulated by intercellular ethylene concentration and low oxygen. Aminocyclopropane carboxylic acid (ACC) oxidase gene expression does the synthesis of ethylene.

NITROGEN

Nitrogen forms the main bulk of the atmosphere (79%). This substance is very important for plants and animals being an essential constituent of chlorophyll and all proteins. Despite its great importance and indispensable nature, nitrogen is not directly taken from the atmosphere by animals or higher plants (Fig. 5.4). Atmospheric nitrogen is rather inert and does not readily participate in any reaction. Herbivore animals derive their nitrogen requirements from plants and carnivores from herbivores. Plants derive their nitrogen requirement from soil in the form of ammonia or nitrates. This in soil, is built up by bacteria and some blue-green algae which fix atmospheric nitro-

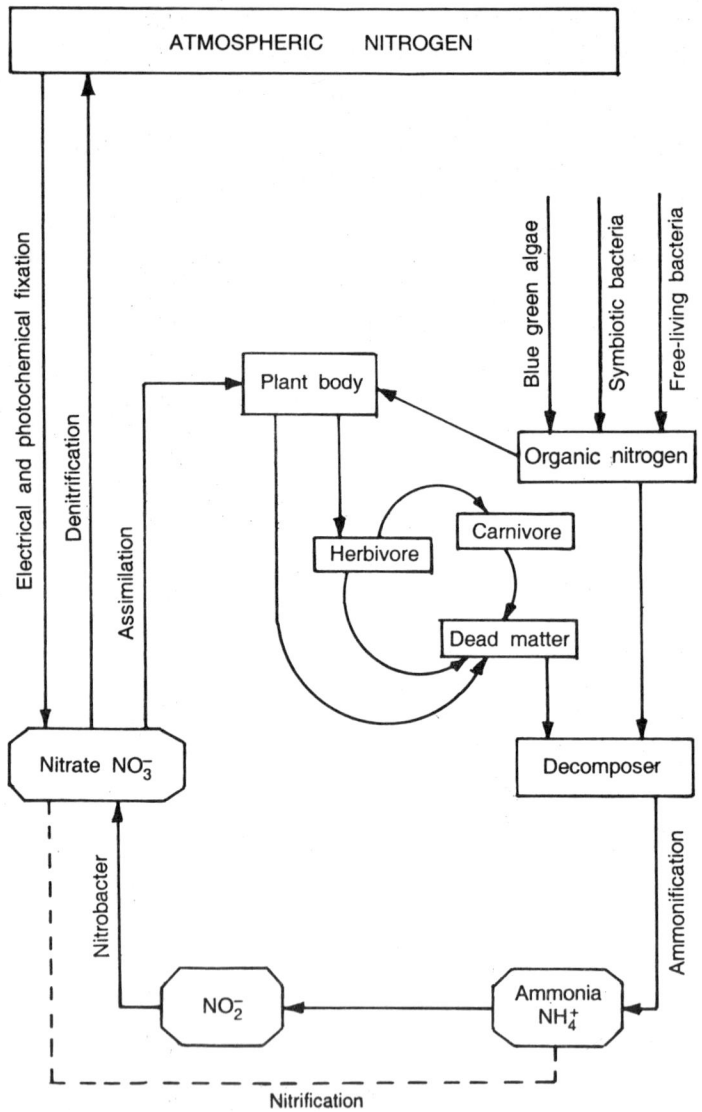

Fig. 5.4. Nitrogen cycle. Stages showing movement of nitrogen from atmosphere to plants, animals, soils and back to the atmosphere.

gen directly. Kormondy (1969) has stated that through electrochemical and photochemical fixation about 35 mg/m²/yr of nitrate is formed and through biological fixation it ranges between 14 to 700 mg/m²/yr. In very fertile areas it may exceed 20,000 mg/m²/yr. Nitrogen fixing organisms are usually grouped into : the *symbiotic* and *non-symbiotic*. Symbiotic nitrogen fixers are chiefly species or strains of the bacterium *Rhizobium*. These are chiefly terrestrial and associated in roots of legumes. Among non-symbionts, some blue green algae like *Anabaena* and *Nostoc*, aerobic bacteria like *Azotobacter* and anaerobic bacteria like *Clostridium* are the commonest. Like *Anabaena* and *Nostoc* certain photosynthetic bacteria like *Rhodospirillum* are also nitrogen fixers (Kamen, 1953).

Some of the prokaryotic organisms like bacteria and bluegreen algae convert atmospheric nitrogen into ammonium form by combining nitrogen and water in presence of nitrogenase enzyme and utilizing energy from ATP and oxygen is released as a byproduct. Symbiotic association of the bacteria *Rhizobium* with legume roots and of algae *Nostoc* and *Anabaena* with roots of cycads and leaf of *Azolla* are commonly found where nitrogen is fixed by the prokaryotes and used by higher plants. Many non-leguminous trees show root nodulation as a result of symbiotic association with antinomycetes particularly in temperate regions. Almost all species of *Alnus, Casuarina, Shepherdia* and *Coriara*, and a few species of *Dryas, Myrica* and *Rubus* have root nodules (Akkermans and van Dijk, 1981). Sharma and Ambasht (1984) have found that considerable quantity of atmospheric nitrogen is fixed by an actinomycetous genus *Frankia* that is present in the root nodules of trees of *Alnus nepalensis* in Darjeeling Himalayan forests.

The locked up nitrogen in the body of dead animals and plants is acted upon by various types of bacteria, actinomycetes and fungi occurring in soil and water. Nitrogen rich organic materials like protein are converted by ammonifying bacteria into an inorganic form like ammonia and the process is called *ammonification*. The ammonium nitrogen is converted by another set of microorganisms like *Nitrosomonas* into nitrites (NO_2). Nitrites are acted upon by bacteria like *Nitrobacter* to produce nitrates. Denitrifying bacteria *Pseudomonas* and certain fungi are responsible for the return of nitrogen to the atmosphere. This inorganic nitrogen is again recycled into the organic system upon absorption by higher plants. Thus, the path of nitrogen cycle as shown diagrammatically in Fig. 5.4 is from the atmosphere to (1) biological fixation by microorganisms, (2) release in soil, (3) absorption by higher plants, (4) absorption by other organisms which feed upon plants, (5) the dead organic material, (6) bacteria and (7) soil. Atmospheric electrical discharges also fix nitrogen that fall to the earth dissolved in rain water.

On the whole in nature, through the process of fixation, amount of nitrogen converted into usable form equals the quantity returned to atmosphere by the process of denitrification. But in recent years there has been exceedingly high quantities of atmospheric nitrogen fixation by industrial processes (Habbers process) and they are not fully and speedily denitrified so as to cause accumulation of nitrates or ammonia in water and soil or NO_2 (from fossil fuels in automobiles) in atmosphere which are polluting the environment.

Kormondy (1976) has given quantitative values of nitrogen in different segments of the earth, based on the data given by Delwiche published in Scientific American (Sept., 1970).

The major storehouse of nitrogen are atmosphere (3,800,000 × 10^6 t), oceans water in dissolved form (20,000 × 10^6 t), ocean sediments (4,000,000 × 10^6 t) and earth crust (14,000,000 × 10^6 t). The quantities on land is 12 × 10^6 t in the living organisms, 760 × 10^6 t in dead organisms, while in ocean organisms living have 1 × 10^6 and dead 900 × 10^6 t.

Nitrogen cycling in tropical deciduous forests dominated by *Anogeissus latifolia* and *Diospyros melanoxylon* trees (Singh and Pandey, 1981) and savanna dominated by *Heteropogon* and *Desmostachya* grasses (Ambasht, 1984; Misra, 1980) have been worked out. In the forest, the storage quantities of nitrogen was 2970 kg/ha and in the body of plants 676 kg/ha. The annual uptake by plants is reported to be 173 kg/ha whereas the return rate to soil was 76 kg/ha via litter and 29 kg via root decay. Rainfall contributed 10 kg/ha/yr of nitrogen to soil pool of which 5 kg/ha/yr was lost as runoff. In the savanna the input through rainfall in the study year was 8.77 kg/ha/yr and loss through runoff was 4 to 6 kg in different vegetation stands. But in a scraped or bare land the runoff loss increased to 110-120 kg/ha/yr, thus depleting the soil of its nitrogen very heavily (Fig. 5.5).

On global basis according to Delwiche's estimates, the total annual N fixation is 92 × 10^6 t of which 54 × 10^6 t is fixed biologically,

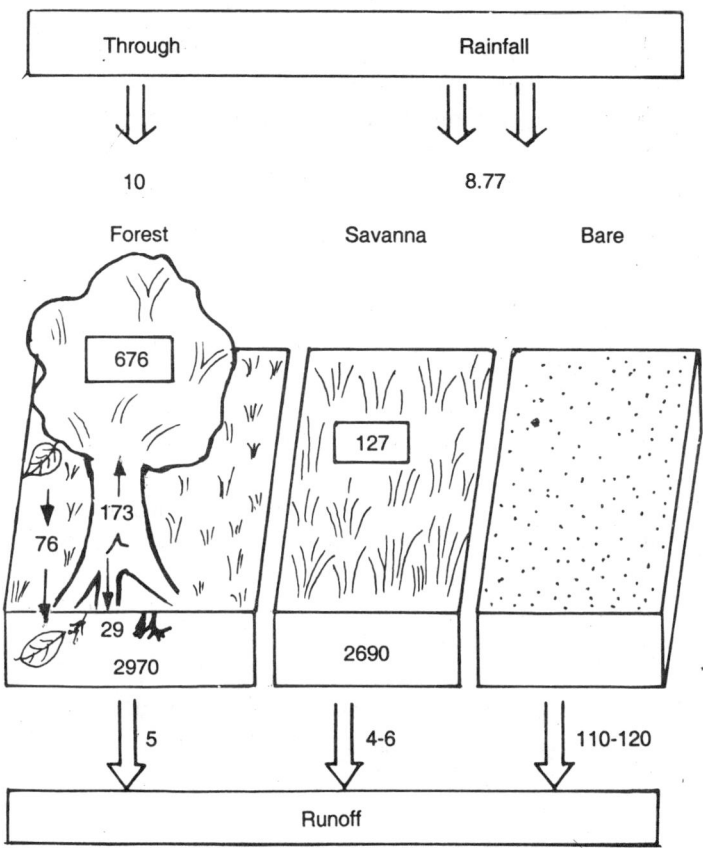

Fig. 5.5. Nitrogen input storage, uptake, release and runoff in forest, savanna and bare land (Data : Singh and Pandey, 1981; Ambasht, 1984 and Misra, 1980). Values are in kg/ha in boxes for storages in soil and plants and on arrows in kg/ha/yr for transfer rates.

30-industrially and 7.6-photochemically and 0.2×10^6 t yr^{-1} by volcanic activity. Denitrifications return only 83×10^9 t yr^{-1} creating an imbalance of 9×10^9 yr^{-1}.

WIND

The strong moving current of air or wind is an ecological factor of great significance. Its movement results from a number of environmental factors such as temperature, atmospheric pressure and geographical features. Wind has played an important role in the cultural evolution of man especially, in influencing the routes of sea voyages, explorations and trade for a long time.

Effects of wind on plant life and plant environment are shown in a wide variety of ways which can be described under the following convenient headings.

1. Dispersal

(a) Micro-organisms and pollen grains

Many micro-organisms, fungal spores and bacteria are freely transported even in moderate wind to long distances. The pollen grains in many species are wind transported from one flower to another and anemophily (wind pollination) is a well known phenomenon prevalent in nature.

Plants adapted to wind pollination necessarily produce pollen grains quite abundantly as the chances for pollen to reach the stigma of flowers of the same species are quite meagre. In certain forests of the Himalaya pollen dust of *Pinus* and *Cedrus* is very abundant in the atmosphere at certain times. Success of wind pollination in such dominant trees becomes certain on account of the numbers of individuals of the species occurring near each other over a wide area. Pollen grains in some genera like *Pinus* are specially adapted to float in air due to the presence of air bladders.

(b) Seeds and fruits

Dispersal of seeds and fruits and other propagules in many plants depends upon wind for their movement from one place to another. Dispersal and migration are very important aspects in ecology. Aggregation of seeds of any species at any one place leads to competition between individuals of the same species whose requirements are similar. Wide dispersal of seeds and propagules leads to better establishment and a more successful and healthy growth.

Many plants produce extremely small seeds which due to their lightness float in air. Others develop wings that help the seed to be carried by wind to long distances. In plants like *Alhagi camelorum* the entire shoot gets broken and entangled with others. This mass of twigs is carried rolling on the ground for long distances in deserts due to wind. In many plants seed are covered with fine white cotton or silky hairs that help them float in air and get wind dispersed.

2. Effect of wind on plant structure

(a) Size of plants

Wind constantly removes water from the tissues of plants. This leads to less expansion of cell walls in the absence of turgidity. The overall effect of smaller size of cells is to bring about decreased total height and girth of plants.

(b) Streamlined shapes

Some plants that grow in areas subjected to strong wind all the year round develop an overall shape that offers least resistance to wind. The mechanisms of development of such a shape are several but most commonly young vegetative buds on the windward side get dried or damaged. Thus growth is restricted on the side where wind effect is greatest. A wind trained-shape is quite common in sea coast plants as there the wind carrying salt particles from sea water is more effective in killing buds and thereby checking the development of branches on the windward side.

(c) Anatomy

The internal anatomy of a stem subjected to constant strong wind leads to greater development of mechanical tissues on one side only. This eccentric growth of secondary xylem tissue prevents the stem from getting easily broken.

(d) Desert plants on shifting sands

Strong wind in deserts moves sand from one place to another. Some plants thus covered with sand show quick elongation of stem apices and development of adventitious roots from the covered stem base. Plants that do not have such adaptations succumb.

(e) Transpiration

The most profound effect of wind on plants as a whole is to increase their rate to transpiration. Wind quickly removes the layer of humid air just above the leaf surface. The rate of transpiration depends upon the dryness or wetness of the atmosphere. Initial transpiration leads to the formation of a moist layer of air above the leaf surface. The moist layer reduces the transpiration rate but wind removes the transpired vapour quickly and therefore, the overall rate of transpiration rises considerably. The plants suffer from water deficiency in their tissues in strong winds. The mechanical to and fro movement of twigs and leaves also helps in rapid moisture loss due to physical contraction and expansion of cells.

3. Effect of wind on plant environment

(a) Expanding deserts

In India as well as in many other parts of the world, due to unecological land management, deserts have expanded. In the absence of trees, wind moves unchecked and therefore, lifts greater quantity of surface sand. Wind erosion naturally leads to deposition of sand on areas where the wind moves unchecked, including the margins of deserts. In India, the Rajasthan desert has, in this way advanced towards western Uttar Pradesh. The deposited sand makes the edaphic environment of cultivated field unsuitable for the good growth of crop plants. In order to check this menace, grasses, weeds, shrubs and trees are being planted. Only such species are used which can withstand the effect of dry wind by their structural and/or physiological adaptation. Blocks of plantation to check wind erosion and wind speed are called *wind breaks* and *shelter belts*. Plant roots bind the soil and protect it from erosion. The shoot canopy acts as a physical barrier for wind and dust and greatly reduces the wind speed. The result is that land and plants on the leeward side of shelter belts escape the ravages of wind, sand and dust.

(b) Rainfall

Wind also has many beneficial effects on the plant environment. For example, most of the rainfall at any given place is regulated by wind movement. In Northern India in early summer the westerly winds are dry and desiccating because they originate on dry land. The easterly in the months of June, July onwards originate in the Bay of Bengal and move from sea to land. These, therefore, are wet and cause widespread monsoon rain in the states of Assam, Bengal, Orissa,

Bihar, Uttar Pradesh, Delhi and Punjab.

4. Measurement of wind speed

Wind movement is usually expressed in terms of miles per hour or kilometres per hour. There are several types of instruments called anemometers that record wind speed. Fig. 5.6 shows one such type in which cups mounted on a pivot revolve at speeds proportionate to the velocity of the wind. In some other types several blades revolve much like the blades of a fan and cause the set of needles to indicate the speed.

In the absence of an instrument the following field observation may help in approximate determination of wind velocity values.

If smoke rises vertically and does not drift in any direction the wind is said to be calm and its velocity is less than 1.5 km per hour, but if the smoke gradually drifts, the speed is likely to be between 2 to

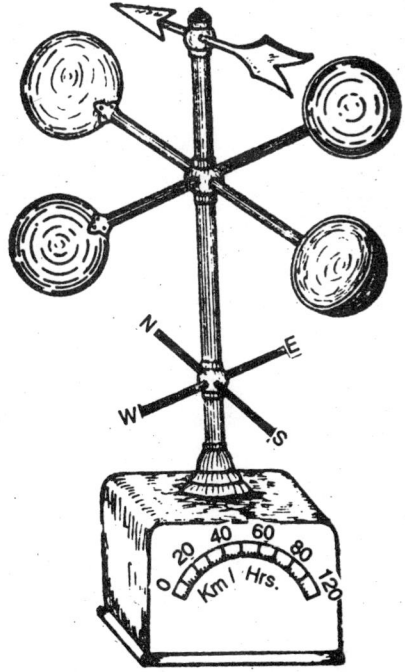

Fig. 5.6. A cup anemometer.

5 km/hr. If the wind is felt on the face and if leaves of trees move slightly in the light breeze the velocity is between 5 to 10 km/hr and if tree leaves the twigs both move to and fro constantly the velocity is between 10-18 km/hr. In a moderate breeze which raises dust and dry leaves and paper pieces from the ground, the velocity may reach up to 25 km/hr. In a yet stronger breeze waves form on ponds and lakes with marked crests and troughs and small trees or branches begin to sway, the velocity may be between 25 to about 40 km/hr and when telegraph wires cause a whistling sound, velocity may be up to 50 km/hr. Gale, storm and hurricane are progressively stronger winds around 60 to 90 km, 90 to 120 km and above 120 km/hr respectively.

Soil and Sedimentary Cycle

The word 'soil' is derived from the Latin word *solum* meaning 'the floor' or the ground surface. Soil is the medium in which roots grow, anchor the plants and from which the plants derive water and nutrients. The thin ramifications of root branches and hairs extend extensively and their total volume is often more than the shoot volume. As such the importance of soil properties in the plant environment is of great significance.

The soil is defined as the weathered (or broken particles) surface of the earth's crusts which is mixed with organic material and in which microorganisms live and plants grow. Soil consists of the inorganic materials derived from parent rocks; the organic materials derived from dead organisms, the air and water occupying the pores between the soil particles which are loosely packed, small organisms like bacteria, fungi, nematodes etc., and the higher plants live in it. Soil is the ultimate source of all food production since plants form the base of all ecological pyramids and plants grow in soil and derive nutrition from them. Soils differ in matter of relative abundance or paucity of certain components and only plants grow in soil and derive nutrition from them. Soils differ in matter of relative abundance or paucity of certain components and only plants adapted to such extremes grow there. On such a basis plants are classified into : (1) *calciphytes* that grow in alkaline soils, (2) *oxylophytes* that grow in acidic soils, (3) *lithophytes* that grow on the surface of rocks, (4) *chasmophytes* that grow in rock crevices, (5) *psammophytes* that grow on sand and (6) *halophytes* that grow on saline soils.

SOIL COMPONENTS

All soils have the following components :

(1) The mineral matter, (2) the soil organisms and the organic matter, (3) the soil solution and (4) the soil atmosphere. Development of soil from parent rock is also called *pedogenesis* and the soil science is called *pedology*. The nature of soil formed at a place is largely dependent upon the nature of rock material, its slope, the prevailing climate, the types of biota of the region and the human activity.

(1) The Mineral matter

Mineral particles such as sand, silt or clay are the primary material that constitute the soil. Bedrocks from which these particles originate are called *parent material*. The process of breaking up of rocks into smaller particles is called *weathering*. There are several causes or agencies of weathering.

Physical weathering

Mechanical forces acting upon the rocks cause physical weathering. Temperature fluctuation is the most important physical factor. Surface rock gets heated while inner part remains cold. This leads to surface chipping or exfoliation leading to peeling off of rock layers. Heating and cooling also leads to cracking and formation of fissures (deep crevices). Freezing temperature causes, expansion of water present in rock crevices during ice formation and the force of expansion causes breaking up of rocks which if roll down the slope further get broken into smaller fragments due to force of gravity. The beating effect of hails and rainwater and scouring effect of fast flowing streams also cause physical weathering. Formation of rounded stone pieces of small to very small sizes are commonly seen on the bed of Ganga river at Rishikesh and Haridwar where fast flowing river reaches the plains from its place of origin at Gangotri in the Himalaya. Wind is also another physical agent of weathering particularly when it carries sand dusts that rub the rocks during their rolling movement and cause abrasion. In the Vindhyan hill forests, it is commonly seen that tree roots often penetrate through the rock crevices and in course of time with girth growth of these roots the rocks gets disintegrated.

Chemical weathering

The rocks while getting disintegrated may also undergo chemical changes in their property. Water is an important agent in bringing about chemical changes due to dissolution or reaction of one or more components of rock material. Presence of dissolved material and warm temperature favour chemical weathering. Some rock components may get dissolved and reprecipitated, thus changing rocky texture to fine dusty texture. Some minerals like felspar and mica readily combine with water through the process of *hydration* and become soft and easily weatherable. Another very important process of chemical weathering is through *hydrolysis* in which water dissociates, particularly in presence of CO_2 and organic acids, into H^+ and OH^- ions which act on silicates like orthoclase and produce silicate clays. Other chemical weathering means are through *oxidation, reduction* and *carbonation*, in which addition of oxygen, removal of oxygen and action of carbonic acids respectively take place in the mineral components of parent rocks.

Biological weathering

On bare rock surfaces patches of lichens are commonly seen flourishing. They weather rock surface through the chemical action of their secretion and carbonic acid formed with action of water on CO_2 liberated in their respiration. Similarly other lithophytes corrode rock surfaces and root systems that enter rock crevices, break them by force of secondary growth and increase in root girth.

Soil transportation

The mineral particles thus formed may remain over the parent rock and this is called *residual* or sedentary soil. When this has moved it is called *transported*. Depending upon the agencies of transport the soils are called *alluvial* (transported by the river or running water), *colluvial* (transported through gravity), *glacial* (transported by ice or glaciers), *eolian* and *loess* (transported by wind) etc. The transect across the *residual* soil at the place of its formation say on a hill top, its drift through gravity of colluvial soil at the hill base, and further away by wind and water transported *eolian* and *alluvial* soil could be observed in what is called a *catena* on a chain of hill top and valley formations. The nature and properties of mineral components of soil greatly depend upon the nature of the parent rocks and process of weathering.

Among the water transported soil, besides the commonest alluvials, there are two more types; *lacustrine* that are deposited in lake beds in form of mineral sediments and marine in which sediments deposited by river into oceans cause upwelling of shore margin or formation of islands. In cold climates, very slow decomposing accumulation of dead plant materials form *peat* where plant parts are still recognisable and *muck* if organic matter is fully broken and plant parts are not recognisable.

In India the principal residual soil types are (i) reddish soil of the Vindhyas and south of it, derived from sedimentary rocks of different geological ages and, (ii) black soil of South West India. The red soils are poor in calcium, magnesium, phosphorus and nitrogen. The black soils, well known for its cultivation for cotton, are rich in clay, devoid of sand and have high contents of calcium, magnesium and aluminium.

Among the transported soil, the *alluvial* soil of the Ganga and Indus plains are highly productive and have been under cultivation for several thousands of years. The parent material of the Gangetic alluvial soils in Uttar Pradesh, Bihar and Bengal is to be found in the Himalayas. On steep slopes due to the force of running water, big boulders of stone move and roll down. This causes fragmentation. The broken stone pieces get further powdered and rounded as they roll and gradually sand is formed. Along the river bank for some distance flood plains are found consisting of alluvial deposits. Huge expanses of alluvial deposits are formed near the confluence of river and sea. This is called a delta. In the alluvial deposits *Acacia arabica, Dalbergia sissoo, Saccharum munja, Alhagi camelorum* are quite characteristic but most of the area is under intensive cultivation for crops. The Sunderban in West Bengal and Bangladesh are good example of delta where a mangrove type of vegetation constitutes the forest.

Glacial deposits are transported by the movement of snow. Here the fine particles are intermixed with pebbles and boulders as against alluvial soil which are uniform in texture. At the base of hills, fine and coarse soils intermixed usually remain in thick deposits. This type of soil mass is called *colluvial* as it is actually formed at the hill top or slope and is transported and deposited below due to the force of gravity and movement of glaciers.

Texture and structure

The weathered minerals are classified on the basis of their size. At present more or less all countries agree to the International Society of Soil Sciences classification on the following lines :

Coarse sand	2.0	mm to	0.2	mm diameter
Fine sand	0.2	mm to	0.02	mm diameter
Silt	0.02	mm to	0.002	mm diameter
Clay		less than	0.002	mm diameter

Particles larger than coarse sand are called gravels. Depending upon the proportion of various size classes in a soil, it is called sandy, silty, loamy or clayey. In fact there are several intermediate names for slight differences. Different size particles have very different properties and as such the texture of soil is of great significance for plant growth. A sandy soil has larger size particles in greater quantity which renders the packing loose. Water percolates through sand quickly leaving the soil dry. Clay is very small and has several additional properties of colloidal particles such as electrical charge on its surface. Clay soils are more compact on account of their small size of particles and therefore offer resistance to root growth and hold a greater quantity of water against gravitational pull, due to capillary force. A mixture of humus (decomposed and mineralized organic matter) further adds the desirable properties to clay. Clays are formed as the final product of weathering and through precipitation of aluminium and silicon salts present in dissolved state in the soil moisture. Small clay particles (less than 0.001 mm) have colloidal properties and form *micelles* in combination with humus particles. The micelles have negative electrical charge on their surface through which positively charged cations get adsorbed on them. The adsorbed cations can be exchanged with other cations and this process is known as *cation exchange capacity*. The relative percentage of coarse sand, fine sand, silt and clay determines the *soil texture*. The relative percentages may be in any possible combinations and the properties differ in soils of different textures. If there is 70-80% sand, we call it *sandy loam*, if silt is so high we call it silty loam and if there is more than 50% clay we call it *clay* or *silty clay* or *clay loam*. Loam refers to soils which have about 20-30% clay, 30-45% silt and the rest of sand. In the warm climate of the tropics there are two main types of clay : (i) *Kaolinite* in which alumina and silica have alternating layers in 1:1 ratio held by hydrogen bonds giving a compact structure due to which water cannot be easily adsorbed and (ii) *Montmorillonite* in which one alumina sheet is held between two silica sheets and it is less rigid and adsorb the water well. This kind of clay expands on wetting and cracks on drying, and has greater cation exchange capacity.

These primary articles of sand, silt and clay aggregate in certain fashion to form large blocks or peds of soils which refers to the soil *structure*. These aggregations pattern into the following different shapes influence the physical properties of soil with respect to aeration, percolation of water and its retention, etc. As the terms indicate the soil structures may be : (i) *spheroidal*, (ii) *crumb* (relatively porous), (iii) *platelike* (overlapping platy layers), (iv) *blocklike* (angular blocky peds), (v) *prismatic* and (vi) *columnar*. The aggregation of primary particles is brought about by the colloidal fractions including sticky material of organic origin and very thin film of water molecules. Soil aggregates are called peds and it offers stability to soil structure (Fig. 6.1).

Water, nitrogen and carbon cycling in the biosphere were described

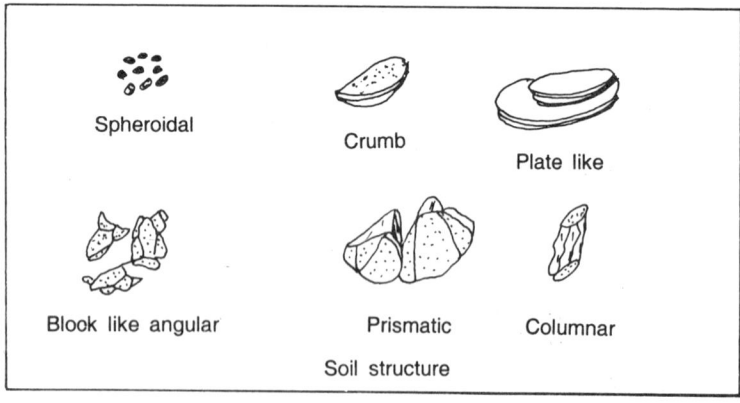

Spheroidal

Crumb

Plate like

Blook like angular

Prismatic

Columnar

Soil structure

Fig. 6.1. Common types of soil aggregates.

earlier under water and atmospheric factors. Phosphorous and calcium are important materials in the soil and it is described here.

PHOSPHORUS

Phosphorus is an essential and important constituent of protoplasm. In all energy transfers and energy fixation, adenosine di- and tri-phosphates (ADP and ATP) are involved in biological system. The major store house of phosphorus are rock deposits. Through erosion and weathering inorganic phosphates become available to plants in dissolved condition. Through hydrologic cycle a good proportion of this available phosphate is washed into the sea. But from the sea only a small fraction is able to return to land system through fishing by man and via certain sea birds. In the past the terrestrial systems were so balanced that the loss of phosphorus through erosion was less and transfer from sea to land by birds was more than what it is today. With greater human activity the process of formation of soluble phosphates, their application to crop fields, runoff and transportation rate to sea has become faster but the opposite direction return to land has not proportionately increased. Thus the phosphorus cycle is not really balanced. From plants, other organisms like animals and on their death microorganisms obtain phosphorus for their own protoplasm. From dead organic matter, excretions and bones, complex organic compounds are reduced by phosphatising bacteria to dissolved phosphates which are partly recycled in biological system (Fig. 6.2) and partly leached away.

On the global basis out of a good fraction of about 1.5 million tons of stock phosphates mined for the manufacture of fertilizers and after use in croplands is lost through runoff to rivers and through them to oceans. Hutchinson's estimates suggest 60,000 tons is obtained back through fishing. Thus ocean is getting richer in phosphorus at the cost of terrestrial mineral stock. This is not a serious problem in this context as the stock of rock phosphate is very large, but the problem created by runoff of phosphorus is serious in receiving freshwater bodies.

In Chakia forests of Vindhyan Hills in India, Ambasht and Misra (1984) have reported input of phosphorus through rainfall as 0.65 kg ha^{-1} yr^{-1} in a savanna land during 1977-78 and its runoff with water and eroded soil from protected vegetation was between 0.28 to 0.39

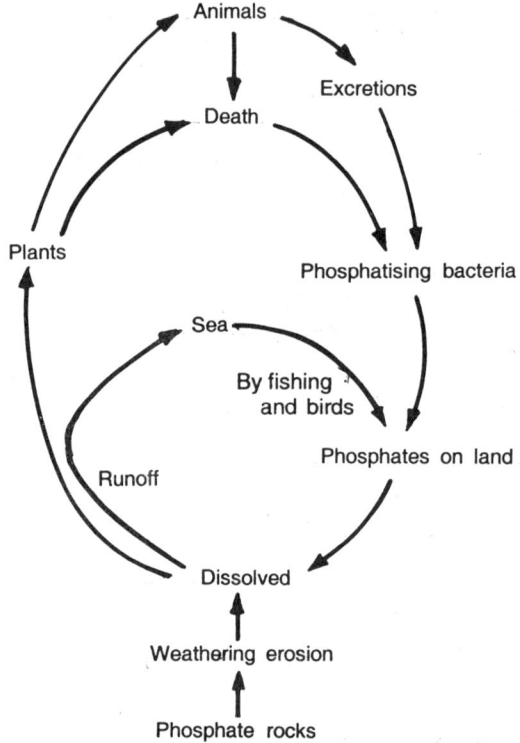

Fig. 6.2. Phosphorus cycle.

kg/ha/yr. Its runoff losses rose to 7.7 kg/ha/yr when the herbage cover was fully scraped. So phosphorus loss increased on grazing.

Ovington (1962) has studied phosphorus budget in *Pinus sylvestris* plantation forest in England. The total uptake by trees in the 55 years period i.e. the plantation age was 413 kg P ha^{-1} and by ground flora 182 kg ha^{-1} so that the plants took 595 kg P ha^{-1}. The total return during the same period to the soil surface in form of litter fall was 545 kg ha^{-1}. Thus a fairly large percentage is returned for recycling purposes while a small fraction is stored in biomass built up.

CALCIUM

Calcium is an important element needed by plants for building their cell walls (calcium pectate) and by animals for bone formation. It is being constantly added to the soil pool through the weathering of rocks and through atmosphere by dust fall and rainfall. A very great proportion of this is kept in a state of cycling by uptake from soil into the biotic pool of plants and animals and their return through litter fall, death and decay. Only a small fraction is lost out of the ecosystem through streamflow and this is replenished by weathering, rain and dustfall.

Bormann and Likens (1967, 1979) have studied nutrient cycling including calcium in Hubbard Brook in New Hampshire and Ovington (1962) has studied calcium movement in a *Pinus* forest in UK. Calcium budget has been worked out by them and many others in different parts of the world. In Hubbard Brook watershed study, the exchange rate between abiotic (soil) pool and biotic (plants) pool was balanced at 50 kg Ca ha/yr. A runoff through stream flow was only 1 kg/ha/yr while the input of calcium through rainfall was 3 kg/ha/yr and the deficit of 5 kg/ha/yr must be met by natural processes of weathering. However, when the vegetation was felled and through aerial application of herbicides the regeneration of vegetation was prevented, the calcium loss through stream flow increased by six times more than what was lost in undisturbed vegetation covered sites. This kind of finding in Indian Vindhyan Hills for nitrogen and phosphorus loss increase on scraping of ground vegetation was reported by Ambasht (1985). After a three year period of devegetation the Hubbard Brook watershed was allowed to recover and on vegetation recovery calcium loss again decreased to normal level over a period of few years. Ovington (1962) has estimated that in a 55 year old stand of *Pinus sylvestris* the total uptake of calcium by plants over the years was 3.04 t/ha of which 0.272 t/ha was in the live biomass of trees at a given time while the total return to litter was 2.56 t Ca/ha.

SULPHUR

Sulphur is an essential nutrient element for plants and animals in trace amounts. In intensive agriculture, therefore, sulphur may be required to be added as fertilizers or SO_2 pollution also compensates this. The largest reservoir of sulphur is the soil, sulphate is the commonest form and it is reduced by green plants in order to get incorporated in certain aminoacids and proteins. Microorganisms in nature perform the specialized roles of conversion of sulphur H_2S to SO_4 by colourless or green and purple sulphur bacteria and from SO_4 to H_2S by the anaerobic *desulphovibiro* bacteria and aerobic *thiobacillus* bacteria. The sulphur taken up by green plants gets incorporated into the organic form and gets returned to soil via litter, grazing animal excreta or through death. Huge quantity of sulphur in SO_2 gas form is discharged into the atmosphere by biomass and coal burning. Pollution effects due to SO_2 and acid rain are described later in the pollution chapter. Sulphur transformations have interacting effects on other mineral nutrients, e.g. in presence of H_2S the insolable form of phosphorus in soil gets converted into soluble form and becomes available to plants for uptake.

Thus the sulphur cycle (Fig. 6.3) starting in the sedimentary stock

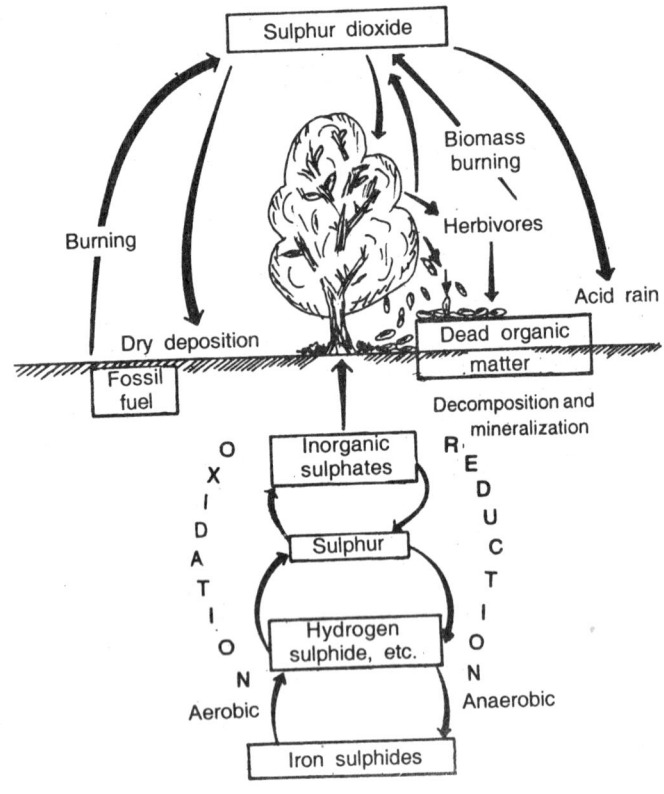

Fig. 6.3. Sulphur cycle.

as iron sulphide gets converted into hydrogen sulphide, then to sulphur and sulphate. Plants as well as microorganisms take up sulphates and then convert them into organic form. Plants also receive SO_2 through rainfall at places where industrial fumes containing SO_2 alongwith CO_2 are being continuously discharged from chimneys. Sulphur cycle cuts across through sedimentary, gaseous and water cycles. The sedimentary part involves ferrous sulphide, ferric sulphide and calcium sulphate. Part of it through combustion of fossil fuel joins gaseous cycle in form of sulphur dioxide, which may join hydrological cycle and get dissolved in rain water to form sulphuric acid and return to land or water bodies. Sedimentary sulphides on oxidation change to sulphur which may get further oxidised to inorganic sulphate, and in this form plants absorb it.

(2) The soil organisms and organic matter

The organic mater in soil is received from the dead bodies of plants and animals of all types and sizes. The dead material cannot as such be utilised but they are acted upon by a series of microorganisms which convert them into humus and ultimately back to minerals. These soil organisms are constantly present causing rapid changes to take place, and so soil is said to be a *living system*. A large variety of soil microflora and fauna abound naturally in soils. Among the microflora are heterotrophic and autotrophic bacteria, fungi like *Rhizopus, Aspergillus, Penicillium*, yeasts, mushrooms etc., algae including nitrogen fixing blue-green algae, and lichens. Common soil animals are amoebae, nematodes, earthworms, snails and slugs, ants, mites, termites, milipedes and centipedes, various kinds of insects, reptiles and burrowing mammals. All these organisms influence the soil properties very much and play important roles in nutrient cycling and aeration.

The earthworms play an important role of breaking up the forest litter, upturning and mixing soil and improving the fertility. The particulate organic materials are further simplified by bacterial action. Protein is reduced to amino acids, amino acids to nitrites and nitrates. Similarly carbohydrates are consumed and CO_2 is released into the soil atmosphere.

The finely broken particles of organic matter, generally of a dark brown colour are called *humus* and the process of its formation is called *humification*. Naturally the property and pH of these soil will depend upon the type of organic material that has humified, e.g., *Pinus* needles on decomposition make the soil acidic while calcium rich leaves of most dicots make the soil alkaline or neutral. The organic materials present in dead plants i.e. litter are celluloses, hemicelluloses, starch, sugars, fats and oils, lignins, proteins, etc. and these are acted upon by different kinds of organisms, principally bacteria, and through a series of stages fine dark coloured powdery colloidal substance called *humus* is formed. Humus remains in soil for different periods in different climatic conditions. For example, in cold European countries humus is not quickly mineralized due to the low activity of microorganisms at low temperatures. In India due to moist and warm conditions humus is quickly reduced to carbon dioxide, water and minerals. This process is called *mineralization* and results in circulation of the minerals from locked up organic compounds. In conditions unfavourable to microorganism the dead remains of organic matter go on accumulating and this results in the formation of *peat.*. In forests the accumulated organic matter on floor is classified on the basis of structure and organisms inhabiting it into *mor* and *mull*. Mor soil is in form of a thick carpet of organic matter usually undifferentiated into layers with a very low percentage of minerals and earthworms do not occur in it. That is to say it is in the form of raw humus. Mull on the other hand is loose porous organic matter in almost equal proportion with inorganic matter and earthworms occur in good numbers.

Organic matter is the source of most of the nutrients for successive crop of plants and organic manuring has been in vogue since time immemorial. Certain elements are however, lost out of the cycle, e.g., nitrogen and this needs refixation by bacteria and other microorganisms, or in agriculture, addition by means of artificial fertilizers. It is of great practical significance to evaluate the amount of carbon relative to nitrogen in soils. This C/N ratio is a good index of fertility of organic manures and it varies depending upon the source and stage of decomposition of the litter

Litter and humus in major biomes

The quantity of litter at a given time in the soils depends upon the percentage of net primary production that falls on ground each year and the rate of its decomposition into humus. Both these factors, i.e., primary production and decomposition rates are dependent largely upon the temperature and moisture regimes of the place. In cooler climates the total disappearance of litter into mineral form is much slower than in the warm climates. Rodin and Bazilevic (1968) have given data on different aspects of litter and humus from cold Tundra to warm deserts in a series of biomass bearing from small plants to large trees. With a contribution of only 4% of the annual productivity in shrubby Tundra falling as litter, there is an addition of 0.1 tons per hectare per year but the total accumulation of litter on ground is recorded to be as high as 24 tons per hectare and of humus as 83.5 tons per hectare. Mineralisation process is so slow in extreme cold place that the ratio of humus accumulated to fresh litter is as high as 92:1. This ratio progressively falls down in less cold Boreal Spruce forest to 15:1, in temperature oak forest to 4:1 and in moist tropical forest 0.1:1. This is on account of very rapid mineralization in warm and moist climates, and therefore, there is a progressive decrease in the humus content from 45 tons per hectare in Boreal Spruce forest to 15 tons per hectare in oak forest to 2 tons per hectare in moist tropical forest. About 23% of the net productivity is returned as litter on the ground in the warm region forest and 28% in temperate oak forest.

(3) Soil water and soil solution

Water is an essential substance for all organisms. It is a medium of reaction and of cytoplasmic materials. Water is itself a reactant in photosynthesis. Terrestrial plants meet their water requirement from soil. The quantity and availability of soil water to plants is a great determining factor of the nature, composition and stature of vegetation of any place.

Rain is the principal source of water for the soil. On sloping topography much of the rain runs off and it may adversely influence soil by causing erosion.

Soil structure is an important aspect which regulates the permeability, infiltration and percolation of water. Permeability refers to the relative ease with which water moves within a particular soil system; infiltration refers to downward movement of water from the soil surface, while percolation refers to the water movement in a column of wet soil.

Water flows due to the force of gravity while some of it may be held or even more upwards due to the force of capillary tension of small pores. Water is also held around the surface of clay and humus particles of colloidal dimensions due to soil moisture tension. Thus the proportion of finer particles like clay and humus determines the water retention capacity. Sandy soil consisting of larger particles are loosely packed and water flows down more rapidly through the large pores between the particles.

A mixture of clay, silt and sand on the other hand holds adequate proportion of soil air and soil water. Most of the mesophytes grow better in the type of soil while in sand or in predominantly clayey soil only specialized types of species adapted structurally and/or physiologically to such habitats can grow well.

Soil water is usually classified into :

1. *Hygroscopic water*—held in the form of a very thin film around the surface of soil particles.
2. *Capillary water*—held in capillary pores of soil where it is retained against the gravitational pull.
3. *Gravitation water*—the water present in soil pores of noncapillary sizes.
4. *Water vapour*—in the soil atmosphere.

Another system of soil water classification is based on a different pattern. It is based on the force with which moisture is held by the soil. This system takes into consideration the *force* with which plant roots are required to absorb water from the soil. From about 0.2 atmosphere to 14 atmosphere of *soil moisture tensions*, water is available to most of the land plants. At 15 atmospheres tension, most plants permanently wilt and cannot recover unless the environment surrounding the plant becomes humid. At this tension the moisture level is called *wilting percentage*. Beyond this soil water tension, plants fail to grow and the soil may be regarded as air dry.

Another useful way of expressing soil moisture is in terms of pF or logarithm of capillary potential or the free energy on a compact scale cf 0 to 7 for the entire range of moisture level. pF = 0 represents saturated soil, 2.7 represents the level at moisture equivalent; 4.2 the permanent wilting percentage, about 5.5 the air dry level and 7.0 the oven dry conditions. At the same pF the actual moisture content is higher in clay than in sand.

When gravitational water percolates down and reaches the ground at the parent rock, it is called ground water. Capillary water also moves both downwards and laterally. Capillary water movement takes place from low moisture tension of high moisture tension.

Some soil moisture constants

1. *Moisture content* : Soil moisture is expressed as *moisture percentage.* This figure represents the percentage of water in the soil on oven dry weight basis. The wet sample is weighed, dried in an oven at 105°C for about 24 hours and then reweighed. The difference of the two readings represents water content. This is converted into percentage (of dry soil) as follows :
   (a) 100 g wet sample dried in oven.
   (b) Oven dry weight say 80 g.
   Water content $(100 - 80) = 20$ g.
   80 g (not 100) of soil contains 20 g of water
   Therefore 100 g of dry soil will contain

   $$= \frac{20}{80} \times 100 = 25 \text{ g of water or moisture content } 25\%$$

2. *Sticky point* : In the dry powder of soil sample water is added gradually in small quantities and mixed. More water is added until the dough just begins to stick to fingers. Moisture content at this stage is called sticky point.

3. *Field capacity* : This represents the moisture content which a wet soil can retain against the force of gravity. Fields containing higher clay have higher field capacity on account of greater capillary porosity. This can be determined by saturating a soil sample and allowing the water to drain down for 2 days. The atmosphere above the soil surface however, should be kept humid to avoid loss of soil water due to evaporation. The moisture content as a percentage of oven dry weight of soil is then determined to get the field capacity value.

4. *Moisture equivalent* : This is very much the same as field capacity

but the term refers to the moisture content at a particular soil moisture tension equivalent to a centrifugal force of 1000 times gravity. A saturated soil sample of 1 cm thickness is centrifuged in a perforated cup at the aforesaid force for about half an hour and then the moisture content is determined.

5. *Wilting coefficient* is the moisture content of soil at which a plant fails to get sufficient moisture and begins to show permanent wilting. This depends upon the moisture content and texture of the soil. In coarse soils even at low moisture levels some water is available to plants and therefore, the wilting coefficients of sandy soils is lesser than clay rich soils. The difference between the moisture equivalent and the wilting coefficient gives the quantity of moisture actually available to plants and that is called as growth water.

Besides such weighing methods, soil moisture is measured in terms of tension with the aid of tensiometers, and pressure membranes. A tensiometer functions on the principle of moisture flow equilibrium between a wet porous cup and the surrounding soil in which it is buried. Variation in the moisture content of soil changes the electrical conductance. For this purpose gypsum blocks with electrodes or nylon and fibreglass blocks are used. In recent years neutron probes have been used to determine moisture content. Their action is based on the fact that water (soil moisture) absorbs neutrons from radioactive sources at a faster rate than of scattering back to a detecting device.

SOIL DYNAMICS

The rocks on weathering produce soil. The process of weathering has already been described at the beginning of this chapter. At any place where parent material is weathering over a period of time, there develops layers of soil one over the other in a progressive state of maturity. Such a vertical section of soil from top manure soil to the rock below is called a *soil profile*. Not only do the layers differ in the size of their constituent particles, but very often they can be distinguished due to their colour differences.

In any soil profile the layers are named from top to bottom as A, B and C (Fig. 6.4). Many ecologists distinguish another layer D. The true soil that is the weathered product of parent material is called *solum* and this constitutes horizons A and B. The semi broken region of the parent rock is called horizon C. In the A horizon the top fertile soil is richly mixed with organic material in different stages of disintegration. The top layer of fallen leaves and twigs still intact or only lightly broken is the Aoo layer. Below it is the Ao layer of partly decomposed organic matter where decomposing organs have lost their identity. Fully decomposed organic matter, i.e., humus enriches the soil by becoming mixed in with mineral components to form aggregations or crumbs. This is referred to as A_1 layer and is dark in colour. A_2 and A_3 have a decreasing quantity of humus and are less dark than the A_1 layer.

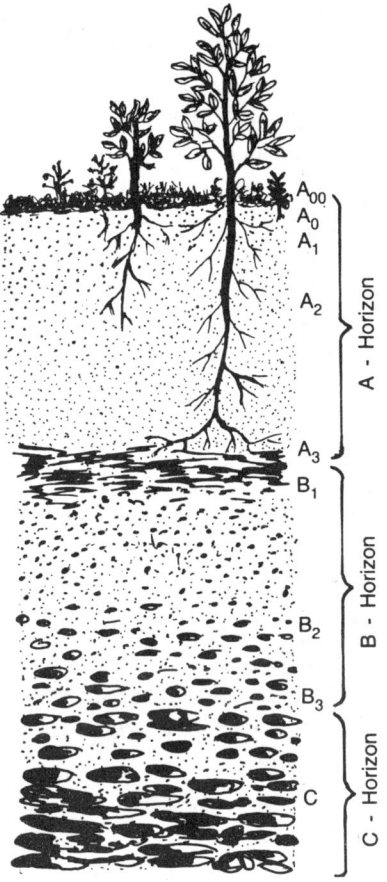

Fig. 6.4. A section of soil profile. Various zones indicate the stages in soil formation.

Horizon A is also known as the *eluviation zone* from which the nutrients released as a result of decomposition are leached downwards to B horizon.

In the B horizon the nutrients received from A horizon are accumulated and it is known as the *illuviation zone.*

The zone below C horizon is made up of the solid bedrock and is called as either D or R horizon. On the basis of profiles, soil is classified into three types (i) *Zonal* soil showing normal profile, (ii) the *intrazonal* with unusual parent material or topography such as the bogs and (iii) azonal with very little or no profile development such as the fresh alluvium consisting of layers of periodic deposition of silt by the river.

Percolation and infiltration of rain water and its upward movement through capillary force to an actively evaporating surface leach down or raise up various materials and thus impart colour and chemical properties to different layers. Variation in climatic conditions and rock properties, therefore, lead to different types of profile and top soil development. The important types of soil development in different climatic conditions are : (1) Laterization, (2) Podsolization, (3) Calcification and (4) Gleization.

Laterite has been derived from the Latin word latus meaning brick. Buchanan in 1807 gave the term laterite for the first time for the soils of Malabar coasts in India. These soils are reddish brown in colour and are found in the warm and humid tropics. In *laterization*, iron and aluminium oxides do not leach down but remain in the surface soil. Silicic acid on the other hand leaches to lower horizons. The Indian lateritic soils are, however, rich in silica as well. Laterites are found in Australia, India, some warm parts of Europe and North America. These soils are comparatively less fertile. Presently almost all reddish tropical soils are referred as laterite or lateritic. In the lower horizons there are lateritic hard pans sometimes referred as ironstone concretion.

Latosol, a wider term than laterite is now used more commonly to refer to soils of warm humid belt of the earth where silica usually leaches to lower horizons while on surface the oxides of iron and aluminium are present. These are poor in bases and have usually low organic content but rich in free quartz grains and kaolinite clays. The other names are oxisol and ferrasol.

Vertisols are the other characteristic tropical soil where there is a long dry season. These are also known as tropical black earths formed from limestones, marl or ferromagnesium rocks and are rich in montmorillonite clays. Deep cracks are formed in the dry summer. These cracks get filled with clay due to dust blow which in the rainy season on wetting swells so much as to form small bumps. Thus the vertisols are characterised by cracks in summer and bumps in rains.

Podsolization is the most extensively studied soil forming process. It takes place in cool humid climate under forest vegetation where leaf litter and other organic matters decompose slowly, chiefly through fungal activity. Percolating water carries down carbonates, sulphates, iron and aluminium compounds often leaving the surface soil acidic. Most of the minerals dissolved in acids are leached down to the B horizon while silica remains on the top. Leaching results into a grey ash like surface which is responsible for the term podsol which in Russian means under ash (*Pod* = under, *Zole* = ash). Litters of conifer forests, which are predominant in cold region on decomposition add to the acidity of surface soil in podsols. But in the Indian conifer forests of the Himalayas true podsols are not met with on account of the alternate phases of wet and dry season. In the dry season rapid evaporation sets in a reverse movement of water and salts thus reversing the podsolic pattern of the profile.

Calcification is a common process found in North India. Here calcium in the form of calcium bicarbonate dissolves in water and leaches down escaping adsorption around dry particles. At depths of about half metre to three metres it precipitates depending upon the quantity of rainfall and depth of infiltration. The precipitated calcium remains in the form of nodules of *Kankar* and form a hard pan. This gradually reduces soil fertility. In Northern India the Kankar nodules are unearthed and are used in the building of roads and in the preparation of lime. Calcification is also a common phenomenon in arid regions of temperate climates.

Gleization takes place in predominantly ice covered regions like arctics where soil is not saline. Peat accumulates on surface. Tundra soils are common example of gleization.

Chernozem is a Russian term and it refers to black earth. These are extremely fertile blackish brown top soil of 3 to 4 feet thickness which bear rich grassland communities.

Under each of the above types of soil development the soil property ultimately depends upon a number of other factors like the nature of parent rocks, the weathering agents, the topography, drainage, erosion, transportation of weathered material and finally the type of vegetation. Man is instrumental in altering the soil characters and appearance of most of the land surface due to his profound all embracing activities. This aspect is dealt with in the chapters on conservation and management and pollution ecology.

A few other characteristic soils are *hydromorphic, halomorphic* and *azonal*. Hydromorphic soils are formed on poorly drained water filled regions with poor aeration, i.e., they are highly reducing and ferric ions are reduced to ferrous. But in dry season the soil becomes aerated and ferrous is again converted into ferric. The hydromorphic soils of tropical paddy fields are also called *gley soil.* *Halomorphic soils* are characterised by the presence of high salt content on the surface. Salinity and alkalinity are commonly found in big patches in the North-Indian states resulting into the formation of *usar* or *reh* soils. These result on account of upward capillary movement of dissolved salts from the rising water table in too frequently irrigated and fertilized regions, or from nutrient rich water received through stream flow from the surroundings during the period of high rainfall. On evaporation, salts in different forms are left on the surface. Grazing animals often meet their part of metabolic salt needs by licking such salt deposits. *Azonal soils* lack horizon B and the horizon A is also thin. Among azonal soils, the common are *lithosols* found on mountain tops where parent rock is quite hard and resistant to weathering. *Regosols* are formed from volcanic deposits. *Alluvials* are soils deposited by water current along the course of rivers.

Catena is the term given to describe regular alternating sequence of thin residual soil on flat hill top with thick deposited soils in the valley followed again by hill tops and valleys in a broad range. The hill top sedentary or residual soil is well drained, with somewhat more acidic reaction and brown to yellow in colour. The valley soil is waterlogged, light grey and less acidic than the soil at the top of the hill. There is a sequential change from top to slope to the bottom of the hillock and each catena is influenced by the one next above it on the slope and influences the one below it due to drainage. Catena or alternating and sequential change is also a characteristic of undulating topography bearing forests or savanna vegetation.

Field and laboratory tests of certain physico-chemical properties of soil

Soil moisture constants have already been described. Other physico-chemical properties are described below :

1. *Colour* : Soil colour is usually expressed in terms of symbols as given in the Munsell colour chart. The soil sample is compared

with standard colour charts and the soil colour is expressed in combinations of *chroma, value* and *depth* or hue. Chroma refers to relative purity of the wavelength, value refers to brilliance that is its value increases from dark to light and depth or hue refers to the dominant wavelength or colour. Tropical soils range from black, grey, slaty, light brown, whitish and reddish and the colour is principally due to dominance of silica, iron, humus, etc.

2. *Soil texture* : This can be estimated in quantitative terms as percentages of clay, silt, fine sand and coarse sand by (i) Beaker method or (ii) International Pipette method. The latter method is more commonly used. A known weight of soil is thoroughly mixed in water. Different size particles settle at different rates and therefore pipetting soil samples from such suspensions at standard depth and interval of time gives a proportional quantity of different textural components. Details can be found in any book on soil analysis (Piper, 1944). For qualitative testing the soil is rubbed between thumb and forefinger and an estimate of its texture as sandy, loamy or clayey is noted.

3. *Soil density* : Soil density is studied for the density of particles and for the entire bulk of soil including the pore space. *Particle density* is the average of densities of different kinds of inorganic and organic particles excluding pore spaces. It is usually regarded as 2.65 g/cm^3. Samples with more of organic particles have lesser density and those with more of heavy minerals have higher density. *Bulk density* is always less than particle density because pore spaces are also included in it. This can be estimated by taking out a core of undisturbed soil by a metal tube or cutting a small pit. The soil is oven dried and weighed. The volume of the soil sample is measured by taking sand in a measuring cylinder and then pouring it down in the pit formed by taking out the sample until the pit is just full. The volume of sand utilized in filling the pit is equal to the volume of soil sampled. The value obtained by dividing the oven dry weight of soil sample in grams by the volume value in cm^3 is the bulk density. Low bulk density is usually better for plant growth and soil aeration.

4. *Soil porosity* : The percentage of space in a soil bulk that is occupied by the interstitial spaces or pores in which air and moisture reside is called as soil porosity. It can be easily estimated by finding out

the bulk density as described earlier and using the following formula :

$$\% \text{ porosity} = 100 - \left(\frac{\text{Bulk density}}{2.65} \times 100 \right)$$

Here 2.65 is regarded as the standard particle density. Size and shape of particles and the compactness determine the porosity. Coarser particles have large pores but the total pore space is less in volume as compared to clay, silt and humus fractions. Clay loam with organic manure has about 55-60% porosity, whereas sandy soil has about 40% porosity. Porosity is an important physical property which determines the soil moisture and air proportions, but for these the capillary (micro) and noncapillary (macro) pores ratio is more important. Capillary pores retain water against the force of gravity.

5. *pH* or hydrogen ion concentration or nature of soil reaction is measured on a scale of 0 to 14. This scale is actually the negative logarithm of the H^+ ion concentration, and therefore, as the pH value increases from 1 to 6.9 the acidity or actual H^+ ion concentration decreases. For example pH 5 means it is equal to 10^{-5}, i.e. from pH 6 to pH 5 there is a ten fold increase in hydrogen ion concentration. pH 7 represents the neutral condition and beyond pH 7 it is alkaline. The soil sample is mixed with water in a 1:5 proportion and thoroughly shaken. The supernatant liquid can be tested for pH by BDH Universal Indicator or pH paper. The colour developed on treatment with indicator or paper is compared to the standard colour chart provided on the bottle of indicator or cover of pH paper booklet to get the pH value directly. Electrical and electronic pH meters are used for more accurate results.

6. *Total soluble salts* : This is usually determined by preparing a soil suspension in distilled water and finding out its electrical conductance with the use of an electrical instrument called a *conductivity meter*. Conductivity is dependent upon the total salt content and its dissociability. The conductivity value of the soil suspension is compared with the conductivity values of known concentration of chemical solutions to get the total salt content in soils.

7. *Carbonate* : In soil, carbonate is mostly in the form of calcium

carbonate which can be detected by adding dilute HCl to a pinch of soil. The acid reacts with carbonate and carbon dioxide gas escapes showing effervescence. Qualitatively, the degree of effervescence is divided in scale of 1 to 5. Low carbonate content shows poor effervescence or CO_3 value 1. Very rapid effervescence indicates very high carbonate content or CO_3 value 5. Intermediate levels are put in class 2, 3 and 4. For exact quantitative estimations the CO_2 gas evolved out of a known quantity of soil on acid treatment is measured in a calcimeter or is absorbed in standard solution of sodium hydroxide and titrated.

8. *Nitrate* content in quick estimations is usually measured colourimetrically. In spot test a small quantity of soil is placed on a white cavity tile and to it is added a few drops of 0.002% solution of diphenylamine in concentrated sulphuric acid. This produces a blue colouration, the depth of which is directly proportional to the quantity of nitrate present. Here also for quantitative purposes the nitrate content is expressed on a scale of 1 to 5. Spot tests for many soil properties have been described by Misra (1945).

9. *Deficiency of exchangeable bases* : Misra (1945) has modified Comber's test to make an empirical estimation of deficiency of bases like Na, K, Ca and Mg. A small quantity of soil is shaken thoroughly in an alcoholic saturated solution of ammonium thiocyanate in a test tube and allowed to settle for a few minutes. A few drops of hydrogen peroxide are then added. Development of a red colour indicates base deficiency and it is noted qualitatively in four degrees of which very light red is degree one and deep red is degree four.

Soil nutrients and plants

Plants need a large variety of elements for their growth and metabolism. Depending upon the quantity needed, the nutrients are classified into *macronutrients* and *micronutrients*. The former is needed in large quantity and the latter in traces. Carbon, hydrogen, nitrogen, oxygen, calcium, potassium, phosphorus, sulphur and magnesium are the main macronutrients and copper, zinc, boron, molybdenum, iron, manganese, chlorine are the common micronutrients. Carbon, hydrogen and oxygen are obtained from air and water while nitrogen is principally stored in air but is routed through soil after its fixation into usable forms by various means as described in Nitrogen-cycle section. The metallic elements reach in the soil from the rocks through weathering and decomposing organic materials and become available in form of either sparingly soluble, inorganic or organic compounds, or in form of actions adsorbed on clay-humus surfaces or in forms of free cations in soil solution. There is a very slow supply of cations from insoluble compounds also. The store-house of available nutrients to plants is the exchange complex of finer soil particles.

Nitrogen is absorbed by higher plants in form of nitrate ions and is used in protein synthesis including all enzymes. Nitrogen deficiency thus reduces growth and metabolic processes, resulting into yellowing and sometimes reddening of leaves. Phosphorus is absorbed in form of phosphates and its supply in soil is both through organic compounds and inorganic sources. It is an essential constituent of DNA and is involved in all energy processes in plants as the energy currency ATP contains phosphate. Its deficiency causes reduced growth, delayed flowering and darkening of leaves. Sulphur is similarly supplied through organic and inorganic sources, absorbed as sulphates and its deficiency causes chlorosis of leaves. Potassium is in form of minerals and is absorbed directly as K^+ ion and its deficiency causes root rotting and water balance of plant is disturbed. Magnesium present as Mg salt is absorbed in ionic form and its deficiency causes chlorosis and reduced growth. This element is the essential constituent of chlorophyll. Calcium is also present in form of salts and is absorbed in ionic form. It is used in the cell wall formation and its deficiency causes leaf deformation and growth is retarded. Calcium also influences the pH of the soil and its aggregation property.

Parent rocks are sources of micronutrient elements. Rocks may be deficient in some micronutrients and rich to the toxic levels in some others. Soil developed from such rocks bear specialized kind of plants that can withstand the deficiency of some and toxic level of others. They may absorb minerals abundantly present and accumulate them in their body, i.e., they act as accumulators. Margaret Vickery (1984) has listed some such plant species as *indicators* of the presence of certain minerals in excess. For example, *Astrgalus* accumulates selenium. In central Africa *Buchnera cupricola, Guttenburgia cupricola* and *Becium homblei* are indicators of copper ore. In Zambia, *Mechovia grandiflora* is a good indicator of manganese and *Acrocephalus robertii* of cobalt. In Zimbabwe chromium ore is indicated by the presence of *Convolvulus ocellatus* and *Pearsonia metallifera*. Srivastava and Ambasht (1990) have reported excessive accumulation of iron in *Polygonum amphibium* in iron enriched polluted river margins.

Plants that grow in calcium rich soil with alkaline pH are called as *calcicoles* or *calciphytes*. These plants cannot tolerate low pH or high concentration of aluminium. As opposed to calcicoles, plants that are adapted to lower pH, low calcium and can withstand high aluminium are known as *calcifuges* or *oxylophytes*. There could also be such an edaphic situation in which the top soil may be acidic on account of leaching down of bases and lower layer becoming alkaline and on such soils shallow rooted calcifuges and deep rooted calcicoles grow together.

There are conditions of 'deficiency' and 'excess' stresses on plants in different habitats. They cause some symptoms on plant parts. Some plant species have *indicator values* with respect to deficiency or excess of mineral nutrients by their presence in such habitats. Schulze et al. (2004) have given examples, and put N_2 indicators into 9 categories from 1, very poor to 9 very high. N1 indicator is *Trifolium arvense*, for N2 and 3 are *Medicago sativa* and *Equisetum arvense*, N4 and 5 is *Primula vulgaris*, N6 and 7 is *Chenopodium album*, N8 *Urtica urens* and N9 *Chenopodium bonushenricus*. Nitrogen deficiency causes yellowing of leave while in N_2 rich conditions leave have high chlorophyll content. Sulphur is a macro nutrient but in nature SO_2 pollution causes damage. Threshold level of SO_2 is 5 $\mu g\ m^{-3}$ in air. Lichens die and in this way indicate SO_2 pollution.

Fire Factor

Fire is of a common occurrence in natural vegetation all over the world, it is more so in dried habitats than the wet. Lightning is the commonest natural cause of fire initiation and dryness helps in its spread. Our earth's surface is hit by lightning every second in one or another part of the globe and many of these are of great magnitude. Other causes of fire are abrasive effects of falling rocks, or dried plant material like bamboos, or spontaneous combustion of very dry and hot material or by volcanic activities. Man of course, has now in recent past become the commonest source of managed as well as accidental fires.

Fire has always been a part of global environment for thousands and thousands of years. It has influenced the course of succession and evolution of ecosystems. Fire arrests the course of succession and modifies the edaphic environment very much. If it takes place at some regular interval of time the seral stage is maintained and, the soil is periodically enriched by the ash. In quite wet areas natural fire and its spread is uncommon. In the Meghalaya region which is one of the wettest region, tribals and even modern farmers practice the *jhum* cultivation by periodically slashing the forest, burning them and then raising crops on the ash enriched soil for a few years and move to another segment of forest as the earlier one becomes eroded and depauperate. After a cycle of 10-15 years when the abandoned plot is again covered with new trees, the tribals return to repeat the slash and burn practice.

Fire in forests are most devastating and difficult to control. Recently during 1997-98, one of the most extensive forest fires have ravaged Malaysia and Indonesia causing widespread misery and pollution. They spread very fast, particularly in the windward direction. In frequent fire zones, trees are adapted to fire in the sense that the while the crown and twigs burn, the main trunk may get damaged in the outer surface only while the inner tissues survive and sprout after the fire is over and conditions become favourable, say in the next rainy season. Fires in natural vegetation is divided into three main types :

1. The *ground fire* usually takes place where there is heavy accumulation of litter. In dry litter, fire is rapid and extinguishes quickly while in somewhat moist litter, the fire is slow and with its heat, the inner parts also get dried and fire continues for a longer period. In ground fire, all herbaceous plants die, but some woody shrubs and trees survive because of their thick protective bark and deep roots.

2. Less destructive are faster sweeping *surface fires*. The flames burn the surface litter and the herbaceous plants. It scorches the tree and shrub by the heat of surface fire. Soil also gets scorched at surface only and after the fire, lots of new seedlings and new shoots from perennating underground parts emerge.

3. The *crown fire* is a forest fire affecting the crown of trees most. It spreads in the top layer from canopy of one tree to the canopy of another and so on. It kills the trees, shrubs and herbs most devastatingly. In moist soil surface, the underground plant parts and buried seeds escape death.

Impact of fire on soil system

Fire intensity is important in determining the quality and totality of impacts. Fire intensity is determined by the heat yield i.e. kcal/g, weight of burning material or fuel g/m^2, and the rate of its spread in metres/second.

Behaviour of fire is determined by a number of factors as the pattern of ignition, the fuel quality, weather condition and topography. For example in windy weather fire spread is fast, in rainy and moist, it gets restricted. Fire first cuts short the normal process of litter breakdown and decomposition. The organic matter gets converted into gaseous CO_2 and other oxides while mineral ash is left on soil surface, which may be rich in Ca, Na and Mg. In normal humification and mineralization process the cation exchange capacity of soil is better than after the fire. Soil bulk density also increases on burning as the organic matter rich soil is more porous or less dense than ash rich burnt soil. Carbon content in soil is inversely proportional to the bulk density. Burnt soil is exposed to greater beating effect of rainfall and therefore, to erosion. In many temperate forests with low pH of account of litter decomposition, burning releases mineral bases and with some increase in pH the activity of soil bacteria including N-fixers increase. During ground fire the surface soil temperature rises but the soil moisture uses most of the heat in evaporation processes and lower layers below 5 cm, therefore, do not get heated too much.

Adaptation against fires

1. Certain trees, particularly conifers like *Pinus* and *Larix* and dicots like *Quercus* develop fire resistant bark with insulating effect against heat. These trees have tall trunk with the crown restricted to upper zone only. This helps in escaping the damages against surface and ground fires.

2. Many trees of frequent fire prone areas like *Eucalyptus* in Australia have dormant buds which escape killing effect of fire and get activated to produce new branches after the fire is over. These buds may be located on the lower part of the trunk and under the soil surface stem and root parts.

3. Latent or dormant buds are also found on a special organ called *lignotuber* in many European and Australian shrubs. The lignotuber is a swollen 'turnip-like' structure which produces new branches after the entire above ground plant is burnt.

4. Some of the plants have fire resistant foliage with lots of water and very little of resin or oil content.

5. Seeds of certain plants have hard coat which effectively protects the embryo against the heat of sub-surface soil during fire. Fire resistant fruits produce new plants which rapidly elongate and produce new fruits quickly. Fire helps the *Pinus* cone to open, to facilitate the release of seeds in favourable season.

6. The indirect effect of fires on some plants is the removal of competition as many nonadapted species vanish from the fired habitats. Fire injury produces open wounds which are easily attacked by parasites. The microclimate is vastly altered due to addition of ash, loss of shade, loss of raindrop interception, accelerated erosion, etc.

Epilobium anguistifolium is a fire indicator species. It normally remains in patches of small scattered form, almost in dormant condition, but when there is a forest fire, these get activated and rapidly grow into full sized plants whereas other plants die due to fire.

Fire is an effective tool in management of habitats for production of useful plants. On slopes, *Saccharum benghalense* are planted to prevent erosion. In late winter the foliage is harvested for thaching and rope making. The remaining stubs are put to fire so as to obtain a much better and vigorously growing foliage during the next rainy season. This improves the rain interception and checking soil erosion and at the same time yield a better biomass. In grazing lands also, fire helps to produce new and soft shoots which are liked by cattle.

Fire alters the soil organic matter, total nutrients, pH and water holding capacity. Soil fungi are reduced while bacteria increase also as a result of post fire changes in soil. Nitrogen loss is due to volatilization.

Role of fire in forests and grasslands has been described by Fox (1998). For Jhum, or slash and burn shifting cultivation a cycle of 20 years (or more if the terrain has greater erosivity) can be followed indefinitely without habitat loss or of biodiversity. Increasing human population accompanied with greater encroachment on forest lands are forcing the farmers and tribals to return to same site at shorter gaps of 10 years of shorter period to slash and burn. This has caused extensive erosion and desertification even in the rainiest of the regions in the North-East India. Fire cycles of shorter duration leads to loss of biodiversity, poor tree regeneration and increase in grasses like *Pennisetum subangustum* (Savill and Fox, 1970), *Andropogon gabunensis* and *Imperata cylindrica* (Fox, 1968).

Trollope (1984) has recorded that the heat generation in grassland is about 17,600 kj/kg. Agarwal and Tewari (1988) have studied the effect of prescribed fires on the productivity of grasslands of Garhwal

Himalaya and found that *Cynodon dactylon* could tolerate short cycles of fire. Short cycle eliminates shrub elements from grazing lands and encourages better grass growth with soft new foliage liked by cattle. Fire is a good tool for managing grasslands. Its frequency alters the natural course of succession. Attiwill (1994) has found that in Australia forest fires kill *Eucalyptus regnans* but regeneration is prolific from seeds, while *E. marginata* resists fire. Edroma (1986) has found that fire eliminates many native grass species and exotics take their place. Fire generated succession is greatly influenced by the intensity and pattern of grazing.

In annually burnt grasslands only a few resistant shrubs persist such as *Lophira lanceolata* trees with thick bark in grasslands of Sierra Leone (Savill and Fox, 1970). While the savanna fire kills the aerial shoot of shrubs of *Prosopis juliflora* it again sprouts from the living basal stem after the fire is over. In some of the American grasslands, fire is reported to help invasion of *Quercus virginiana* tree seedlings, increase in species diversity of forage grasses and forbs (Springer et al., 1987).

Fire is reported to stimulate dormant leguminous seeds buried in soil to germination due to heat and smoke (Brown, 1993). Fire events also influence flowering and increase in biomass production in *Tephrosia virginiana* (Dudley and Lajtha, 1993). Howe (1995) has differentiated species into flowering guilds in relation to burning responses as compared to unburnt controls. In Serangeti National Park (Tanzania), grazing and burning are reported to increase species diversity, particularly of annuals.

More and more application of fire has been in vogue in modern times in grassland management as it helps in regeneration of soft new foliage liked by cattle. However, in large animal reserves such as for tigers, fire is prohibited.

CHAPTER

8

Biotic Interrelationships

Green plants are independent and produce their own food upon which the rest of the biological world depends. This sweeping remark appears wholly correct but when we look closely at ecological balances and interdependencies we see that a variety of direct and indirect biotic interrelationships and interdependencies of all types exist between plants and animals. In fact no organism, not even the simplest ones, can live completely isolated without experiencing the influences of some other organisms. The CO_2 used in photosynthesis by plants is being continuously supplemented by animals through respiration. Nitrogen used by higher plants is fixed by the bacteria, actinomycetes and some blue green algae. Besides competition, there are many more types of close inter-relationships between organisms for example *symbiosis, parasitism, epiphytes* and *lianas, grazers, carnivores, zoophily, entomophily, zoochory, ornithochiory,* multifarious role of man and so on and so forth.

1. *Symbiosis and related phenomena* : The association of two different species for mutual benefit is called *symbiosis*. The term is derived from the Greek words *sym* (= together) and *bios* (= life). Among animals there is some slightly different association where the benefit, derived by weaker partner is less clear it is called *commensalism*, while specialised associations like of algae and fungi in *lichens* is called *mutualism*. But in plant ecology symbiosis is a fairly broad term to include a variety of associations such as *disjunctive* and *conjunctive symbiosis* where associated plants are not in constant contact with each other or they are in contact,

139

respectively. Among disjunctives the relationship may be *trophic* (related to food) such as grazing of plants by animals, of insects deriving food (honey) and in return bringing about pollination or *atrophic* (not related to food) such as shade and humid cool air by a large tree to others living under it. The conjunctive symbiosis may also be trophic such as nitrogen fixing bacteria living in the root nodules and atrophic such as lianas and epiphytes which live in association for support or space. Many ecologists however regard symbiosis as a narrow term to include not all associations but only those where it is strictly for the mutual benefit of the associated species.

The symbiosis of bacteria and leguminous roots for example is advantageous for both the organisms and in fact it modifies suitably the edaphic environment for the succeeding crop community in the next season. Bacteria receive a protective space to live in and derive prepared food from the roots. Legumes on the other hand receive fixed nitrogen in usable form and manufacture protein from it (legumes are rich in protein). Otherwise atmospheric nitrogen cannot be taken by these plants. After the legume crop is harvested or the plant dies, the nitrogen fixed in roots is converted by the decomposition into nitrates and ammonium ions which are of significant nutritive benefit to the next crop of plants.

2. *Mycorrhizal and other similar associations* : In mycorrhizal associations, tree roots become infested with fungal hyphae. The fungi derive their food from the tree roots. But it is not a case of parasitism as the fungal hyphae in return supply water and minerals that they absorb from the soil, much like root hairs of trees. It is now estimated that in about eighty percent of the families of seed plants some species have mycorrhizal associations. Much experimental evidence indicates that the fungi derive much more than they give to be the higher plants but, somehow, these associations are found to be essential for a good and successful growth of some conifers. It is believed that the fungus regulates the pH and sugar content level for a good growth of roots. Mycorrhiazae may be on the surface of roots (*ectotrophic*) or inside between the cells of the roots (*endotrophic*).

Ectomycorrhizae are common in temperate forest trees like

Quercus, Pinus, Salix, etc. and are commonly basidiomycetous, and a few ascomycetous fungi. Endomycorrhizae are of wide varieties and have been grouped in four categories by Smith (1980). There are (1) *Vesicular Arbuscular Mycorrhizae* (VAM) found on tropical as well as temperate herbs and deciduous trees and the fungi involved is *Glomus* and a few others. (2) *Ericoid Mycorrhizae* are common on sclerophyllous shrubs, particularly of the family Ericaceae and a few others and the fungi involved are *Clavaria* and *Pezizella.* (3) *Arbutoid Mycorrhizae* are found on autotrophic in Mediterranean climates as on *Arbutus* and heterotrophic herbs like *Monotropa* with *Arimilaria* fungus, some of the fungi are on surface but grouped in *endo* category because their hyphae penetrate the cortical cells. (4) *Orchidaceous Mycorrhizae* are found on autotrophic and heterotrophic orchids particularly by *Rhizoctonia* as saprophyte or epiparasite. VAM are reported to connect several plant species in a herbaceous community while in trees, ectomycorrhizae do not connect different plants. VAM are efficient in releasing phosphorus from the soil. From herbaceous stage, as the succession progresses to woody stage, there is an increase in leaf litter on the ground surface. Ecotomycorrhizae then become predominant and remobilize phosphorus and nitrogen from the organic mass (Read, 1994).

The other similar root associations are with actinomycetes (*actinorhizal*) in form of nodules and with some blue-green algae (or cyanobacteria) forming coralloid roots. Both these kinds of associations are connected with nitrogen fixation by the microscopic associated organisms present in the roots.

3. *Lichens:* Some particular types of algae & fungi associate together to give rise to a peculiar type of life form called *lichens.* The algae manufacture food which the fungal component also uses. Fungus on the other hand protects the algae from drying up and so these are able to colonize dry rocks and trees barks. This type of association is also termed as *mutualism.* In lichens the alga-fungus association is also present into reproductive bodies called *soredia.*

4. *Parasitic relationship* between plants and plants, or animals and plants, are of very widespread occurrence. In fact all plants and animals have some kind of parasite which impoverishes the host.

Parasites greatly regulate the population level of plants and animals. There are quite a few vascular plants too, which have in the course of evolution become parasites. Many have the green pigment in their leaves and these are called partial parasites in the sense that they derive the raw materials from the host. *Loranthus* and *Viscum* are common partial parasites infesting the stems of hundreds of tree species in India and elsewhere. The host trees are at times so extensively covered with the parasites that they have a peculiarly different look. The parasites derive their foods from the host through haustoria entering the body of the host. Other parasites like *Cuscuta* on stem and *Orobanche* on roots have lost chlorophyll and are complete parasites. The *Cuscuta* stem twines around a large variety of hosts and at contact points pierces their haustoria into the host tissue. In *Orobanche* the seeds germinate near the host plant and soon get attached to the host root. The host and parasite appear as separate plants but the contact is there within the soil *Orobanche* appreciably affects the yield of mustard, brinjal and some other members of the Solanaceae, Cruciferae and Gramineae families. However, an interesting fact is that the host root of mustard grows much longer to come into contact with the root of *Orobanche* (Rawat and Ambasht, 1958). Such mustard plants in association with *Orobanche* have a more robust growth; an anomaly still remaining unexplained.

Several fungi, like rusts and smuts, parasitise several crop plants and affect their production appreciably. Some are complete parasites in the sense that they cannot live as saprophytes even for a short period after the death of their host. These are called *obligate parasites*. Other parasites, however, for brief periods can live as saprophytes on the dead remains in their host or other organic materials. These are called *facultative saprophytes.*

5. *Prey-predator relationship* : This is a kind of relationship of eating and being eaten of one species by another found universally among animals of all kinds of ecosystem. There are predatory organisms like snake or hawk that prey upon rats, frogs or small birds. Both the predators and prey regulate the population size of each other and in ultimate effect regulate their own populations through the medium of the other. For instance if the population of predators

like tiger gradually increases at the cost of prey population, say of deers. With the rise to some extent in tiger population there would be a decrease in deer population. This leads to reduction in predator population for the want of sufficient food and as it decreases the prey population rises with decreased predation. Thus there is an oscillation of rise in predator and fall in prey population on one hand and fall in predator and rise in prey population on the other. But as it happens in all biological dynamic phenomena, here also there is some *time lag* difference in the periodicity of rise of one and fall in another population. They do not synchronise at exactly the same time.

6. *Epiphytism* is still another type of biotic association where the epiphytes grow on other plants but do not derive their food from the supporting plant. These also grow on a variety of plants occupying different positions, but mostly they are to be found on the stems of higher plants. Epiphytes are more common in the warm and humid tropics. They derive their supply of moisture and nutrient from the frequent rains and debris accumulation in bark crevices. Dust is also a source of the nutrients. Drought (dryness) is a limitation to the epiphytic mode of life and therefore epiphytes occur in humid climates. Roots of epiphytes often store water in a special tissue called *velamen*. A very remarkable type of adaptation for meeting water requirements is found in *Dischidia* where the leaves fold and form a jug like structure with a narrow mouth where water accumulates. Roots of the plant enter and grow in these specialised cavities of the leaf and so the plant escapes drought. Many epiphytes produce a thick network of roots upon which wind blown dust accumulates and provides the necessary edaphic environment.

7. *Lianas* are also more common in tropical rain forests when light at ground level is scarce because of the dense and multistoreyed growth of vegetation. Lianas are woody plants rooted in the ground but they climb up with the support of other trees and reach almost to the top of the forest canopy. The woody stem is closely attached to the supporting tree but it is not involved with it in a nutritional relationship. In Indian tropical deciduous forests *Bauhinia vahlii* is a common liana.

8. *Carnivorous plants* : A number of highly specialized plants like

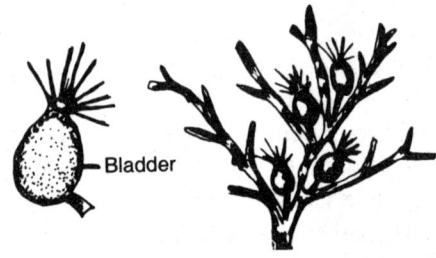

Fig. 8.1. An aquatic insectivorous plant *Utricularia* or bladderwort. The tiny insects enter the bladder along with water current but fail to come out due to valve action. They die and get digested.

Utricularia, Nepenthes, Drosera Dionaea (Fig. 8.1 to 8.4) etc. are dependent for part of their nitrogen requirements upon small animals like insects. These insectivorous or insect eating plants, have some specialized structure and mechanism to trap insects and then secrete some enzymes to 'digest' the protein contents of the insect body. In *Nepenthes*, a pitcher plant commonly growing in the humid forest of Assam, the leaf lamina is folded

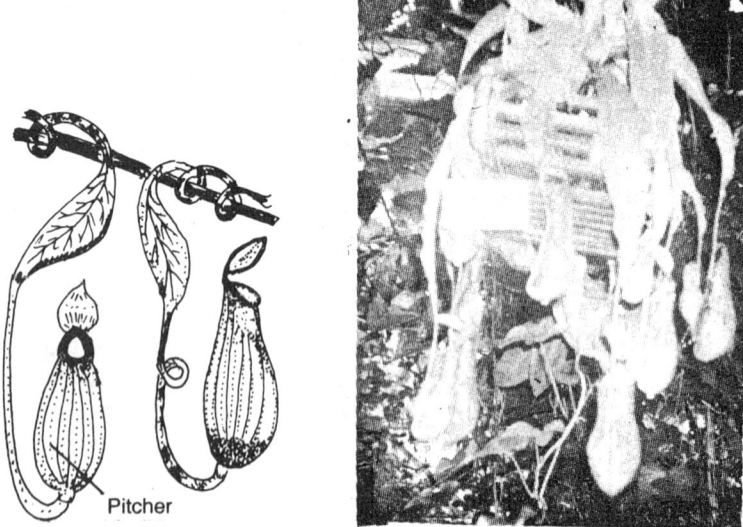

Fig. 8.2. *Nepenthes* or pitcher plant showing pitchers.

Fig. 8.3. *Dionea* or venus fly-trap. The jaw like trap of trigger hairs close tightly and trap insects at the slightest pressure.

Fig. 8.4. *Drosera* an insectivorous plant. The bright coloured leaf hairs attract insects. When they sit on it the hairs close and trap them.

and modified into a pitcher like structure. A lid regulates entry and checks the exit of small insects. The entrapped organism is digested by some proteolitic enzymes. In *Drosera* and *Dionaea* the leaves are covered with hairs. Their coloured leaves attract small flies and other insects.

Utricularia an aquatic insectivorous plant commonly found in India, is covered with 'bladders'. The morphology of these bladders is variously interpreted but it is obviously formed from a leaf. Small planktonic animals floating in water current enter the bladder where they are ultimately digested by the plant.

9. *Competitions* : In all ecosystems there is an element of interaction and interference between individuals of same as well as of different species. When individuals interfere one another in order to get more of space to live, or light, or nutrients or water from soil or any other of their requirements, particularly when the resource is in short supply to meet full requirements of all individuals, there develops *competition* among them. If the individuals of the same species exert interference on one another, we call it as *intraspecific competition* and between different species as *interspecific competition*. Competition is therefore dependent upon the resource level or carrying capacity of ecosystem and the number of individuals present there are the extent and kind of requirement of each individual. Competition results into elimination of weaker species, population regulation, density controls and adjustment of different kinds. Population controls and related aspects are described later in a separate chapter.

In competition studies between two or more species, first the *nature of association* is studied. Three types of associations are commonly found : (i) the *positive association* in which the individuals of two species occur together in the given space, (ii) the *negative association* in which the individuals of the two species occur in separate stands in the given space and (iii) *random association* in which individuals of both the species occur at random without any preference for a positive or negative association.

In sampling a given area for competition between two species, say A and B, there are four possibilities of their occurrence in sampled quadrats. *First*, the number of quadrats in which A and B occur together; *second*, the number of quadrats in which only A is present, *third* in which only B is present and *fourth* the number of quadrats in which neither A nor B is present. These four values can be converted into percentage i.e. out of 100 quadrats.

Competition is one reason for *endemism* i.e. restricted occurrence of species to a small exclusive habitat. Many species adapted to otherwise inhospitable situations like halophytes in highly saline areas escape competition from *glycophytes* or plants that cannot tolerate high salinity. Even epiphytism is regarded as a specialized adaptation to escape competition from other terrestrial organisms. Competition is among associated species only. Plants with similar size, shape and rooting habit compete more severely for the common resources. Individuals of the same species have the same or overlapping *niche* dimensions. So *intraspecific competition* is more serious as it exploits only one zone of habitat and same kind of resources while other zones remain unutilized. But when there are individuals of different species and of different sizes, the exploitation zones and resource use get divided leading to some adjustment among competing species due to less or not overlapping niche dimensions. *Interspecific competition* is a better exploitation of overall resources. So mixed cropping in agriculture is recommended for higher net yield over a longer period from the cropfields.

10. *Amensalism* is a comprehensive term in biotic interrelationships to denote the adverse or depressing effect on one of the competing population or species while the other remains stable. Among lower organisms the production of harmful substances for the life of another organism is called *antibiosis* while in higher plants it is called *allelochemics*. This offers a better competitive ability for the survival of one species over another, but it must not be confused with the phenomenon of competition. Competition is for successful removal of water, nutrients or use of light or space, etc. whereas *amensalism* including *antibiosis, allelochemics* like *allelopathy* are addition (production) of chemicals to harm the competing species. Within the overall allelochemics, if both the interacting species are plants, then it is called *allelopathy*. This term is derived from two root words *Allelon* meaning 'each other' and *pathy* meaning 'suffering'.

There is a very definite kind of impact of the nature of litter producing species on the chemical nature, pH and microbial population in the soil surface. For example, conifers uaually decompose to produce acid pH humus favouring fungal growth, while most dicots produce neutral to alkaline pH favouring bacterial growth. Among the conifers *Pinus* is more acid producing than other conifers like spruce.

Rainwater reaching through the canopy of trees also contain appreciable quantities of nutrients, much more than directly falling raindrops. Water received as stemflow i.e. after canopy interception, vary in nature and concentration or chemicals under different tree species.

11. *Allelopathy* or chemical control of distribution among plants has been quite extensively worked out. Muller (1966) reported that in Southern California, *Salvia leucophylla* leaves emit some volatile oils which reach the soil surface and inhibit the germination of seeds of other species and the inhibitory effects persist for several months. Allolopathy term is derived from Greek words : *Allelos* = another and *pathos* = suffering, i.e., inhibitory or suffering effect of one species on its another neighbour. Similar allelopathic effects are also reported to be exerted by *Adenostoma fasciculatum* (Christensen and Muller, 1975).

The chemical excretions from one kind of organisms into the inorganic medium like soil, water or air and their reaching to and influencing other kinds of organisms have also been called by the terms *ectocrines, exocrines* or *environmental hormones*. The excretions of one species may be harmful to certain other species or it may be useful and growth promoting. Antibiosis and allelopathy are very similar phenomena termed differently by different scientists. Allelopathy is not confined to root exudates only, but it has been found that even aromatic leaves fallen on ground on decomposition release chemicals that inhibit growth of many herbaceous weeds around the plant from which leaves had fallen. Antibiosis is more commonly referred to secretions by microorganisms that check the growth of others. For instance *Penicillium*, a fungus found in soil produces antibiotic substances that check the growth of a large variety of bacteria. All harmful interactions between different species through the production of chemical byproducts are collectively known as *allelochemic interactions*. Allelopathy and antibiosis can both be regarded as types of allelochemism or amensalism. Datta and Chakrabarti (1982) prepared weed plant extracts such as of *Croton bonplandianum* and *Clerodendrum viscosum* in water and tested their effects under different dilutions on the seed germination and growth of crops as pea, mustard, rice and lettuce. *C. bonplandianum* inhibits germination of the above crops except of pea at high concentration. The inhibitory effect of younger leaf extract is stronger than of mature leaves and the effect is also found to vary with the season. Besides seed germination, the root and hypocotyl lengths are also reported to depress. They find that the active ingredient in *C. bonplandianum* leaves showing allelopathic potential are abscisic and phaseic acids while in *C. viscosum* inhibiting mustard seed germination may be a terpene. *Clerodendrum* is also reported to exert allelopathic potential on several weeds like *Abutilon indicum, Amaranthus spinosus, Cassia tora*, etc.

Bar et al. (2000) have shown that in *Kalanchoe* the mother plant excretes from its peripheral roots chemicals that prevent growth of daughter plants lying too close to the mother plants. This is an extreme case of intraspecific allelopathy. In Walnut, *Juglans regia* the decomposing leaves under the tree canopy produce a chemical called *juglone* (under the influence of bacteria) which check, the growth of grasses (Schulze et al., 2004).

Grazing and scraping

A most obvious biotic factor in forests and grasslands is grazing by cattle and wild herbivores. This has a profound effect on the composition, structure and physiognomy of vegetation. Heavy grazing reduces the photosynthetic parts so much that many plants succumb or their population decreases. This is true for palatable species. The unpalatable species avoided by grazing animals multiply and increase in number in absence of competition due to the decrease in the population of palatable species. There are certain species in every grassland or forest floor whose populations are not appreciably altered due to grazing. Thus in a grazed land the species composition and population structure are altered differently in different component species. In *Dichanthium annulatum* there is a considerable loss of seeds and herbage as a result of grazing. The reduced reproductive capacity due to loss of seeds is compensated by a remarkable adaptation of increased reproductive buds on rhizomes (Ambasht and Maurya, 1970). Thus the species is successful even in well grazed fields.

Another indirect effect of grazing is on the rate of soil erosion. This is certainly the greatest menace of grazing aspect of biotic factor. The decreased shoot cover and root growth due to heavy grazing may expose the land surface to eroding forces such as beating effect of rainfall, cutting or scouring effect of water run off and to wind erosion.

The rain drops falling directly on the soil surface in the absence of plant cover splash the finer particles selectively to a greater height. Thus slowly the coarse particles remain below while clay and humus are splashed up and settle on the top. Once the soil is saturated the excess water begins to run down the slope and the clay and humus laid on the surface are washed down first. Depending upon the topography, erosion leads to the formation of gullies and ravines as the bulk of running water increases. The whole process impoverishes the soil and in several instances completely exposes the underlying rock. Thus the soil formed very slowly over thousands of years and stabilized by plant cover is quickly lost through erosion due to the indirect effect of grazing. In the outskirts of Indian forests, villagers scrape grasses to feed their cattle in summer. Scraping is usually done right at the ground level. Completely exposed and loosened top soil is blown by strong summer wind. This is followed by rainy season and much damage is usually done to these bare ground in the first few heavy showers before annuals have grown.

Another aspect of even moderate grazing is the large scale consumption of seeds due to their high nutritive value. Seed loss results in poor appearance of successive crop of their species, unless the plant is adapted to perennate and propagate vegetatively. Rats selectively remove seeds from grasses and store them in their burrows in large quantities.

Grazing or eating of fruits by other animals and birds have some advantageous effects also. Fruits and seeds of some species get dispersed by grazing animals. Some seeds fail to germinate unless they have passed through the gut of birds or other animals. The seed coat in these seeds are hard and impermeable to water or oxygen. The seed coat is softened (scarified) and often digested by the animals and the rest of the seeds come out and germinate quickly.

Role of animals in pollination and dispersal of seed and fruit

In a large variety of species the pollen grains are carried from the anthers of one flower to the stigma of another of the same species. Entomophily and zoophily are the two terms for pollination through insects and other animals respectively.

Bright coloured, or highly scented or honey producing flowers attract insects for these reasons. The visiting insects gets dusted with pollen at the slight touch of the anthers. Pollen grains reach the stigma of flowers of the same species by such visiting flies and insects. The phenomenon of cross pollination by insects is the only means of pollination in a very large number of species.

Honey bees are one of the most efficient cross pollinating insects as they visit one variety of flower in one trip of nectar collection. In the course of evolution of ecosystems in many species the flowers have evolved specialized structures or shapes that suit specialized types of visiting birds or insects of the ecosystem in bringing about cross pollination.

A large variety of seeds are dispersed from one place to another through animals eating them. *Ficus* species are seen growing on top of building or on another tree where seeds reach through birds' droppings. Seeds pass through the intestine unharmed. Such is the case with seeds of partial stem parasite *Loranthus*. Grazing animals carry a large variety of hooked seeds or fruits as in *Xanthium*, *Achyranthes* and *Setaria* species. These possess hooks or barbs and cling to the coat of animals and get dispersed. Many seeds without any specialised clinging device also get transported on account of their habitat. For example the pond heron and other marsh-visiting birds carry spores and seeds from one pond to another in the mud clinging to their feet.

Man is the greatest biotic factor and he regulates plant life in hundreds of ways. Much of the land has been brought under crop cultivation by him. Forests are being managed, lakes, rivers and all profoundly influenced by the activity of man. The role of man in the management of resources is described in a separate chapter on Conservation and Management.

9

Autecology and Population Dynamics

Autecology is the study of environmental interrelationships of a species, population or a set of populations at all stages of its life cycle. A person who has some experience of raising plants even in garden or in pots, understands the species requirement with regard to light, temperature, water (climate), soil condition and nutrients (edaphic conditions).

The effect of the biotic environment such as presence of weeds (interspecific competition), or close or wide spacing between individuals of the same species (intraspecific competition) or still other biotic effects on the germination, growth, development and seed output, etc. are also easily understood to a certain extent. Agriculture (regarding crop plants), silviculture (regarding forest trees), horticulture (regarding ornamental plants) etc. are based on detailed aspect of the autecology of economically important species. Besides detailed ecological life history studies, autecology further aims towards an understanding of natural distribution, adaptation, differentiation of population and speciation. For a proper study and understanding of autecology some basic knowledge about (i) taxonomy, (ii) morphology, (iii) cytology, (iv) physiology, etc. and about (v) environmental complex and measurement of its components is necessary.

ECOLOGICAL LIFE CYCLE

Each step in life cycle of a plant is greatly influenced by a number of environmental factors.

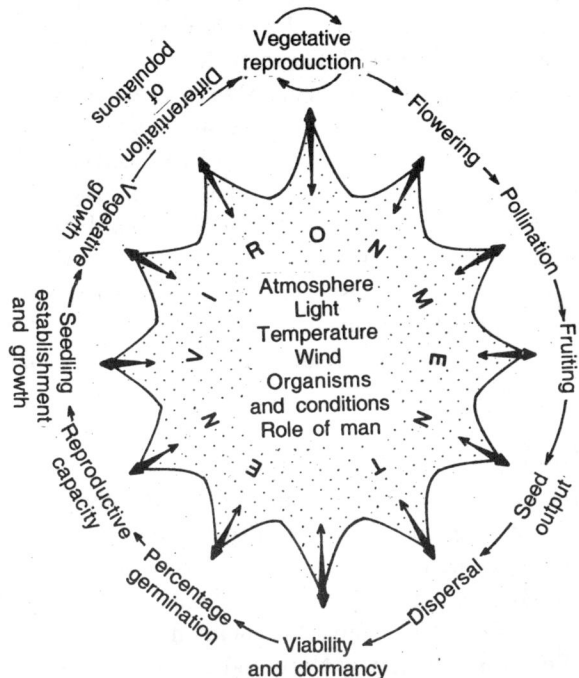

Fig. 9.1. An Ecological clock, phenological stages in the ecological life cycle of plants. The environmental complex made up of several factors in various combinations affects each stage of the life cycle. Plants also influence the environmental complex.

In nature no species grows singly. There is always a large variety of species found at any locality growing in a certain pattern as to give the area a particular appearance. Study of all stages of the life cycle of a particular species in nature in association with other species of the community, completely bathed in environment both above ground and underground is the core of autecology. Each stage of the life cycle is studied in detail, both in the field and in cultures grown under controlled or known environmental conditions and then the whole picture of autecology is built by putting observed and interpreted facts in proper sequence and perspective. The biotic and abiotic aspects of the environment of the species are measured quantitatively at different stages of growth at regular intervals of time to study dynamic changes including phenological events like germination leaf fall, floral bud initiation, etc. With the environmental changes through the seasons. Ambasht and Lal (1979), Lal and Ambasht (1973, 1979) and Ambasht and Chakhaiyar (1979) have made interesting observations on the autecological behaviours of many weeds, particularly *Scoparia dulcis, Chrozophora rottleri* and cropland weeds particularly with respect to growth and biomass production rates. Silvertown (1987) has considered life tables and divided populations into age classes and age specific mortality risk. *Cohorts* or batches of seedling from their birth to death of the last individual of the cohort are studied to obtain a *dynamic life table*. These aspects will be considered later in this chapter. We can presently divide important phenological events in the ecological life cycle of higher plants and discuss the fundamentals of environmental impact and control on them, one by one.

1. Flowering

In terrestrial ecosystems, higher plants predominantly reproduce by sexual means of formation of flowers and then seeds. However, aquatic plant communities are equally efficient in reproduction through vegetative methods. Environment is greatly responsible for bringing about the initiation of flower formation. We commonly observe that mango begins to flower in February in North India while it flowers in other months in south India. Similarly many other species flower in different months at different latitudes or at different temperatures. It is quite evident that temperature and light (day length or photoperiod) are very important factors of environment which induce and regulate flowering. The mechanism of regulation of production uf auxins responsible for the formation of flowers in response to light conditions and temperatures are described in many modern books of plant physiology. In nature population of the same species like *Xanthium strumarium*, are differentiated into photoperiodic ecotypes (Kaul, 1965). *Ecotypes* are genetical races of the same species differentiated due to interplay of certain environmental conditions. The characters so *developed always go hand in hand with the specific environmental factor*. Alternating range of maximum and minimum temperatures also regulate flowering in many plants (e.g. tomato) and the phenomenon is called *thermoperiodism*. A certain variety of tomato does not flower despite all other optimum condition of water, nutrition and light unless the diurnal maximum and minimum temperatures reach a certain level. Further, the temperature factor hastens flowering in certain varieties of wheat grown in temperate regions. Unless seed is chilled at freezing temperatures over a period of time, the plant fails to flower until in the vegetative phase the plant is subjected to one season of cold temperature. The phenomenon related to this aspect of flowering physiology is called *vernalization*.

Besides light and temperature, large number of other environmental factors influence the plant at the flowering stage. Biotic factors may be of great importance. In very closely cultivated crops flowering may be less because of rapid vegetative growth and large scale sharing of limited nutritional resources in the soil. Man and his domesticated animals are further responsible for selective removal of flower of certain species such as of *Nymphaea* and *Nelumbium*, or grazing of many grass inflorescence by cattle.

2. Pollination

It is largely governed by the morphology of flower with regard to the position and time of maturity of the stamens and the carpels, the colour, scent, season of flowering, wind movement etc. In a community of several species, for example, a honey bee visits selectively the flowers of the same species on each nectar collecting trip and as such it is more effective in pollination. Unfortunately in most of the autecological work this aspect has not been studied as critically as it should have been.

3. Fruiting

The structure and number of fruits, number of seeds per fruit, the seasons of fruit formation, etc. are important aspects in the ecological life cycle that greatly influence the success of a species among the members of the community in regeneration and establishments, generation after generation.

Even after effective pollination and fertilization the environment affects fruit formation to certain extent. Biotic factors especially pathogens like certain fungi and insects, damage fruits a great deal. Smuts may damage the inside of fruits, as do several insect species. Fruits are richest in food value and as such many animals including man, eat the fruits selectively. Under such stresses of the biotic environment, many species have evolved to produce fruits in much larger quantities than needed and the small quantity left by consumers are still sufficient to perpetuate the species. In several species the seeds are effectively protected in strong fruit coat, or covered by pulp which attracts potential seed disseminating animals, etc.

4. Seed output

Plants always produce seeds in much larger number than the habitat could sustain if all of them successfully germinate and establish. This is necessary in view of heavy wastage, destruction or consumption by herbivores and heavy seedling mortality. The number of seeds produced in each flush (i.e. the *seed output*) is of considerable autecological significance. Annuals normally produce seeds once in their life time, perennial shrubs and trees produce seeds usually once in a year. Variations like species producing seeds only once in several years or continuously throughout the year are also commonly found. The seed output of some Indian species has been calculated. Every species has its own range. Some produce only a few hundred (e.g. many legumes) and some others produce thousands (e.g. several grasses). Some members of the Orchidaceae with exceptionally small seed size, may produce much greater numbers. In several large trees like *Ficus spp.* the seed output is also very large. Several environmental factors influence the seed output of an individual plant but biotic factors are the most important. Seed infecting fungi and insects damage the embryos at a very young stage leading to inhibition or malformation of seeds. Herbivores mostly remove the top of grassland plants giving little chance for the inflorescence to develop, and grazing cattle also trample the plants. Man himself is a great force in influencing the seed output of many species chiefly crop plants. His efforts to eradicate some species and cultivate others, raising the yield of their seeds, etc. are well known. Competition between individuals of the same species (intraspecific) and between different species (interspecific) also influence seed output. For example, in a closely cultivated field the average seed output (per plant) is a lower than in plants spaced at intervals wide enough to avoid competition for nutrients, light and other resources. In some species the seeds are produced in several flushes which regulate germination and development of seedlings at intervals. This ensures continuity of populations even when there ar e adverse situations in which one set of seedlings perish due to drought, disease, grazing, etc. In *Alysicarpus monilifer*, a herbaceous legume, Maurya and Ambasht (1973) have found the formation of dimorphic seeds which show a variety of difference and which are responsible for differentiation of its populations.

Plants have evolved mechanisms to retain stocks of dormant buds rhizomes or seeds in soil all the time to await suitable period for germination in batches as shown by Ambasht (1962) in *Alhagi camelorum* and by Tripathi and Harper (1973) in *Elymus repens*.

5. Dispersal of seeds

If all the seeds fall at one place around the mother plant the natural consequence will be poor regeneration, for there will be much competition for space and nutrients. Movement of plant populations to wider areas ensures success in establishment of the species in increasing numbers of their individuals. Studies of the dispersal of seeds and other propagules are important aspects of autecology. Many species of the Orchidaceae family produce extremely small seeds that are blown by the wind to great distances because of their lightness. A large variety of plants in widely different families produce flat and compressed seeds able to float in air and they are carried to great distance by the wind. Many others, mostly fruits of Compositae and Asclepiadaceae, have fine hairs covering the small seeds which also enable them to float in air and get transported. For many species growing of the sea shores, on river banks or in aquatic bodies water currents may carry the seeds and fruits to great distances. Some members of the Acanthaceae like *Ruellia tuberosa* and *Andrographis* sp. have a mechanism of dispersing the seeds by explosion of fruit when moisture is available. Biotic agencies like man and animals are also important agents of dispersal of seeds, fruits and diaspores. Many wild plants have hooks, stiff hairs or other projections on their seeds through which they get attached to the bodies of animals. Man transports the seeds of crop plants, and inadvertently the associated weed seeds also, to different parts of the globe for agricultural purposes.

Dispersal of seeds and other propagules can be through the same species (*autochory*) or by other agencies (*allochory*). In autochory, the dispersal may be by (i) runners or creeping nature, (ii) rolling down by gravity and (iii) seeds dispersed in bursts of explosions of fruit due to moisture or other agencies. Allochory may be by wind (*anemochory*) rolling on ground surface (*chamaechory*) or in air current above in the air (*meteochory*). Other transports are by water (*hydrochory*) because of float mechanism in flowing water (*bythisochory*) or stagnant water (*nautochory*); by animals (*zoochory*). Schulze et al. (2004) have elaborated the classification in some detail. Animals ingest the seeds and pass through their intestine (*endochory*), by ants (*myrmechory*) by birds (*ornithochory*), or by animals not involved in eating but mechanically transporting the seeds that get attached on its skin or fur (*epichory*). Some monkeys eat the fruit pulp and then spit the hard seeds from their mouth (*stomatochory*). Human role in seed transfers (*anthropochory* or *hemerochory*) either intentionally e.g. foodgrain export (*ethelochory*) or unintentionally like the weed seeds along with grains (*speirochory*).

6. Viability

Seeds are capable of withstanding adverse environmental conditions for long periods and they usually germinate when conditions become suitable. Therefore, the seeds have a life span of their germination. The period between seed production and the time when they begin to lose their capacity to germinate is called the *viability* period and it varies from species to species. It is usually longer in seeds that cannot germinate immediately on formation. Viability of seeds for longer periods ensures regeneration even when favourable condition occurs after long intervals and, therefore the study of this aspect is important for understanding the ecological equipment of the species for regeneration of populations. Viability is largely governed by environmental factors especially the conditions during the period when seeds lie dormant. Conditions that reduce the metabolic activity of seeds are usually responsible for increased longevity; for instance low temperature, low oxygen and high carbon dioxide contents.

Viability period may be just a few weeks as in *Shorea robusta* (Sal tree). Some crop plants germinate even after a storage of 5-10 years. Seeds obtained from herbarium specimens collected over a hundred years back could also germinate in *Mimosa glomerata*, *Astragalus massilienses* and *Cassia bicapsularis* (Bacquerel, 1932, 1934). Some species show seed viability upto over 50 years such as *Lens esculenta, Medicago orbicularis, Nelumbium luteum, Trifoleum arvense* etc. In one extreme case Indian lotus (*Nelumbo nucifera*) seed obtained from mud of a lake bed in Manchuria germinated on breaking the seed coat and from radio-carbon dating the seed was estimated to be one thousand years old (Mayer and Poljakoff Mayber, 1963).

7. Dormanncy

In many species viable seeds do not germinate for some time despite favourable environmental conditions. Such seeds are said to be *dormant*. Dormancy can be broken artificially, but under natural condition it gradually breaks by natural agencies in the course of time. The 'time' is called the *dormancy period*. Dormancy is regarded as an important ecological adaptation for staggered germination. If the seeds germinate all at once as soon as they have been formed, the complete population would be at risk in the event of some catastrophe like severe drought, or epidemic disease etc. Dormancy may be due to the physiological immaturity of the seed for germination or due to impermeability of the seed coat to water, gases, specific light requirements or even to presence of some substances inhibiting germination. There are yet other species where seeds can readily germinate if conditions are favourable or else enter a phase of *secondary dormancy*. Certain conditions of seed storage especially light or dark and low or high temperatures and oxygen concentrations, etc. also induce secondary dormancy. The most common type of dormancy occurring in seeds result from the impermeability of a hard seed coat to water. In the absence of water, oxygen and requisite light the embryo fails to respond. Enzymatic changes necessary for germination do not take place in the absence of water and dormancy is retained. Such a dormancy can be broken by mechanical abrasion, or scarification, damage, or puncture of the seed coat. Treatment of hard seeds in conc. sulphuric acid for a few minutes, followed by washing under running water also renders the seed coat permeable and germination may be brought about as in seeds of *Alhagi camelorum* (Ambasht, 1963). In this case the percentage of germination increases considerably when seeds are treated with sulphuric acid for 2 to 5 minutes. In *Coriandrum sativum* the seeds are mechanically pounded before sowing to ensure a higher percentage germination.

Dormancy of one kind or another, and the availability of natural agencies to break it in different seasons, are responsible for the development of different population of the same species in different seasons. For example the two seeds in each fruit of *Xanthium* often germinate in different seasons due to dormancy of one on account of its higher oxygen requirement. Under certain water-logged condition, however, in certain populations both the seeds germinate simultaneously. All this leads to flushes of populations one after the other. Light is another important factor which is responsible for inducing dormancy in some, and breaking dormancy in others. In some species the percentage of germination increases and in some it decreases when placed in varying light conditions. Day length or photoperiod is also responsible for inducing germination among seeds of certain species. In *Anagallis arvensis* the seed coat is coloured reddish. The light reaching, embryo, therefore, is only red and this has a great effect in bringing about germination (Pandey, 1968). In such seeds, highly adapted to the colour of light, the thickness and colour of the seed coat is responsible for regulating the dormancy and germination. As with other phenological aspects, in dormancy and its breaking, the interacting effect of light and temperature may be highly important. In a certain variety of lettuce 95% of the seeds germinate when kept in the dark at 20°C, but if kept at 35°C, about 90% of the seeds become dormant. This dormancy can be broken by first keeping the seeds in red light and then bringing down the temperature to 20°C. Exposure of seeds to red light breaks dormancy and increases germination percentage in many species.

In many plants of cold regions the seeds remain dormant and germinate only after an exposure to low temperature for certain periods. This treatment of seeds to moisture and cold temperature to break their dormancy is called *stratification*. During this period a variety of changes take place in the seeds. Nitrogen and phosphorus contents change slightly and certain chemicals that check germination are also affected.

In quite a large variety of species the dormant seeds have some chemicals which at a particular concentration prevent or inhibit germination. These are called *germination inhibitors*. Sharma and Lavania (1978) have shown that germination of seeds and seedling growth of *Vicia* species are affected by some hormones. Similarly Sahai, Kaur and Roy (1977) have made such investigation of *Lathyrus aphaca*. In nature the dormancy of such seeds is broken (i) by slow chemical changes in the inhibitors themselves, (ii) by the decay of seed coat which contains the inhibitor or (iii) by the washing away of the inhibitors by rain or running water, etc. Certain inhibitors act on account of their high osmotic pressure and so prevent the absorption of water so necessary to germination process. Certain other inhibitors bring about dormancy due to their interference in the normal metabolism of the seed germination. In agriculture a large variety of chemicals are artificially sprinkled to check the germination of undesirable weeds. These, however, inhibit germination permantnely and do not necessarily induce dormancy. Coumarin and its derivatives are highly successful germination inhibitors in a very wide range of seeds. This is widely used artificially and it is found to be naturally present in dormant seeds of some species, e.g., in *Trigonella arabica* (Lerner et al., 1959).

8. Seed germination and reproductive capacity

Not all seeds produced germinate. Many of them never receive suitable conditions and some others are incapable of germination for a reason like imperfect development of the embryo. Of the average number of seeds produced per plant the mean number that can normally germinate is referred to as the *reproductive capacity* of the species. Salisbury (1942) in his book '*The Reproductive Capacity of Plants*' has given the following formula to calculate it :

$$\text{Reproductive capacity} = \frac{\text{Average seed output} \times \%\text{age germination}}{100}$$

This fact, together with percentage seedling establishmnt, is of great autecological significance for physiognomy and sociological structure of populations in subsequent generations. A number of Indian weeds, grasses and forest trees have been investigated for their seed output and percentage germination which give the reproductive capacity figures. However, this figure differs under different environmental conditions of storage and germination for the same species to a certain extent. For germination of seeds almost all environmental factors are of some significance. For example, light and temperature are very important factors. Some seeds germinate more in open sun, others in shade and still others in darkness, some need alternating dark and light phases of specific duration and others are neutral to light conditions. In fact in course of ecological evolution and adaptation to the many different types of temperature and light combinations available in the natural habitats of a species, seed germination has been so adjusted as to be of maximum utility for the continued survival of population. Some seeds germinate immediately on formation, while still attached to parent plant, some germinate quickly on reaching the soil some a few months or years after, and some germinate periodically in batches.

Water is by far the most important factor in the germination of seeds as with its absorption the seed metabolism is activated and germination process begins. Proteins present in seeds are largely responsible for the inhibition of water. Cellulose and pectic materials also swell to a certain extent, whereas starch accounts least for swelling of seeds. Temperature is also somehow related to absorption of water, and over a certain range, increase in temperature accelerates water intake in seeds.

Levels of concentration of oxygen and carbon dioxide around seeds also affect the germination. Generally higher oxygen content favours better germination and high carbon dioxide retards it. Temperature is of less importance as long as the seeds are not soaked. But once soaked, the germination process begins and the temperature of the surroundings assumes great significance. This regulates the speed of many chemical reactions necessary for germination. Many species show a narrow range of optimum temperature while in many others seeds germinate best at alternating temperature of low and high levels in a diurnal cycle. Among the angiosperms Orchidaceae family has the smallest seeds. It is indeed a very weak point in the life cycle of any species to make a successful autotrophic start with almost no food reserve. A very high seed output has been regarded as the main asset to meet successfully the challenge of heavy seed or seedling mortality. It is reported that these impoverished seeds in fact heavily depend upon some fungi for their initial nutrient requirement and form an underground part called *mycorrhizome* with no leaves or aerial shoot. This is a parasite on fungus and after growing for some time and accumulating food, the autotrophic plant grows and mycorrhizome disappears. In some orchids, after producing flowers and seeds the aerial shoot dies and underground part reverts to parasitic mycorrhizome stage and remains dormant for years together (Summerhayes, 1968).

9. Seddling growth

Shortly after germination the seedling may be exposed to extremes of many environmental factors. In the seed stage the embryo is protected, biological activity is at a minimum level and it can withstand the extremes of environment easily. But once germination is over, in the young seedling biological activities like cell division, differentiation of tissues, biochemical reactions, etc. proceed at a rapid pace. The young root is soft and delicate and has to bear the hard and granular soil through which it makes its way. A large variety of soil fauna and microorganism also find it easy to damage soft roots. Even a temporary dryness around roots may cause to them permanent damage. On germination the reserve food in seeds is quickly exhausted, and the shoot has to meet the high food requirements for a rapid growth. Formation of leaves, development of chlorophyll, adequate sunlight, proper temperature etc. are the basic requirements for the successful establishment of the seedling. Birds and certain mammals do harm to seedlings specially in cropfields. Grazing animals also do maximum damage to seedlings both by grazing and tampling. The oxygen and carbon dioxide contents in root atmosphere are important ecological aspects.

In forests the seedling establishment of trees usually only occurs in areas where open sunlight is available for at least a few hours each day such as on forest margins and in spaces left open due to the death of old trees. Seeds fall everywhere and germinate in large numbers, but only a few that are in suitable situations survive, the rest perish. But in many species of annuals, perennial shrubs or woody climbers (lianas) etc., the seedlings survive in shade because of their ability to carry on photosynthesis at low light intensities. Similarly, biotic influences also regulate seedling establishment. In heavily grazed forest, tree seedlings are seen to come up to sapling or pole stage if thorny bushes protect them from grazing animals. If seedlings are absent in open areas it usually indicates a heavy grazing pressure. Rainfall and moisture status also largely determine the seedling establishment or seedling mortality in any area. Seedling establishment and growth, further more depend upon topography and drainage. On sloping surface the direction and degree of the slope are important aspects. On slopes due to runoff and erosion, seeds and young seedlings may be washed off and regeneration of plants is usually through vegetative means. In *Capparis sepiaria* and *Alhagi camelorum* on eroded habitats regeneration through seedling is rare and due to soil erosion produce new plantlets on the newly exposed roots (Ambasht, 1962).

A phenomenon known a *Oskar syndrome* has been described by Silvertown (1987) which refers to a prolonged arrest of tree seedlings at a stunted growth stage awaiting the death of older trees especially where there are closed canopy stands such as in temperate belts in *Tsuga canadensis, Picea abies* and *Quercus alba*. The seedlings remain more or less dormant and wherever the capony becomes open at a place the seedlings present there rapidly elongate, produce branches, flowers and seeds. In some herbs, in absence of dormant seeds, there develop perennating tubers which become active when there is ploughing of field or removal of population of other species.

In any autecological investigation study of seedling establishment is important, and its knowledge helps in climate and site selection for the maximum success in seedling establishment of desirable species.

10. Vegetative growth

The intensity, duration and quality of different environmental factors like temperature, light, soil conditions and water influence the vegetative growth of any species. Well established seedlings have already a well grown root system to meet the requirements of salts and water and a well developed shoot manufactures food for growth of new tissues, increase in size and biomass and for metabolic activities. Every species has its own *ecological amplitude* of tolerance to an extent of fluctuation towards higher and lower side from the optimum. For instance if a plant grows best at 25°C, in loam soil with 15% soil moisture then these levels of temperature, soil and moisture are called *optimum* conditions, and the ecological amplitude would be the range of temperature above or below 25°C, the range of soil texture from loam towards clay or sand and the range of soil moisture between excessively wet or waterlogged to dry condition which limit the growth of the species. In vegetative growth of a species it is of great ecological significance to have an understanding of the optimum environmental condition and the ecological amplitude of the species.

In population studies of a number of weeds and grasses it has been found that the vegetative growth, e.g., length of shoot, depth of root, number of nodes, length of internodes, number and size of leaves, stomatal frequency, thickness of cuticle on leaf, etc. are all affected by varying environmental factors. In nature such variations in habitat condition have often led to the development of different types of population, each suited to its own habitat. Such populations within a species in response to diverse habitat condition lead to the production of ecological races.

The effect of environmental factors are both direct and indirect. Each factor affects singly as well as in combination with other factors. Rainfall affects the vegetative growth as a direct source of water and as water vapour in the atmosphere, and indirectly by replacing soil atmosphere around roots.

Temperature regulates the rate of growth by influencing the rate of biochemical reactions and enzymatic activities necessary for growth. Every species has its own cardinal temperatures of maximum, optimum and minimum. Some grow in cold temperatures of arctic, others grow well in temperate countries and still others grow best in tropics. Infact at different stages of growth a plant has different cardinal points. Temperature regulates the rate of net production of vegetative parts by appreciably affecting the rate of respiration. Out of gross production the quantity lost by way of respiration will be about twice at 30°C than at 20°C, while in gross primary production rate the temperature rise from 20°C to 30°C may not have a corresponding increase. It may even decrease some times. Another way in which temperature growth is by rhythmic alternation of high and low temperatures in the 24 hours cycle. A tomato plant grown at constant optimum temperature of 25.5°C shows good vegetative growth but the stems show much better elongation if the day temperature 25.5°C is alternated with a night temperature of 18°C or so. Such a physiological response of alternating high and low temperatures in diurnal cycle on the phenology of plants is called *thermoperiodism*. Exceedingly low temperatures cause freezing of the water in plant tissues. This stops vegetative growth and may damage the tissues. Frost is also influenced by low temperature and many species however, have mechanism to resist the potentially damaging effect of frost. In the summer months on the North Indian plains high temperatures combined with low moisture also causes injury to vegetative parts. High temperatures result in evaporation of moisture and under limited moisture supply desiccation of air and soil results and causes wilting. But the direct effect of heat does not adversely affect vegetative growth if water is not limiting and in the hot but rainy months of July and August plant growth is at its best. The majority of annuals die and perennials just survive in a dry hot summer, and vegetative growth is suddenly activated with the first few showers of the monsoon.

Light is one of the most important factors regulating the pace of vegetative growth. With adequate light availability, chlorophyll production and the gross productions of stem, roots and leaves increase. Light also influences the quantity of water lost through leaves by its regulatory effect on the opening and closing of stomata. Unidirectional light rays induce differential growth and consequent phototropic curvatures in stems. Light retards the rate of elongation and darkness promotes it. Plants growing in shade, therefore, becomes elongate and weak as against normal and tough in light. In overcrowded populations where the ground does not receive light on account of dense shade, individuals grow taller than others of the same species and age growing in the open, but the net growth and biomass are always higher in the latter. Duration of light period is by far the most important ecological factor determining a variety of vegetative growth aspects such as leaf expansion, size, leaf fall period, vegetative phase etc.

Soil conditions affect the roots and shoot growth in several ways. Soil moisture and soil solution of minerals are important aspects. In loose sandy soils, the root extension is greater on account reduced mechanical resistance but total biomass may not necessarily be increased. In heavy soil with high moisture, reduced soil aeration limits the rate of respiration and affects internal metabolism to the extent that vegetative growth is curtailed. The availability of soil minerals like calcium, nitrogenous compounds, phosphate, etc. affect vegetative growth tremendously as they are essential constituents of plant tissues.

The influence of other biota on vegetative growth may also be there in a variety of ways. Organisms like bacteria, fungi and insects infect roots, stems and leaves and may adversely affect the tissues by arresting the normal growth, deformation of organs and even killing the entire plant. Many angiospermic plants grow parasitically on others drawing their food from the host. Some biotic associations are of advantage to hosts as well, for example root nodule bacteria in legumes are responsible for nitrogen availability. In climbers and lianas vegetative growth is profoundly influenced by the availability of some support to climb upon. In the development of communities competition for a limited supply of resources among plants growing together may result in stunted growth, sometimes elongation of stems, sometimes stratification and adjustments, etc. In economically important plants where abundant growth of vegetative parts is desired such as in various leafy vegetables and sugarcane, man is responsible for manipulating environmental and genetical factors in order to increase vegetative growth and check reproductive phases.

ECOTYPIC DIFFERENTIATION

Barbour, Burk and Pitts (1980) have defined a taxonomic species as "groups of morphologically and ecologically similar natural populations which may or may not be interbreeding, but which are reproductively isolated from other such groups". The taxonomic species is a heterogeneous group of plants adapted to a given ecological condition. Traditionally it has been regarded that individuals of same species are interfertile, but in many cases crossing between individuals of different population kept together in green house have failed to reproduce fertile offsprings. So, such populations are seggregated to distinct *biosystematic species*. However, such seggregation or splitting of species has to be taken with due caution as green house condition results may not hold true in nature. There are partial level of interfertility in natural populations. Out of say 100 trials only 10 to 15 may turn out to be fertile and 85 to 90% may fail. Numerical taxonomists divide species into several *operational units* and the degree of similarity is divided on a scale of 0 to 100, zero representing no similarity and 100 as total similarity. In numerical taxonomy a very large number of traits are taken into consideration.

The concept of *ecological species* began with the pioneering works of the Swedish genecologist Gote Turesson (1922, 1930) who hypothesized that many of the variations within a species which go hand in hand with habitat variation are heritable and the variations have adaptive values. Turesson called such ecological races with heritable variations which are associated with habitat conditions as *ecotypes*. These are the products of genetic response of populations (within the framework of taxonomic species) with the environmental conditions. *Genecotypes* and *ecological races* are more or less synonymous with ecotypes. As against such heritable variants, there are a number of cases where populations show different morphological appearance in different habitats, but on transplantation to an identical habitat the variations disappear i.e. the variations are not genetically fixed and these are called as *Ecophenes* or *ecads*.

Ecads are plants of the same species which differ in appearance such as size, erect or prostrate nature, reproductive vigour etc. in differing environmental conditions. These variations are not genetically fixed, and when transplanted to neutral conditions the variation vanishes. For example plants of *Euphorbia hirta* growing in disturbed conditions are *prostrate* and cushion-like whereas in undisturbed conditions erect. These morphological variations are not permanently fixed and upon transplantation the prostrate form may become erect.

Turesson collected vegetatively propagating plant parts or seeds of populations of many species from all over Europe and grew them in his garden at Akarp in Sweden to test whether the original variations of traits are retained or not. In cases where the traits are lost, the populations were regarded as genetically not fixed i.e. *ecads*, and where traits were retained in such a neutral habitats of the garden, the traits were regarded as genetically fixed and the population was classed as *ecotype*. Turesson identified a few important characteristics for ecotypes such as :

1. Variations are *morphological, physiological* or *phenological* or combination of two or all the three.
2. Variations are associated with certain distinctive habitat types.
3. Ecotypes are distinct and discrete entities.
4. They are genetically based i.e. genetically fixed.
5. They are interfertile or potentially so.

However, the modern concept of ecotypes has undergone much changes. While, Turesson was working on ecotypes in Europe, at about the same time, in America (1922) a cytogeneticist J. Clausen, a taxonomist D. Keck and a physiological ecologist W. Hiesey carried out population variation in plants on a line transect of 323 km in

California from near the sea bed at Stanford University to mountain range timberline upto 3000 metre altitude. The selected species were also raised in green house and gardens at Stanford i.e. at sea level (30 m alt.) at Mather in Sierra Nevada at mid elevation of 1400 m alt. and at the timberline height of 3050 m alt. Clausen, Keck and Hiesey (1940) recognised four ecotypes in *Potentilla glandulosa* based on morphological and phenological behaviours associated with altitudinal i.e. temperature variations : (1) *P. glandulosa* ecotype *nevadensis* was of shortest height, flowered early in the season and it was frost tolerant, (2) ecotype *reflexa* was taller and occurred at mid-elevations in dry habitats, (3) ecotype *hanseni* was also tall and occurred at mid-elevations but on moist or wet meadowlands, and (4) the ecotype *typica* was shorter than (2) and (3) in height but grew best in low elevation at Stanford and poorly at mid-elevations and absent at higher elevations. These three workers made such studies on a number of species and came to the conclusion that almost all species showing a wide ecological ranges of distribution are composed of several ecotypes.

The concept of ecotypes as distinct and discrete units of species i.e. *stepwise change*, like in the staircase, *each step representing a distinct ecotype* leading to the next-say in a transect across ecotones was given by Turesson in Europe and Clausen, Keck and Hiesey in America.

Another concept of gradual changes between ecotypes not like the steps of a staircase but like the ramp showing a continuous change in successive ecotypes and the habitats has now been recognised and called as *Ecocline*. Works of Gregor (1946) on *Plantago maritima* in Scotland and Langlet (1959) on *Pinus sylvestris* in Sweden prove that there are no discrete and easily distinguishable boundaries between different ecotypes as there are so many integrading varieties between the two extremes of habitats. It is difficult to distinguish (on the habitat transect) where the first ecotype ends and the second begins. That is to say that there is an *ecocline* of continuous and gradual change in ecological races and habitat types (Fig. 9.2).

There is still another term the *genoecoclinodeme* given by Gilmour and Heslop-Harrison (1954) which almost means the same as ecotypes. Their system combines several kinds of population variations which are given a name ending in-*demes*, like *topodeme* (location), *gamodeme*

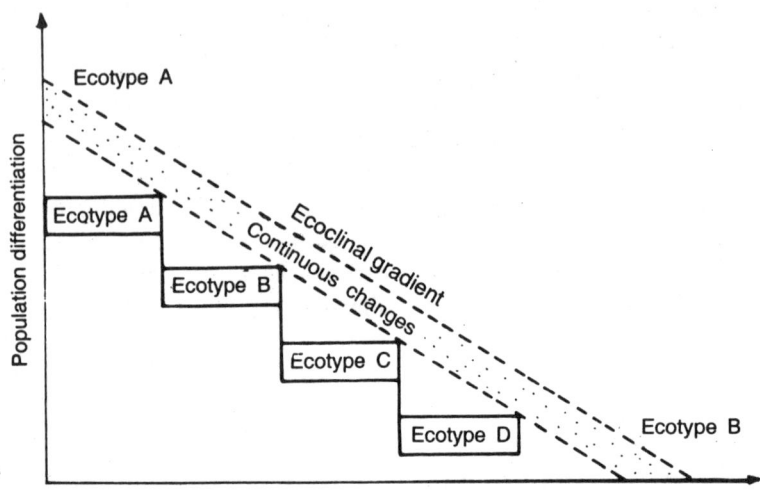

Fig. 9.2. Ecotypic differentiation. Turesson's stair-step concept of discontinuous differences in ecotypes A, B, C and D; and Gregor's continuous ecoclinal changes between ecotypes A and B on the ramp-like slope pattern.

(interbreeding), *ecodeme* (habitat related), *phenodeme* (non-genetic differences), *genoecoclinodeme* (continuous clinal gradation of ecotypes). Some of the researches on physiological aspect of ecotypes have revealed that in *Oxyria digyna* even when their arctic and alpine forms are artificially grown together in controlled chambers they showed to retain their adaptational metabolism (Mooney and Billings, 1961). Even the photosynthetic efficiency is reported to be double in a high elevation ecotype of *Typha latifolia* than that of low lands (Mc-Naughton, 1966). Bjorkman (1968) through a series of experiments on ecotypic differentiation in *Salidago virgaurea* herbs has distinguished *Sun* ecotype growing in open with a high light saturation point and higher photosynthesis and *Shade* ecotype with low light saturation level and the adaptation in this case lies at the enzyme level with 2 to 5 time more of ribulose diphosphate carboxylase activity in sun ecotype than in shade ecotype.

In India, Kaul (1959 has shown the occurrence of photoperiodic ecotypes, i.e. climatic ecotypes) in *Xanthium strumarium*. Ramakrishnan (1959) has shown edaphic differentiation in *Euphor-*

bia thymifolia where red colour of leaf and stem is associated with soil calcium content. Biotic ecotypes develop in a variety of ways, as for instance under constant browsing pressure in pastures. Singh (1972a) has distinguished two ecotypes in *Portulaca oleracea* showing differential responses to photoperiods, temperature and light intensities on their growth, seed germination and flowering. He has also found ecoclinal variations in seed samples from a gradient of population.

From the foregoing account it is clear that ecotypes are of wide occurrence particularly in widely distributed species as a result of adaptation to long range environmental impacts. The responses are possible at several levels, such as morphological, phenological, physiological, biochemical or in combinations of these. The responses when temporary lead to formation of ecads or ecophene and when genetically fixed to ecotypes or genoecoclinodemes. The species undergoes a process of aclimation or aclimatization as a result of variation in new environment to which the species moves out from the centre of its origin.

POPULATION CHARACTERISTICS AND DYNAMICS

At any given plane a large number of individuals of different species are found growing together forming a community. The group of individuals of each species forms its own *population*. Population is therefore, an assemblage of individuals of same species growing or living together in one habitat and showing many important group properties such as density, rates of birth (natality), death (mortality), competitions and association among members of same species (intraspecific) or different species (interspecific), etc. Their group attributes of population is different from those of an individual for instance an individual cannot have a density, it cannot have a rate of birth or death, although it occupies a place, is born and dies. Their group attributes or population characteristics with respect to rates of birth, death, reproduction, etc. is a science by itself known as *demography* and when ecologically studied, i.e., in relation to environmental conditions, it is called *population ecology* or *democology*. We can discuss these characteristics separately.

The seeds population present in the soil for different species are referred as *seed bank* or sometimes as the *seed pool* (Silvertown,

1987). All the seeds do not germinate or all the seedlings do not establish, some die due to environmental stresses and it is referred as *environmental sieve* which allows only the stronger individuals to survive. The seeds in most cases germinate in batches and seedlings of one lot is known as *cohort*. Thus from a huge seed bank through ecological selection, cohorts are formed and these in turn result into adult population. This process is called as *recruitment*. A plant may originate from a vegetative part called *ramete* or from seed called *genet*. *Clone* is the term used to designate the population derived from rametes of the same parent plant.

The *Population size* (N) at any given place is determined by the processes of birth (B), death (D), new arrivals from outside or *immigration* (I) and going out or *emigration* (E). Therefore, the change in population size between an interval of time Nt + 1 is Nt (initial stage) + B − D + I − E.

Density

Density refers to the number of individual per unit area of space or in case of plankton in unit volume of water. This alone is a static value and does not give its changing character with passage of time, and it is important to study the dynamic aspects in order to understand population behaviour. For measuring density we usually use quadrat of an appropriate size and sample about 1 to 5% of the total area to be studied. The methods are described later in the chapter on plant communities. The measured density value is also called as *crude density* whereas the number by which the given habitat can be colonised is called as *ecological density* or *specific density*, sometimes, instead of numbers the biomass value of a given area is also taken in density study. The population is a growing and changing entity, and a record of its density at daily, weekly or monthly intervals gives its dynamics. If we divide the change in any measured parameter by the time elapsed during which the change has taken place, then we get the rate. For instance if the number of individuals of a species in a given area increases from 50 to 100 in 10 days then the population growth rate will be 100 − 50/10 = 5 per day. A changing condition such as mentioned above is denoted by the symbol D (delta) and D N/Dt represents the rate of change in the number of organisms (D N) over a period of time (D t). For instantaneous rate of change at any one time in the growth of population the calculus notation dn/dt is used. In most cases the growth curve is S-shaped (if the time is plotted on the horizontal and density on the vertical axes).

Natality is the ability of any population to increase through birth rate. It is the opposite of mortality which means death rate. Although a plant species may produce a large number of seeds or even seedlings, not all of them, survive. Only a few grow into adult plants. The actual increase in population is called the *ecological* or *realized natality*, whereas the theoretical maximum increase that could be achieved under ideal conditions is called the maximum *natality*.

Natality rate tells us the number by which a population is increasing over a given period of time. It takes into considerations the new individuals added and not simply the difference in density level at certain intervals. For instance if in a population of 100 individuals 10 new ones are born in one month and three die, the net increase is only seven but natality rate will be 10 per month in a population of 100 individuals. Thus natality rate is always positive and in absence of birth may drop to zero level but it is never in negative. It does not take into account the change in numbers on account of mortality or incoming or outgoing members of population. In plant communities some members of a population produce more seeds and some less. The natality of the population is calculated on the basis of average reproductive capacity of different individuals. As already explained reproductive capacity of plants refers to the numbers of seeds produced by an individual plant that actually germinate.

Mortality means the rate of death among members of a population. Like population growth and natality, mortality is also expressed in terms of mathematical notations. Two aspects of mortality may be distinguished—*ecological mortality* which is the actual loss of individuals in a given ecological condition, and *minimum mortality* representing the least loss under ideal conditions. All organisms have a *physiological* longevity representing the age upto which the organism can live under ideal condition without stresses or disease and death occurs purely due to ageing.

Mortality age differs for individuals of different age groups. A large number of seedlings die at the stage of their establishment. In nature there is always over production of seeds so that even after heavy consumption of seeds by animals and seedling mortality, there remains enough plants to develop to a desired population level. But all organisms that are born must die even though it has survived natural hazards, by the process of ageing. There is a total life span divisible into three main periods, the (i) *juvenile* or *dependent phase*, (ii) *reproductive* or *adult* phase and (iii) *post reproductive* or *old age*. Mortality at these phases differ from species to species and under habitat to habitat conditions.

Survival

If we follow the pattern of death and number of survivors at the above mentioned three phases we find three main types of survivorship curves (Fig. 9.3). In many large size plants of perennial habit mostly the plants die after reaching post reproductive phase within a narrow period of old age. Not much mortality is noticed at young and adult stage and the pattern I of the Fig. 9.3 is followed. Some other species show a more even death rate among individuals of all ages and the survivorship curve follows the II pattern. But in many short lived weedy annuals, there is a heavy mortality at the young seedling stage and the III Pattern of survivorship curve is obtained. There can be many other intermediate types of survivorship curves possible in different species or different situations.

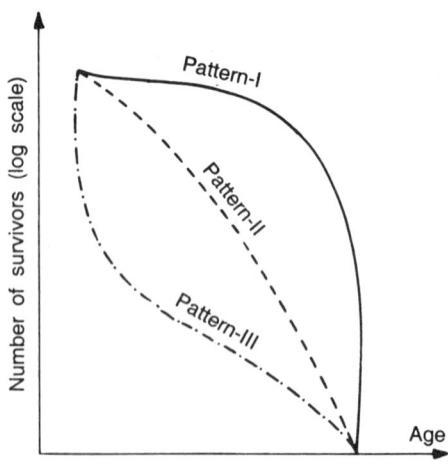

Fig. 9.3. Curves showing three main patterns of survivorship.

r and K selections

McArthur (1965) gave a concept of r and K-selection for population growth. Such organisms which are short lived annuals that usually follow up the pattern III of survivorship curve rapidly grow and utilize most of their production products (photosynthates in plants) in the production of flowers, seeds and fruits are called *r-selection* types. These plants are usually seral in nature and greatly affected by droughts and floods (Fig. 9.3).

Those plants which live for long periods as shrubs and trees and which produce massive vegetative parts follow I and II patterns of survivorship curve and remain in climax or nearly stable balance with the habitat, etc. are called K-selection populations. Among K and r-selection populations there may be some with more r-selection than another r-selection or more K-selecton than another. There may be some with intermediate selection between r and K. Therefore r and K are relative terms and taken from logistic equation of McArthur given by Barbour, Burk and Pitts (1980). Change in population size with time is represented by the formula :

$$\frac{dn}{dt} = rN = \frac{(K - N)}{K}$$

where r = innate capacity of population to increase (birth rate— death rate without resource limitation),

   N = population size and

   K = highest population density that can be maintained in real environment, i.e., at carrying capacity.

Pianka (1970) has given the important differences in the traits of r- and K-selection. These can be briefly summarised as follows :

   (i) The climate is very variable and uncertain in r-selection and nearly constant and more certain in K-selection.

   (ii) Population size is variable in course of time and not in equilibrium stage in r-selection at a much below the carrying capacity. In K-selection the population size is fairly stable and nearly at the carrying capacity of the habitat.

   (iii) Population needs repeated recolonization on the habitat in r-selection since it is not in climax stage, but in K-selection, because of equilibrium stage, no recoloniation is required.

   (iv) Competition in *r*-selection is not so keen but in *K*-it is quite keen.

   (v) Survivorship curve in *K*-selection is of type I and II while in *r*-selection it is of type III.

  (vi) Mortality in *r*-selection is more due to catastrophic events and less on account of competition, but in *K*-selection competition being keen, it is the main cause of mortality.

  (vii) Longevity in *r*-selection plants is short, usually one season or less than one year, while *K*-selection plants are perennial like the shrubs and trees.

 (viii) The body size is small and reproduction is usually only once in the life time of an individual in *r*-selection while there are repeated reproduction in *K*-selection population.

  (ix) A greater share of the primary production goes into reproductive process in *r*-selection than in *K*-selection.

   (x) Seed longevity is long in *r*-selection.

  (xi) *K*-selection plants are more efficient and better adapted and fit than the *r*-selection population.

R, C and S selections

Grime (1979) has given three population selections viz. *R* for *ruderal*, or resource abundant temporary habitats. *C* for *competitive* or resource abundant predictable habitat and *S* for *stress tolerant*, resource stressed habitats. Here *R* is of small plants with short life cycle, early flowering, greater production efforts allocated to reproductive strategy and long seed dormancy. Grime's *R* selection is more or less similar to *r*-selection of Mac Arthur. *C* is largely of *K*-type, but also includes herbs and shrubs of competitive populations. S is an additional category to include such population as of lichens, herbs, shrubs and trees specially adapted to survive in stressed conditions. They may have small, leathery leaves with storage tissues in roots, stems and leaves.

    Biotic potential of any population is towards a geometric increase in numbers (density) but in nature the ideal condition for limitless growth is absent and the population growth is restricted due to ecological resistance prevailing at the given habitat (Fig. 9.4).

    After a period of time the population growth reaches a mature stage when $\Delta N/\Delta t$ approaches zero. At this point the growth curve,

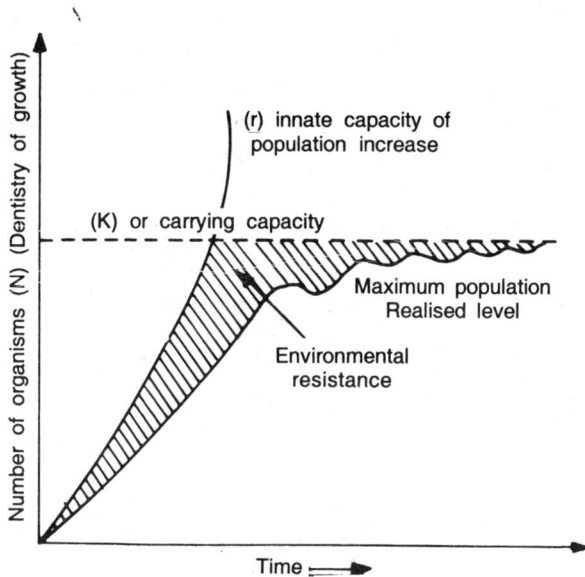

Fig. 9.4. Curves showing the intrinsic and realised population growth.

instead of just levelling off, assumes a wavy appearance indicating regular or cyclic oscillations of rise and fall within a narrow range. In organisms with short life cycle there is often a considerable increase and decrease in population structure governed by seasons.

There are two main types of population growth forms : (1) *J*-shaped and (2) *S*-shaped or sigmoid forms. There could be intermediate shapes also. The growth forms are due to the nature of species and prevailing environmental conditions. In *J*-shaped, there is a rapid increase in density with the passage of time. The density values when plotted against time give a *J*-shaped growth curve and at the peak the population growth ceases abruptly due to environmental resistance. But this *J*-shaped arithmatic growth curve when redrawn on a logarithmic scale for density values it assumes a linear shape. The population growth curve in many others shows an initial slow rate and then it accelerates and finally slows giving the growth curve a sigmoid or *S*-shape. The peak constant level represented by *K* or upper asymptote of the sigmoid curve is called the *maximum carrying capacity*.

Another aspect that affects population density of a habitat is the going out or migration and coming in or immigration, i.e., population dispersal properties. Emigration or outward movement is a net loss or it simulates mortality effects and immigration or inward movement is a net gain or simulates natality effects. Migration could also be periodically both out-and inwards.

In many mammals and birds there is a cyclic oscillation of large scale migrations from one to another region and the cycles need not be just seasonal or annual, it could be at gaps of 3 to 5 years or more. Still another type of oscillation in populations of plants and insects thriving on them could be conceived. There is a rise in population of trees in a forest for a few years followed by an outbreak of large scale population growth of moths eating the leaves. There is a thinning effect on the forests and in absence of sufficient foliage the moth populations crash. Individual trees then get a chance for a more robust growth in absence of too high a density. Thus there is a natural forest management effect due to such periodical burst of moth or leaf feeding insect populations. Predator-prey population is also commonly seen in many kinds of habitats.

The population structure or pattern of distribution is usually classified into three main types : (1) *random*, (2) *uniform* and (3) *clumped* (Fig. 9.5). There is no special pattern of distribution of individuals in a given space in the random type. In the uniform type the individuals

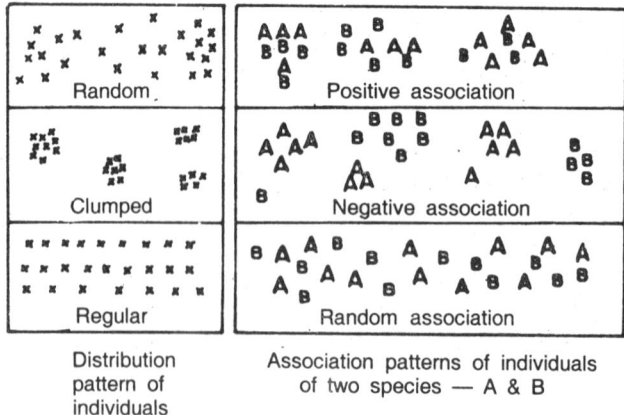

Distribution
pattern of
individuals

Association patterns of individuals
of two species — A & B

Fig. 9.5. Patterns of distribution and association of individuals.

are evenly dispersed. In the clumped type there are several small groups or more or less equal numbers either dispersed randomly or uniformly. The pattern of dispersion is largely controlled by the biotic, edaphic and climatic conditions of the area. As already mentioned in an earlier chapter (under biotic factor) the interactions between individuals of different species may be *negative* or *positive*. The negative effects obviously result in severe competition of all types where there is a tendency for competitive exclusion among species of similar types. In positive effects the populations in the course of evolution have evolved towards their mutual benefit, and they are often interdependent.

Species interaction and competition

Plant communities are made up of several populations that have developed as a result of their own biotic potentials, carrying capacity of the habitat and interactions among members of the populations and communities over a period of time. The interactions may basically be of either *positive* (helpful) or negative (harmful) types. There may be many different intermediate variations and Haskell (1949) and Burkholder (1952) have given a series of nine possible interactions as given by Odum (1971) and Barbour, Burk and Pitts (1980). These are: (i) *Neutralism* in which the two populations do not affect each other; (ii) *directly inhibiting competition* in which both the populations negatively affect each other; (iii) *resource use competition* in which both populations compete for common natural resources like water or nutrients in short supply; (iv) *amensalism* where one population adversely influences the other by chemical exudates, etc. whereas the other does not influence the first; (v) *Parasitism*, (vi) *predation*, (vii) *commensalism*, (viii) *protocooperation* in which both positively benefit each other although the association is not obligatory, i.e. even if one partner or population is missing the other continues to grow, and (ix) *mutualism* where both partners are beneficially associated in an obligatory fashion, i.e. for survival, association of both is essential as in lichens.

The chapter of Biotic Interrelationships gives details of all these category of competitive and non-competitive relationships.

Gadgil and Bossert (1970) and Gadgil and Solbrig (1972) and Stearns (1976) have discussed various ideas of life history tactics shown by different kinds of populations and have given a few generalizations. In a population where juvenile mortality is very low, lesser than that of adults, the species needs to reproduce only once in its life cycle while if juvenile mortality is high, the reproductive phase has to be extended several times. Another idea is that in expanding population maturation stage for reproduction reaches quite early and in stabilized population level the reproductive process needs to be delayed as is applicable in human population (delayed marriage for stabilized or low population growth). In populations facing severe predation or food shortage at birth time, the young one has to be large in size. A large portion of energy is routed to reproduction efforts in an expanding population from a young age itself when food is in abundance while in a stabilized population much less portion of energy and at delayed age is to go in reproductive efforts when the food is a limiting factor. In complex ecosystems different populations with short or long life cycles adjust themselves according to the resource limitations or abundance in different seasons.

KEYSTONE AND FLAGSHIP SPECIES

Paine (1966) first gave the term Keystone species to denote those species which have a disproportionate or far reaching effect on the persistence of all other associated species and Bond (1994) suggested for their conservation. Bond has described different types of Keystone species such as: (1) *Predators* like sea otters which prey on sea urchins and sea urchins eat kelp like *Laminaria* thus the otters in a way have far reaching effect on the population of a number of other species. (2) Keystone *herbivores* have numerous examples like the elephants which easily transform the closed woodland into an open savanna; (3) Keystone *pathogens* and parasites like tsetse flies in African forests are avoided by elephants and thus tsetse control is indirectly responsible for destruction of woodlands. Another category is of (4) *competitors* like some weeds on forest floor through better competitive ability prevent tree regeneration while weed control leads to invasion of many tree species. (5) *Mutualistic* Keystone species are many dispersers, pollinators and plant resource species.

Even though there is a fair degree of interdependence among the community constituent species, the keystone species have far greater influence on the existence and survival of other species. Species whose elimination causes elimination of a chain of species are known as *keystone* species. They by their presence, cause the richness of diversity and therefore need care and conservation.

Not necessarily the keystone species, certain other species attract much greater attention of governments and nature conservationists like the whales, lions, tigers, panda, rhinoceros among animals and many orchids in plants and these are called *Flagship species*. Some species occupy very large areas and provide protection to (understorey) other species are known as *Umbrella species*. In conservation of such species, the best strategy is to conserve the habitat so that the keystone, flagship and umbrella species along with their weaker partner species are conserved in entirety.

10

Synecology and Community Dynamics

The overall combinations of different plant and animal populations sharing a common habitat result in the development of *biotic communities* where organisms and populations of different species interact between themselves and with their physical environment to form an ecosystem. Community is a larger unit than the population and it acquires many characteristics that are not found in its constituents, i.e., the organisms and the populations. Communities can be conceived in a wide range of sizes, ranging from a very small patch of land or water-body to extensive forests. Minor communities are greatly influenced by inputs from adjacent communities while major communities are relatively independent and self sufficient on their habitat. Communities differ from the place to place and at the same place at different times. Quite a few basic ecological concepts have been developed to explain and describe the nature of plant communities in relation to space and time. Despite the possibility that a large variety of seeds and diaspores may reach great distances, we find that all species are not found at all places. The biological potentiality of each species determines a tolerance range towards environmental conditions (including the physical) in which the species or ecological race can successfully grow. The range of environment that the taxa can tolerate is called its *ecological amplitude*. Thus, the nature of the plant community at any place is determined by the species content, their ecological amplitude, the climate, soil and biotic influences available there. In fact, the ecological amplitude of several species usually overlaps over a certain range. Depending upon the degree of overlapping environment, the best suited species grow with higher ecological importance values in association with less conspicuous ones. One important feature of biotic community is that it is an orderly arrangement of organisms and not a group of haphazardly put together individuals. The organisms have always certain food or trophic relationships amongst themselves, communities merge with each other in adjacent areas sometimes sharply and sometimes gradually.

In any plant community the influence of associated animals can scarcely be ignored. Many varieties of insects, worms, nematodes, or higher animals always go together with types of plant communities and as such the term *biotic community* is gaining greater popularity over the terms plant or animal community. In most cases biological activities (including periodicity, phenology etc.) in plants are remarkably interwoven with animal activities and vice versa.

As already indicated, plant and animal aggregation organizing into communities are strongly dependent upon habitat character and other condition of the environment. In a pond the abundance of overlying water etc. determines the region where communities of *Nymphaea, Nelumbium* or *Salvinia, Azolla, Lemna* or *Eleocharis, Cyperus*, and *Scirpus* will develop.

In a broadly similar environment, different components of the community occupy different zones, strata or layers to suit their ecological requirements. Wholly diverse types of life froms like tall trees may grow mixed with less tall shrubs, herbs and lianas. There may be many epiphytes, bark dwelling lichens, and saprophytic fungi on decomposing organic material. Each of these has a different way of life and may have a different role to play in the community. The way of life and the role of each species is referred as its *niche*. The members of communities compete as well as adjust the resources locally available. This results in many kinds of interactions among members of the communities. For example a tall tree casts shade and also provides support for climbers, its fallen leaves on decomposition renew soil nutrients that may have been drawn from deep soil layers by its deep roots. There may be many other ways in which dominant plants influence other members of the community. There is, therefore, an intimate interrelationship of all organisms in a community, some influencing profoundly, others to a lesser extent. The mutual influences are both *vertical* and *horizontal* in communities. For example, light intensity goes on decreasing from the top of the canopy to the ground in a multilayered closed forest. The organisms influence others in a community to different extents and, therefore, each species in a community has its own ecological importance with respect to the rest of its members.

The vertical distribution of different species occupying different levels is called *stratification* or *layering*. These layers may often bring together very different life forms; for example a tree and a climber may both be represented in the top layer. To distinguish between morphologically distinct groups of species the term *synusiae* is used. Plants of similar morphological forms are put together as for instance we may speak of synusia of trees or a synusia of climbers or of epiphytes.

As already stated several communities may merge into each other. Sometimes, as on the margin of a lake, the boundary between two types of communities is quite sharp, while in some other cases with gradual differences in community structures the boundaries are less clearly defined. Thus every community has its spatial limits. The zone of vegetation which demarcates the transition of one type of community into another is called an *ecotone*. Another fact about community structure is that if, we study the vegetation at the same place for several years we find it changes due to several external as well as internal factors of the community. Thus, another very important ecological concept regarding community character is that it is dynamic and it changes in form and composition in the course of time. This is called *ecological succession*. A series of community changes ultimately result in the development of a relatively stable community that retains its shape and composition indefinitely and virtually gains control over the area. This is called the climax community. A *climax community* regenerates and perpetuates over and over again. In order to study community structure a large number of methods have been developed in different parts of the world. The vegetation is sampled in small unit areas only and through these the picture of the entire community structure is formed in qualitative or precise quantitative terms.

Larger units representing an area of similar vegetation are often called as biomes, provinces, biochores or formations. Latitude, altitude (through temperature) and water play key role in determining the broad nature of the biome through their control on climate. In tropical belt between 0 to 20° latitude on plains upto 1000 metres altitude the forests are classed as tropical according to the Kormondy (1978). In the same latitudes between 1000 to 2000 m alt. the forests are sub-tropical, between 2000 to 4000 m alt. they are temperate and between 4000 to 6000 m alt. it is arctic or alpine. In the subtropical belt (20-40°C lat.) there is paucity of tropical vegetation and on altitudinal sequence the subtropical forests of the plains are replaced by temperate between 1000 to 2000 m alt and by arctic/alpine beyond 2000 m alt. Like-wise in temperate belt (40-60° lat.) beyond 1000 m alt. itself the alpine vegetation is found while beyond 2000 m snow line begins. On the Arctics and the Antarctics (60 to 80° lat.) the arctic/alpine kind of vegetation is found on the plains.

Just as in the study of populations we noticed a number of emergent group properties; here too, in communities there are several emergent characteristics like (i) the community architecture, (ii) foliage geometry or leaf orientation, etc., (iii) leaf area index i.e. leaf overlapping in the vertical space, (iv) phenological events of constituent species in relation to seasons, (v) relative phytosociological combination among individuals of the constituent species, (vi) niche specializations and niche adjustments and overlaps, (vii) diversity, richness, eveness, etc., (viii) nutrient movement, storage, release, decomposition rate, (ix) succession, stabilization (climax), and (x) community biomass and energy productivity, and then (xi) there is a mechanism of controls of all the above aspects developed as a result of ecological evolution, information contents and feedback mechanisms.

METHODS AND PURPOSE OF STUDYING PLANT COMMUNITIES

Classification

Communities have been classified by different ecologists from different view points.

According to the amount of water in the habitat, communities may be divided into : (1) *hydrophytic* in predominantly aquatic habitats, (2) *mesophytic* in moderately moist soils and (3) *xerophytic* in arid or dry conditions. Communities growing in conditions of abundant light are called *heliophytic* and those growing in shade *sciophytic*.

Communities may be similarly classified on the basis of other environmental considerations. In terms of their general shape and growth form with a particular environment, communities have also been classified into *forest communities, desert communities*, etc. Clements (1916) recognized the fact that plant communities are not always the same at any place and he classified the communities on two parallel lines— one in the process of change which is called seral communities and the others *stable* or *climax* communities. In both the changing (seral) as well as stable (climax) ones, he recognized communities on the basis of the relative dominance of the species in the vegetations. Seral communities are generally indicated by the suffix ending—*ies* and climax communities by the ending—*ation* (Table 10.1).

Table 10.1. Important units of communities

Community character	Seral communities	Climax communities
(a) The dominant species with maximum canopy coverage of two or more species	Associes	Association
(b) The community represented by a single dominant species	Consocies	Consociation
(c) Some local variation in dominant and sub-dominant communities	Locies	Lociation
(d) Community for sub-dominant species belonging to lower level of life forms than the dominants	Socies	Society

QUALITATIVE CHARACTERISTICS OF COMMUNITIES

The community structures, composition and other characteristics can be readily described by visual observation without actual measurement. This is a qualitative approach which is easier than the quantitative vegetation analysis where measurements are actually made.

The first and foremost prerequisite of a vegetation study is the study of flora or kinds of species present. This necessitates a knowledge of taxonomy which may be obtained by periodic observation of the flora, collection of plant material, systematic studies, herbarium preparation, etc. for the whole year. Although, floristic study is so essential, yet it is only a prerequisite and not an end in itself. Description of species content provides little insight into community structure, the appearance of the landscape, the number of individuals, coverage, height, etc.

Therefore, other characteristics like stratification, and life forms like herbs, shrubs, trees climbers etc. should be recorded for each species. With change in season the appearance of vegetation also changes. This is termed aspection and for this the periodicity and , phenology should be recorded for different seasons and years. The *sociability* or the nature of groupings of individuals of different species whereby they grow singly, in patches, in colonies or evenly intermixed is also an important qualitative character of community study. Rough visual record of number of individuals in terms of (1) rare, (2) seldom present, (3) often present, (4) mostly present or in terms of *rare, less frequent, frequent* and *abundant* are also good qualitative expressions.

Constance and *Presence* are two terms used to refer about the level of occurrence of a species in a community spread over several stands. If a species is present in 7 stands out of a total of 10 stands, its presence is 70% and if one quadrat is placed in each of the ten stands, the same species may not be present in all the seven quadrats; it may be present in only 5. Then the *constance is 50%*. *Sociability* refers to the dispersion of individuals of a species and expressed in a scale of 1 to 5, 1 meaning growing singly, 2 for small but dense clumps, 3 for small patches or cushions, 4 for small colonies or carpets and 5 for growing in large almost pure stands. *Vitality* refers to the general vigour of individuals of a species in a stand and *periodicity* refers to the seasonal importance.

Depending upon the general appearance and growth the species are grouped into different life forms. Raunkiaer's (1934) classification of life forms in plants is most widely understood and used by ecologists throughout the world.

Raunkiaer's life-forms

The famous Danish Botanist Christen Raunkiaer in 1903 (English translated book in 1934) considered that it is the unfavourable environmental condition that really matters as it limits the growth form. Much depends upon how different species overcome adverse temperature conditions, i.e., cold or heat. For this he considered the *position* and the *degree* of *protection* to *perennating bud* during the adverse season as the principal feature of plant adaptation to climate. On this basis he classified higher plants into five major life-form classes viz. (1) Phanerophytes (P), (2) Chamaephytes (Ch), (3) Hemicryptophytes (H), (4) Cryptophytes (Cr) and (5) Therophytes (Th) (Fig. 10.1).

1. Phanerophytes

These are trees, shrubs and climbers where the growing buds are located on the upright shoot much above the ground surface and they are the least protected. These are most abundant in regions like the tropics where tree growth is most favoured, and their number progressively decreases from the tropics to temperate and polar regions. Phanerophytes are further classified into four sub life-forms depending upon the height of mature plants. These are : (i) *Megaphanerophytes* for trees over 30 metres tall, (ii) *Mesophanerophytes* for trees between 8 metres and 30 metres height, (iii) *Microphanerophytes* for trees between 2 and 8 metres and (iv) *Nanophanerophytes* for shrubs smaller than 2 metres.

2. Chamaephytes

These plants are more commonly found in cold regions at high alti-

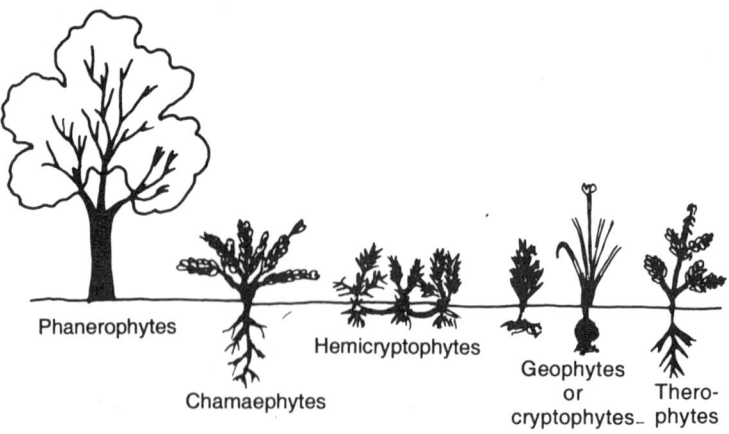

Fig. 10.1. Different life forms. The regions shown in solid black continue to live in adverse season.

tudes or high latitudes and in these the buds are located close to the ground surface or up to a maximum height of 25 cm e.g. *Trifolium repens* and *Buchloe dactyloides* found in temperate America. Sometimes in chamaephytes the rest of the aerial parts die in the cold season and the dead shoots cover and protect the perennating bud. Fresh vegetative growth takes place every spring.

3. Hemicryptophytes

This group of species is again predominantly present in cold climatic regions. Here the perennating buds are present just under the surface soil and remain protected there. Mostly these are biennial or perennial herbs whose vegetative growth and aerial parts are conspicuous in warm seasons only. Buds may also be present at the soil surface but they are never exposed, they remain concealed under dead leaves and twigs.

4. Cryptophytes

These are also called *genophytes* and in plants this category of lifeforms the buds are invariably buried in the soil or the substratum such as the bulbs and rhizomes. These are adapted to withstand long periods of adverse climatic condition where storage of food in the perennating organ is also an important aspect.

5. Therophytes

These are plants which survive the adverse season in the form of seeds. The plants produce flowers and seeds in the favourable season. They are annuals, predominantly found in extremes of dry, hot or cold conditions.

All the species of higher plants of any community can be classified in one or the other life forms. The ratio of the life forms of different species in terms of numbers or percentages in any floristic community is called the *biological spectrum* or *phytoclimatic spectrum*. Since the life form is related to the environment around the plants, the biological spectrum is also regarded as indicative of the prevailing environment, e.g., higher percentages of therophytes indicate long dry seasons, of chamaephytes indicate an extremely cold climate, of hemicryptophytes indicate conditions suited for the development of extensive grasslands, etc. On the other hand occurrence of a similar biological spectrum in different regions indicates similar climatic conditions.

Raunkiaer (1934), who first put forward idea of life forms and biological spectrum, prepared a normal biological spectrum for the phanerogamic flora of the whole world. The percentage values of different life forms in this normal world biological spectrum are Phanerophytes 46, Chamaephytes 9, Hemicryptophytes 26, Cryptophytes (or Geophytes and Hydrophytes) 6 and Therophytes 13. Generally the biological spectrum of a region is worked out and compared with this normal. In most spectra there is at least one life form whose percentage value is much higher than that of the same life form in a normal spectrum. This indicates the predominance of a particular type of climate that favours the development of that life form in higher proportion. For instance, Raunkiaer found that the percentage of therophytes is 42 to 50 (as against 13 in the normal spectrum) in some deserts and the percentage of hemicryptophytes is 49 to 51 in the temperate climates of central Switzerland (as against 26 in normal spectrum). The percentage of phanerophytes is also much higher (upto 74%) in the tropical West Indies than the normal 46%.

But the usefulness of the Raunkiaer's biological spectrum as indicative of climatic condition is limited as it does not take into consideration other ecological factors such as biotic disturbances. For example, the climate of the Gangetic plains in India is suited to forest development and therefore phanerophytes should be dominant. But actually therophytes like annual weeds and crops are in higher proportion due to the agricultural practices of man. Thus biotic influence like agricultural practices, grazing, scraping etc. materially alter the biological spectrum, and its uncritical comparison with the normal world spectrum may lead to misleading or erroneous conclusions about the climate of area.

Another system of community classification, based on leaf size, was also devised by Raunkiaer. He classified plants into the following six *leaf size classes*. The leaf size of each class is nine times larger than the preceding class.

1. Leptophyll	Leaf size		25 sq. mm
2. Nanophyll	Leaf size	$25 \times 9 =$	225 sq. mm
3. Microphyll	Leaf size	$225 \times 9 =$	2025 sq. mm

4. Mesophyll	Leaf size	$2025 \times 9 =$	18225 sq. mm
5. Macrophyll	Leaf size	$18225 \times 9 =$	164025 sq. mm
6. Megaphyll	Leaf size		Larger than class 5

The size of leaf has been correlated to climatic conditions. In recent times this type of study has been profitably correlated with the active productive area of trees.

Pierre Dansereau (1957) has recognised six major categories for structural description of plant community components. These are : 1. life form, 2. size of plants, 3. function, 4. leaf shape, 5. leaf texture and 6. cover of the canopy. In each of the above six categories, Dansereau has recognised several subgroups with their respective notation and symbol. In the life forms, there are six categories : **T** for trees, **F** for shrubs, **H** for herbs, **M** for bryophytic forms, **E** for epiphytes and **L** for lianas. For size of plants, there are three main categories : **t** for tall, **m** for medium and **l** for low heights. For functions there are four categories : **d** for deciduous, **s** for semi deciduous, **e** for evergreen, **j** for evergreen succulent or evergreen leafless. 'The leaf shape has six types : **n** for needle or spine, **g** for graminoids, **a** for medium or small, **h** for broad, **v** for compound and **q** for thalloid. For leaf texture the three types are **f** for filmy, **z** for membranous, **x** for sclerophylls and **k** for succulents. The canopy cover is recognised into **b** for barren, **i** for discontinuous, **p** for in tufts or bunches and **c** for continuous. These notations have also been given graphic notation so that with those diagrams and the above letter symbols a community profile could be drawn on paper to give a clear total perspective.

QUANTITATIVE CHARACTERISTICS OF COMMUNITIES

In considering qualitative characteristics we have noticed that floristic enumeration, aspection, sociability, visual rough estimate of the number of individuals of different species, life forms, biological spectrum, leaf size, classification etc. could be studied in field without the necessity of any special sampling technique.

European ecologists especially under the leadership of Braun Blanquet, have developed systems of description and classification of communities and this aspect of ecology is known as *phytosociology*. The structure of sociological order in any community or set of communities cannot be studied by observing each and every individual in an area. Some sort of vegetation sampling has to be done. The sampling may be made by studying in a line or lines across the area of the study and recording the species that occur along this line. Such a sampling technique is called a *transect*. A transect may have a wide breadth useful for instance in a large tract of forest, and is called a *belt transect* or it may be just a line laid across an area of vegetation when it is called a *line transect*. The transect study is usually employed when studying the pattern of changes between different communities and it is therefore laid across the ecotone. Hence, the length of a line transect is governed by the extent of sharp or gradual change in vegetation. It is important to determine the appropriate direction in which the line transect is laid for obtaining meaningful information. On hills, the transect is usually laid between two points at different altitudes.

A study of plant communities may also be made in relation to vertical distribution of canopy and root system of different individuals and species along a line. This is called a *bisect* study. Bisects include measurements of stratification of plant parts both in air and in soil. This type of study is useful in understanding competition and adjustment among members of communities for light, space, nutrients etc.

In community studies, quantitative estimation of community structure, and composition are necessary. This needs precise sampling and accurate measurements. Besides transects other sampling units may be confined to small *areas* or *quadrats* or to *arealess-points*. Sampling in small areas saves considerable time and energy, otherwise needed to make a really comprehensive survey of the entire study site, and at the same time yields quite significant results. Through both quadrat and point sampling techniques, dispersion, numerical strength coverage, etc. can be studied in quantitative terms. For different purposes quadrat and point methods may be of different types.

Kinds of quadrats

For the purpose of listing the occurrence of a species in sampling area the quadrat used is called a 1. *list quadrat*. Simple sampling areas of predetermined size are marked out and the names of all the species occurring within the area are listed. This is especially for the study of degree of dispersion or frequency of different species. Numerical counts of individuals of each species are also made for the study of their abundance and density using quadrats called 2. *Count quadrats* or *list-count quadrats.*

For true to scale study of species distribution in space the individuals are recorded on a miniature quadrat on graph paper at suitable reduction. The quadrat is called a 3. *Chart quadrat* and is laid at random or at predetermined points and every individual is faithfully recorded from it on a graph paper—often with the aid of an instrument called a pantograph. A pantograph has two pointers and when one is moved on the actual plants in the quadrat the other automatically maps them at proper reduction on the graph paper fixed nearby. The utility of the chart quadrat is seen especially in the study of changes in community structure at the same place at intervals of time. If the quadrat is left undisturbed after charting vegetation and the same site is repeatedly studied over a long period of time, the quadrat is called a 4. *Permanent quadrat.* Periodic charting is also done accurately and conveniently by photographing the quadrat through a camera mounted on a stand vertically above it. This type of study yields quite a valuable picture of vegetation changes with seasons and years at any place. On a cleared area, if this procedure is followed, a sequence of invading communities can be recorded which invade the area, grow, yield the place to new set of species and so on. So the phenomenon of succession could be studied by this method.

The shape of a quadrat is usually square, as the term denotes but it may be of different shapes depending upon convenience and use-fulness. It may be circular or rectangular. The size of a quadrat varies with the type of vegetation to be studied. For small plants like mosses, lichens and liverworts growing in patches, small quadrats of 20 cm × 20 cm size may be quite useful, in grasslands if the stand is of relatively pure type, 50 cm × 50 cm size quadrat may serve the purpose, in grassland with great diversity 1 metre × 1 metre or more may be needed. Many workers find 1 m × 0.5 m rectangular quadrats for grassland studies most suitable. In forests the quadrats may be quite large. 10 metres × 10 metres or even bigger. The number of sampling quadrats should be so much as to cover about 5-10% of the area under study.

There is a method of determining the minimal size of a quadrat for studying vegetation. The idea is that the sample area should be on the one hand small enough for ease of study and on the other sufficiently large to cover about 5-10% of the area under study.

The procedure is to lay a quadrat of small area say 20 cm × 20 cm (0.04 sq. metre), on a plot of vegetation and count the number of species occurring within it (Fig. 10.2). This is plotted on a graph paper with number of species on the *y-axis* and the area of quadrat on the *x-axis* (Fig. 10.3). The size is increased to say 30 cm × 30 (0.09 sq.m.) when some additional species are likely to occur in the enlarged quadrat. The number of accumulated species and the enlarged area are again plotted. Likewise, the area is increased to say 40 cm × 40 cm (0.16 sq.m.), 50 cm × 50 cm (0.25 sq.m.), 60 cm × 60 cm (0.36 sq.m.), 70 cm × 70 cm (0.49 sq.m.), 80 cm × 80 cm (0.64 sq.m.), 90 cm × 90 cm (0.81 sq.m), 100 cm × 100 cm (1.0 sq.m.) (Fig. 10.2) etc. When the number of accumulated species versus the area is plotted (Fig. 10.3), it will be noted that the increase in the number of species is rapid at initial stages but gradually the rate of increase in the number of species fall appreciably and the curve tends to become horizontal.

From this *species-area curve*, the desirable minimal area of quadrat is determined by various techniques, but the easiest is to take the point on the curve where it flattens, or where the increase in species

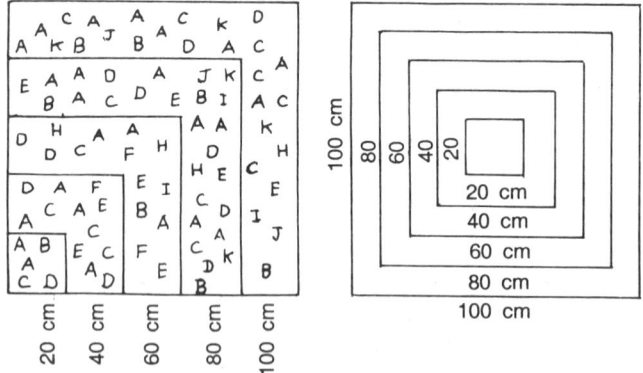

Fig. 10.2. Diagrammatic representation of increasing size of quadrats and the numbers, of species (represented by latters A, B, ... K in determination of the minimal requisite size of a quadrat by species area curve method.

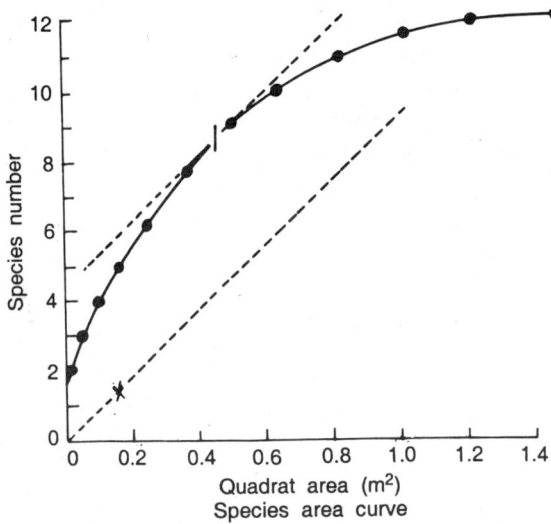

Fig. 10.3. Species-area curve showing an initial rapid increase in number of species with increase in a area of the quadrat. Gradually the rate of increase in number of species declines. A tangent is drawn on this curve parallel to a line drawn at 10% value joining the axis and extending upwards.

number is very little, and join this point to the x-axis denoting the area. This gives us the size of the quadrat where with minimum size the sampling efficiency is maximum.

Braun Blanquet's method of sampling of vegetation by such a *minimal* area representing the smallest area in which adequate sampling of species is possible is called a *releve* and the method is denoted as (i) releve, (ii) SIGMA (or Station Internationale de Geobotanique Maditerraneene et Alpine) or (iii) Z-M (Zurich Montpellier School) or (iv) simply Braun-Blanquet methods. Under this method a number of aspects of vegetation character is measured, such as the name of species occurring within the releve (or quadrat) the number of individuals (density and abundance) the pattern of dispersion (frequency), the percentage basal and/or canopy cover (dominance) in relation to releve area or the value of one species in respect to the sum of the same values of all species present in the study rise (relative percentage of frequency, density and dominance).

In a mixed community with a number of layers or strata of vegetation the quadrat size differ for each of the life forms at the same place. At ground level small quadrats should be laid but for shrubs and for trees larger sizes of quadrats are taken. Each stratum is separately sampled in its own way and such superimposed quadrats of different sizes are called *nested quadrats*.

Quadrats in the field may be laid at random or in a set pattern but the aim should be to cover the entire range of vegetation. Even in random sampling it is desirable to divide sampling areas broadly into different stands and then in each stand to lay quadrats at random. Sometimes quadrats are laid on several intersecting lines at suitable intervals to ensure a wide sampling at equal distances.

PLOTLESS OR POINT METHODS

In point method sampling is done at several points on the ground where a nail or set of nails touches on grids, lines or at random places in the study. There are principally two main types of point sampling methods : 1. *point frame* and 2. a *point centre quarter* (distance).

Point frame method

The method described by Levy and Madden (1933) essentially consists of a scale-like frame with usually ten holes. The frame ends are mounted on a pair of legs and in holes are fitted about 60 cm long nails (Fig. 10.4). The frame is consecutively positioned at several places in the field and the species that nails impinge upon and the number of times they are hit, are recorded. From these values community structure is quantitatively estimated as explained later.

Distance or point centred quarter method

The method described by Cotton and Curtis (1956) eliminates the use of any frame. A brass needle or a nail fitted on the top with a rubber

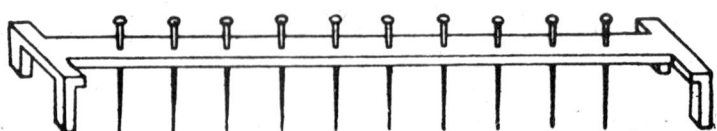

Fig. 10.4. A point frame with ten holes, each with a needle or nail used in sampling vegetation by point frame methods.

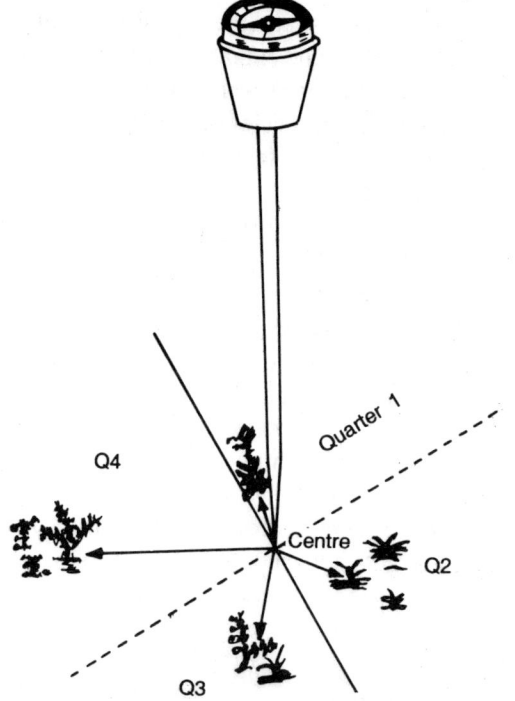

Fig. 10.5. A nail mounted with rubber cap and magnetic needle. It is used in analysis of vegetation. Four quarters are prepared taking the needle point as centre to the nearest plant and its diameter are measured. From these values various sociological characters are calculated.

cork and compass needle (Fig. 10.5) is fixed at random or in a desired direction at desired distances at several consecutive points. Taking the point as the centre, four quarters on four sides are taken and the name, distance from needle and diameter on one plant in each quarter nearest to the needle (centre) are recorded. If the needle hits a plant then that plant's diameter is only recorded and distance is put as zero. Through the aid of observation of the distance and diameter of species at several points a variety of community structure can be calculated.

QUANTITATIVE STRUCTURE OF PLANT COMMUNITIES

With the aid of sampling techniques like transects, bisects, quadrats, point centre or point quarters, etc. the organisation and structure of communities can be studied and expressed and quantitatively both in absolute terms for each species as well as in relative terms of species with respect to all other plant species of the area. The special field of the study of communities with respect to their components, structure and classification, forms the basis of a division of ecology called *phytosociology*. Of course, phytosociology embraces many more aspects than community structure and classification.

What are the species that grow at a place to form a community? How are different species dispersed in space? How many individuals of each species are there and how much of the area is covered by each species? What are relative positions of different species (in terms of distribution, number and coverage) in comparison to the rest? These are some of the more important parameters for describing community structure in precise quantitative terms. Singh, Misra and Ambasht (1980) have made extensive studies of most of these, parameters in a series of grasslands on Vindhyan hills, following the quadrat method of sampling vegetation. Such a study has helped in establishing the relative significance of different grass and weed species that naturally grow there and has important bearing on the management of grasslands. Let us take these aspects one by one in order to understand the methods of their analysis.

Frequency

This term refers to the degree of dispersion of individual species in an area and is usually expressed in terms of percentage occurrence. It can be defined as the chance or probability of an individual of a given species to be present in a randomly placed quadrat. This can be studied by sampling the study area at several places at random, or in a desired pattern to cover the site adequately, and recording the names of the species that occur in each sampling. For instance if a species occurs in five quadrats out of a total 20 quadrats studied, then its frequency is 25%, i.e.

$$F = \frac{\text{Number of quadrats in which a species occurs}}{\text{Total number of quadrats sampled}} \times 100$$

A species most abundantly spread all over the area will have chance of occurring in all the samplings (quadrats or points) and, therefore, its frequency will be 100%. A poorly spread species (even with large number of individuals in one corner) will have a chance of occurrence in only of few quadrats and its frequency value will be low. Thus a higher frequency value shows a greater uniformity of its spread or dispersion. In recording data for frequency simply the presence or absence of a species in the quadrats studied is recorded and not the number of individuals of each species.

Frequency can also be studied on a line transect or on a belt transect in suitable segments of sampling at suitable interception or gaps. Each segment of the line or of belt recorded is taken as equivalent to a quadrat for the purpose of calculation of frequency. Through the point frame method, frequency is calculated from the following formula :

$$F = \frac{\text{Total number of hits made on the species}}{\text{Total number of hits made}} \times 100$$

Raunkiaer (1934) made extensive frequency studies and grouped species into five frequency classes. At any site the species contents show a wide variation in frequency value. Some species have low, some have high and others have intermediate values. Raunkiaer's five frequency classes are :

A	class	with	1-20%	frequency value
B	class	with	21-40%	frequency value
C	class	with	41-60%	frequency value
D	class	with	61-80%	frequency value
E	class	with	81-100%	frequency value

Raunkiaer further propounded a 'Law of Frequency' on the basis of study of over eight thousand quadrats. The number of species in frequency class A is greater than that of B, B is greater than C, C is greater or equal to or lesser than D and D is less than E.

This can also be written as A > B > C > = < D < E and it means that the species with poor dispersion of frequencies are higher in number than the number of species with higher frequency values. On the basis of the averages of a very large number of quadrats Raunkiaer prepared a *Normal Frequency Diagram* in which frequency classes A included 53% of the species, B 14%, C 9%, D 8% and E 16% (Fig. 10.6). This law of frequency led to many frequency studies in many parts of the world and is is generally found that the number of species poorly dispersed (of class A) are larger in number in most natural communities. In biotically disturbed sites the frequency structure is different from the normal.

In frequency studies, the size of the quadrat is of great significance as an increase or decrease in the size alters the result. Generally values of class A and B increase if the quadrat size is enlarged and therefore, values of C, D and E decrease. While comparing frequency values of different sites or at the same site at different times, the same size of quadrat should be used. For instance, if we take the entire study area as one quadrat then all the species, irrespective of their levels of dispersion will show

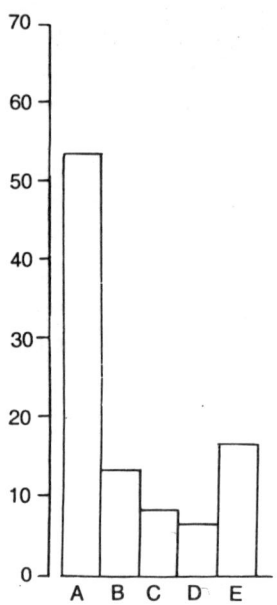

Fig. 10.6. Histogram showing the normal frequency diagram of Raunkiaer. Percentage of species under frequency classes A, B, C, D and E.

100% frequency. On the other hand if the quadrat size is reduced to extremely small size, rare species are likely to escape sampling, i.e. a zero percent frequency will be obtained. Therefore, correct quadrat size and number of samplings are necessary for obtaining reliable data.

One may also study the frequency of a species in terms of its dispersion relative to that of all the rest of the species.

Relative frequency is determined by the use of the following formula using the data obtained by the quadrat methods :

$$\text{R.F. of a species} = \frac{\text{Number of occurrence of a species}}{\text{Number of occurrence of all species}} \times 100$$

Similarly with the use of point method the relative frequency is calculated as follows :

$$\text{R.F.} = \frac{\text{Number of points of occurrence of a species}}{\text{Number of points taken for all species}} \times 100$$

The relative values are more useful than absolute ones in computing the ecological importance of individual species in community of plants.

Abundance

This is the study of number of individual of different species is the community per unit area (e.g. 1 square metre in grassland conditions). By quadrats or by other methods, samplings are made at random at several places and the number of individuals of each species is summed for all the quadrats. The total number of individuals of a species if divided by the number of quadrats where that species occurs. Abundance, therefore, does not give a total picture of the numerical strength of a species in an area because only the quadrats of occurrence are taken into consideration and not all the quadrats studied (Table 10.2).

$$\text{Abundance} = \frac{\text{Total no. of individuals of a species in all quadrats}}{\text{Total no. of quadrats in which the species occurred}} \times 100$$

Density is also an expression of the numerical strength of a species where the total number of individuals of each species is divided by the total number of quadrats studied as illustrated by Table 10.2 and expressed in numbers per sq.m.

$$\text{D} = \frac{\text{Total no. of individuals of a species in all quadrats}}{\text{Total no. of quadrats sampled}} \times 100$$

The above formula would give the density value per quadrat and it should be converted into per square metre. For instance if it comes to 20 plants per quadrat of 50 cm × 50 cm size (0.25 sq.m.), the density would be 20/0.25 or 80 plants per square metre.

Abundance and density may also be calculated by counting the number of individuals on a line or belt transect at regular intervals for 1 metre length. The term *individual* sometimes creates difficulty in the case of grasses or other plants that propagate vegetatively. In such cases each tiller or aerial shoot arising out of soil may be regarded as an individual. Relative density is the study of numerical strength of

Table 10.2. Method of recording data and calculation of frequency, abundance and density

No. of species	Number of individuals in each quadrat (of 1 sq. metre size) Quadrat Numbers										Total no. of individuals of each species	No. of quadrats of occurrence	Total number of quadrats studied	% Frequency = $\frac{4}{5} \times 100$	Abundance = column 3/4	Density = column 3/5
	1	2	3	4	5	6	7	8	9	10						
1. *Cynodon dactylon*	3	7	6	1	17	2	2	2	1	2	43	10	10	$\frac{10}{10} \times 100$ = 100%	$\frac{43}{10}$ = 4.3	$\frac{43}{10}$ = 4.3
2. *Euphorbia hirta*	1	19	2	8	2	×	×	×	×	1	33	6	10	$\frac{6}{10} \times 100$ = 100%	$\frac{33}{6}$ = 5.5	$\frac{33}{10}$ = 3.3
3. *Indigofera* sp.	6	×	12	×	×	1	×	×	×	×	19	3	10	$\frac{3}{10} \times 100$ = 30%	$\frac{19}{3}$ = 6.3	$\frac{19}{10}$ = 1.9
4.																
5.																

a species in relation to total number of individuals of all species and can be calculated as :

Relative density of a species =

$$\frac{\text{No. of individuals of the species in all quadrats}}{\text{No. of individuals of all spp. in all quadrats}} \times 100$$

Cover

This is an expression of the area covered or occupied by different species and is usually given as percentage. The cover may be studied both at the canopy level and at ground level. Cover is of great ecological significance because although the frequency and density of trees or shrubs may be lower than those of smaller plants, yet the dominating influence of trees may be greater in the community because of their more extensive canopy coverage. In grassland species the total coverage of ground by stems and leaves is called herbage cover. The cover can be expressed in terms of percentage of area covered and can be determined through actual measurements by quadrat, line intercept or point methods of sampling.

For the study of grassland types of vegetation, quadrat is divided into 100 segments each measuring 10 cm × 10 cm. The herbage cover is charted on graph paper or alternatively the area is determined by planimetry. For basal cover, the canopy may be clipped in herbaceous species and the diameter of the stems emerging from ground is measured with the help of scales or calipers. The results may be expressed as percentage basal cover or as area covered per square metre.

The basal area is regarded as an index of dominance of a species. The higher the basal area, the greater is the dominance. The average basal area, is calculated from the average diameter of emerging stems. This area of an average individual when multiplied by the density (number/sq metre) gives the basal cover per square metre. In trees the basal cover is usually measured at breast height (1.5 m) by the formula πr^2, but in many tropical trees with butt flares, measurement is taken at ground surface. For measuring crown cover the diameter is measured at ground surface with a metre tape across the canopy perimeter (approximately) passing by the main stem. Since canopy is not full circular several diameter readings in different radii may be measured and averaged. In case of herbaceous vegetation pantograph is used.

Relative dominance is coverage value of a species with respect to the sum of coverage of the rest of the species in the area (and not with respect to ground area) :

$$\text{R.D.} = \frac{\text{Total basal area of the species in all the quadrats}}{\text{Total basal area of all the species in all the quadrats}} \times 100$$

IMPORTANCE VALUE INDEX (IVI)

In any community structure, the quantitative value of each of the frequency, density, abundance and cover has its own importance. But the total picture of ecological importance cannot be obtained by any one of these. For instance frequency gives an idea as to how a species is dispersed in the area but we do not get any idea about its number or the area covered. Density on the other hand gives the numerical strength and nothing about the spread or cover. Dominance gives the basal cover only. Therefore, in order to have a really overall picture of ecological importance of a species with respect to the community structure, the percentage values of the relative frequency, relative density and relative dominance are added together and this value out of 300 is called the *Importance Value Index* or IVI of the species. Usually after the quantitative estimation of relative values of density, dominance and frequency the species are listed in order of decreasing importance.

Besides the comparison between species of a community the data collected on dispersion, number and cover can also be profitably used in comparing the vegetational structure of two or more stands, or of the same stand over a period of time. Vegetation structure in respect of varying environmental factors can also be studied through such studies in sets of varying environmental conditions. The IVI as such, gives the total picture of sociological structure of a species in a community but it does not give the dimension or share of relative values of frequency, density and dominance. There are various polygraphic methods called *phytographs* to show the individual as well as combined aspects of the position of each species in the community structure. One method is to draw a circle and divide it into four equal segments by drawing two lines at right angles to each other passing through the centre. Each of the three radii is divided into 100 parts from the centre to the circumference and the fourth into 300. On the 0-100 scales are marked the values of relative frequency on A, relative density on B, relative dominance on C and the IVI value on the 0-300 scale on D. All these points are joined as shown in Fig. 10.7. Such illustrations give the sociological characters and the IVI of a species at a glance.

Phytographs

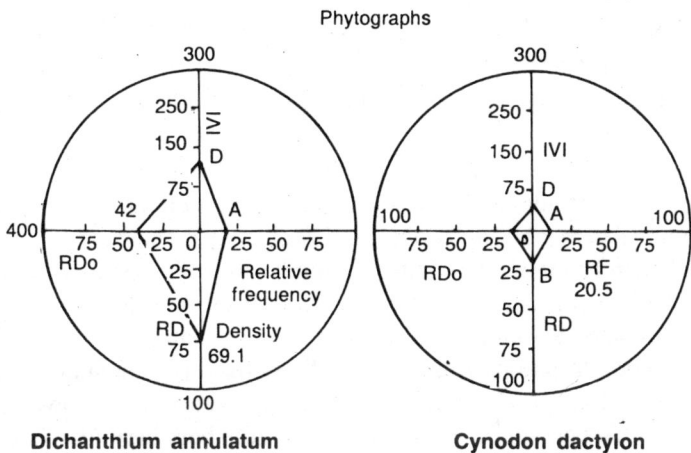

Dichanthium annulatum   Cynodon dactylon

Fig. 10.7. Polygraphic method of showing sociological characters of individual species (the diagrams are based on data from Ambasht and Maurya, 1969).

ECOTONE

The zone of vegetation separating two different types of communities is called *ecotone* or *tension zone*. These are the marginal zones sometimes easily recognisable. The development of a community is intimately related to the prevailing environment and as such difference in community structure and physiognomy (appearance) reflects a difference in their environments such as the moisture condition, biotic influence or any other factor. Therefore, an ecotone is a region where the influence of two different patterns of environment work together and hence the vegetation of ecotones is somewhat specialized. The width of an ecotone may be narrow or wide. For instance the ecotones between adjacent plots—one fenced and protected from grazing and other openly exposed to grazing, or between a pond and an adjacent upland, are quite sharp and narrow whereas among many other types of communities, ecotones are very wide and community differentiation is vague and continuous. Usually in ecotones the variety of species is greater than in any of the adjacent communities. The phenomenon of increased variety and of plants at the community junctions is called the *edge effect* and is essentially due to a wider range of suitable environmental conditions.

ZONAL AND GRADIENT APPROACHES OF COMMUNITY ORGANIZATION

It has been indicated above that communities are well identifiable entities made up of several populations occurring on a prescribed area or physical habitat. They constitute the biotic component of ecosystems. Communities have several group characteristics which are additional to those of individuals and populations. There are two principal approaches to identify the structural entity of communities. One is called 'zonal' approach in which it is argued that since a community occupies a well defined area of physical habitat, it must have boundaries around it. With a well defined zone, the community as a whole exhibits group characteristics of its organization, growth and maturity (succession). This zonal or discrete boundary idea is also known as 'organismic' approach because the behaviour of the community is likened to an organised entity. The concept is supported by Clements (1916). Braun-Blanquet (1932, 1951) and Daubenmire (1966).

Another approach is that the constituent populations of a community respond independently with the changes in environmental gradient. It means that the group behaviour is more strongly operative at population level. Thus within the prescribed physical space there are gradations of changes in the populations. When the idea is extended to adjacent communities, there could be overlapping populations between them also. Thus there could be overlappings among the populations of different community stands in a continuous manner so as to make the boundaries meaningless. The idea of *continuum* in ecology is gaining popularity among quite a few synecologists who advocate that there are nothing like well defined communities in plants with recognizable boundaries around or between them. Vegetation structure changes continuously and gradually both from place to place, and in the course of time at the same place. This concept of gradual change in vegetation structure does not recognise the existence of other clear cut communities with ecotones between them. There is an arrangement of pòpulation in a series of environmental gradients in which plant species or their communities are arranged in an order. Such arrangement of species or communities along a slowly changing gradient is called *ordination* and because of gradual change the term *continuum* is applied. If the change in environment is rapid as for instance from the base to the top of a mountain, the separate communities may be distinguished within a short space. The continuum may not be easily recognised in such cases. But Whittaker (1954) in his classical studies of the Great Smoky Mountains in the USA has prepared distribution curves of dominant trees along a gradient of changing habitat conditions. He found that each population had its own bell shaped curve with their peaks of relative abundance with overlappings across the main community types. Some tree species had a narrow range of distribution and some others had a wide range. Thus Whittaker supported the continuum concept of Gleason (1926) and Curtis and McIntosh (1967). Goodall (1963) has also supported this concept of '*individualistic*' view of distribution of populations as against the '*organismic*' or discrete community concept. Lieth (1968a) has made a critical evaluation of the earlier works and reviews concerning the continuity and discontinuity in ecological gradients and plant communities.

In ordination studies a number of stands in a community are identified and the *ID* or *dissimilarity* index and *CC* or *similarity index* between different stands are computed. Then the highest and lowest ID (pairs of stands) are plotted at the two extreme ends of a graph plot and other pair values are appropriately plotted between these two extreme values. This method of arranging the stands is known as *polar ordination*. Instead of a graph of the values can also be arranged in a table.

HABITAT AND NICHE

Individual organisms of a community occupy certain physical spaces and perform some special functions, and the totality of all these get organized into a balanced ecosystem. The place occupied by an organism is its habitat, and its habitat plus functions performed there by the organism is its *ecological niche*. In functional attributes, there could be a large variety of interactions on a multidimensional scale or on a *hyper volume*. Different species of a habitat may have overlapping niches with its neighbours. Different species in different biogeographical regions may perform similar functions, i.e., they are regarded as ecological equivalents. At first niche was used to designate the place of occurrence or spatial niche, later with Elton's publications the functional property was added to it. Hutchinson (1965, 1978) gave the concept of a multidimensional or hyper volume niche in which different factors are shown on different axes of a multidimensional structure. The functional niche is the maximum 'abstractly inhabitated hyper volume' and realized niche is the smaller hyper volume achieved for an organism under the prevailing ecological constraints or stresses. *Fundamental niche* is the hyper volume that a population can fill in when there is no competition from any other population. But in nature no population is free from competitions and as such the *realized niche* is always less than its fundamental niche. If the niche of one species completely overlaps the niche of a very similar another species, i.e., their requirements along the niche dimensions are the same, then the chances are that one species will be eliminated from that place. But in cases of partial overlaps the possibilities are that either one of the species will fully occupy its own fundamental niche leaving the other weaker species to occupy the remaining unoverlapped space, or both the species will have smaller realized niche so as to accommodate the other's niche dimension. Therefore, niche theory is an extension of Gause's hypothesis which says that different species with same resource requirements cannot co-exist for a very long period of time as the better competitor will sooner or later gain control and the weaker will be eliminated from the place.

Odum (1983) regards *address* of an organism as the habitat and its *profession* as the niche. Thus the "activities especially its nutrition, energy sources and their partitioning; the relevant population attribute, such as intrinsic rate of increase, fitness and so on; and finally the organism's effects on other organisms with which it comes into contact and the extent to which it modifies or can modify important operations in the ecosystem" are the attributes of niche. A term *guild*, first proposed by Root (1967) is now being used for groups or clusters of species with similar role in a community.

DOMINANCE AND DIVERSITY

Lewis (1970) has discussed different concepts regarding species diversity in plant communities and says "species diversity is a statistical abstraction with two components." These are the number of species or *richness* and *evenness* or *equitability*. If there are fifty species in a stand, its richness is fifty. If all the species have equal number of individuals, the *evenness* or equitability is high and if some species are represented by only a few individuals, the evenness is low. The product of richness and evenness is referred as species diversity. Generally the community richness is associated with higher diversity, but it is not always true. For instance, in a community of 15 species if the number of individuals are very uneven, then the diversity may not be as high as in another community of 14 species with an almost equal member of individuals of the 14 constituent species. Evenness is generally referred in terms of number of individuals. Sometimes, the canopy cover or biomass is taken into consideration. So while comparing the diversity indices of different communities one should take into account that the parameters of calculation are same.

The inverse of evenness is concentration of (1) *individuals*, (2) *cover* or (3) *biomass* of a few species. Whittaker (1965) has described three different types of diversity depending upon the range of environment and the community. The *alpha index of diversity* is the diversity with a single community; *beta index of diversity* pertains to diversity in plant composition in several communities occurring in one range of environment and *gamma* diversity deals with a number of samples of community taken from a range of environments. Delta diversity is regarded as dimensionless at a much higher organisational level of landscape than the beta diversity. Levels of organisation in the scale of time and space are from genes to organisms to species to ecosystems to landscapes and finally to biomes, both from structural and functional levels. Temporal diversity is important and keeps changing with succession and/or disturbance. It is commonly found that during succession there is a progression in variety of species or diversity increases to a certain extent. Finally towards maturity or relatively stable state a few species assume dominance. It is often found that with increase in dominance the diversity decreases. Pielou (1966) recognizing two types of diversity, the *species diversity* and *pattern diversity* says that during autogenic progression (succession) the species diversity decreases but pattern diversity increases. Lewis (1970) has quoted McNaughton (1967) thus: (i) diversity is principally a mechanism which generates community stability; (ii) dominance is principally a mechanism which generates community productivity; and (iii) increasing the number of species in stand, rather than enhancing efficiency through more efficient exploitation of site resources decreases efficiency, perhaps through competition. In fact greater diversity provides a number of alternative pathways in the ecosystem functioning which gives stability to the ecosystem whereas single species crops are least stable.

A number of indices to indicate the dominance, similarity and dissimilarity and species diversity have been given by ecologists from time to time.

Index of dominance : Those species which have strongest control over energy flow and the environment in given habitat are known as *ecological dominants*. Simpson (1949) has given the following formula to estimate the Index of dominance (c) :

$$c = \Sigma \left(\frac{ni}{N} \right)^2$$

where, ni is the importance value of the species in terms of number of individuals, or biomass or productivity of each species over a unit area; N is the total of corresponding importance values of all the component species in the same area and period; and Σ (sigma) refers to summation. Suppose in a community of ten species are in all 100 individuals. Now let us take two extreme possibilities of high dominance and low dominance. In situation 1, let us presume that one species is of very strong influence with 91 individuals and the rest nine species are represented by 1 individual each. Let us apply these data to the above formula :

c = $(91/100)^2$ + $(1/100)^2$ + $(1/100)^2$ + ... upto 9 times for 9 unimportant species with one individual in a community of 100. The c value comes to 0.829.

Now let us take situation II where there is no dominance of any one species and each of all the 10 spp. has 10 individuals, c value would be :

$(10/100)^2 + (10/100)^2 + \ldots\ldots$ for 10 times or 0.1.

Thus in communities where one or two species contribute very highly the dominance is quite high showing values more than 0.5 and when all or most of the constituent species share the numbers or biomass almost equally and the dominance values are low.

Similarity index : Sorensen (1984) has given a simple formula to establish the index of similarity between two stands of vegetation :

$$S = \frac{2C}{(A + B)}$$

where S is the similarity index, A and B are number of species on ' stand A and stand B whereas C is the number of species common to both the stands.

If there are 10 species on site A and 10 on site B and 7 spp. are common for both the stands, apply the data to the above formula :

$$S = \frac{14}{20} \text{ or } 0.7$$

Dissimilarity index : Dissimilarity index is $D = 1 - S$ and in this case $D = 1 - 0.7 = 0.3$.

Species diversity index : Different kinds of indices for species diversity are given by a number of workers.

Odum, Cantlon and Kornicker (1969) have regarded diversity (d) as the number of species (S) per one thousand individuals :

$$d = \frac{s}{1000}$$

Margalef (1968) has given the following formula :

$$d = \frac{S - 1}{\log N}$$

and Menhinick (1964) has given the following formula :

$$d = \frac{S}{\sqrt{N}}$$

where d = diversity index, S = number of species and N = number of individuals (or any other measured parameter like biomass or cover).

Shannon index of diversity, also referred as *Shannon-Weaver* and *Shannon-Wiener* index given by Shannon and Weaver (1949) is

$$\overline{H} = -\Sigma \left(\frac{ni}{N}\right) \log \left(\frac{ni}{N}\right)$$

or
$$\overline{H} = \Sigma Pi \log_2 Pi$$

or
$$\overline{H} = -3.3219 \, \Sigma \left(\frac{ni}{N}\right) \log \left(\frac{ni}{N}\right)$$

where $\overline{H}$ = Shannon's index, ni = number of species, N = total number of individuals (or any other parameter of importance). Pi = importance probability for each species.

$\left(\dfrac{ni}{N}\right)$ or the proportion of all samples which belong to species *i*.

COMMUNITY DYNAMICS

Communities are dynamic systems constantly interacting with another system, the environment, which is equally dynamic. It is a common experience that in neglected fields or cleared or burnt forest-lands a series of plant communities make their appearance in the course of time and are themselves eventually replaced by another patch of plants. The changes are gradual and imperceptible at any time but easily recognisable if observed at regular intervals over a long period of time. Seasonal changes in plant communities always occur at every place. It is more characteristic where seasonal changes are rapid, and over a wide range such as in temperate climates. Year after year very similar cycles of vegetation and environment recur but, at many places there are small changes every year. Thus changes in vegetation, animals and environment keep on taking place all the time at all levels in most of the situation in the biosphere. At some places these changes are fast and at others very slow. In course of very long period of time at many places the communities have reached a peak stage and attained a dynamic balance with the environment at one place in the course of time is called *ecological succession*. The processions of communities that arise and disappear yielding place to others are called *seres* and the relatively stable communities in dynamic equilibrium with the environment are called *climax communities*.

The origin and development of communities of any piece of bare area takes place through a series of processes and stages. The first stage is (i) *nudation* or exposure of new surface followed by the arrival of diaspores (that is seeds, fruits or propagule) to an area from neighbouring regions and this process is called (ii) *migration*. This process is present in all communities at all times, but it can be easily recognised on bare areas because of a contrasting background. The next stage is the (iii) *germination* of the propagules which takes place in batches over a period of time depending upon environmental requirements and the availability of optimum conditions. Only a few establish successfully as the seedling or plantlet stage is the weakest point in the ecological life-cycle. The successful establishment is called (iv) *ecesis*. On ecesis the site no longer looks bare and the process of (v) *colonization* continues. Colonization by more and more individuals of the successive offspring and new migrants initially does not create any problem of paucity of resources and the number of individuals increases. This process is called (vi) *aggregation*. The plants that initially colonise and aggregate are called *pioneers*.

The number of individuals increases so much that space and other resources like nutrients, water, light etc. become limiting and (vii) *competition* sets in. Competition leads to the elimination of weaker elements on the one hand and increases the population of better adjusted ones. All these processes not only modify the plant communities but also the rest of the environment as well by casting of shade, selective removal of some minerals from soil, alteration in water balance and addition of organic matter through death and decay of plants and plant parts etc. The modified environment gives ever changing patterns of competition. (viii) *Invasion* of newer and more aggressive species, often better suited to changed environmental conditions, also takes place all the time. The difference is that the first arrivals or pioneer are said to *colonize* whereas the later and more aggressive arrivals are said to *invade*. Invasion naturally results in *reaction* or series of *reactions* on the part of the existing community to interaction among individuals and population of the community and effect of plants on the habitats.

The whole process is at a fast pace initially but as the complexities increase the interactions lead to some sort of (Fig. 10.8) *stabilization.*

ECOLOGICAL SUCCESSION

Succession in general refers to the act of repeated following up of one by another in order of time at a given space. In ecology succession means an orderly sequence of communities of plants (and animals) which occurs over a period of time at the same place. In general the communities or plants at any place modify the environment because of their presence and biological activities, and this modified environment is usually less suitable to the existing community. As such better suited community of plants replaces it. This new com-

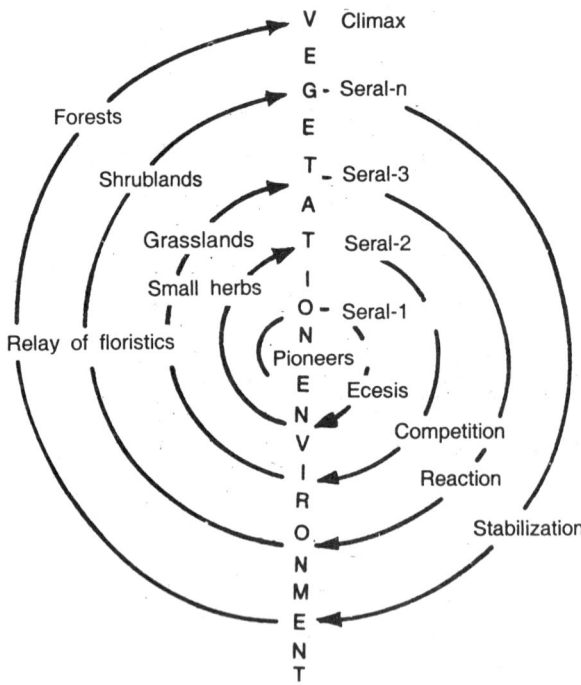

Fig. 10.8. Diagrammatic representation of succession through autogenic processes. Seral communities modify the environment complex which in turn changes the vegetation and the process continues until the dynamic equilibrium is reached.

munity again modifies the environment in such a way as to make conditions less suitable to itself. Thus, changes take place continuously in the community structure, organization, physiognomy, the associated animals and the environment at a place in the course of time and this phenomenon is called *ecological succession*. The rate of successional changes is rapid initially and gradually it slows. The pioneers and early species are of r-*selection* with short life cycle and most of the photosynthates go into the reproductive efforts, whereas plants of late seral stages are of *K-selection* type perennials. With succession there is a convergence of changing communities towards a common, relatively stable climax community in a broad climatic belt.

Seral communities are large in number depending on minor differences in edaphic, topographic or other local conditions. These communities are open to a large scale invasion of other species. Seral communities react to the prevailing environment in such a way that the environment is modified sufficiently to become less favourable to the existing sere. Seral communities are therefore short lived and quick in development. As against these the climax communities are only a few and similar in identical climatic conditions. These are more or less closed to invasion of outside floristic elements. Climax communities attain a dynamic equilibrium with their environment. They cannot change their environment to such an extent as to make it less suited to the existing communities. Climax communities are formed after a long period of time and once formed they maintain their structure and composition.

Successional changes are orderly and develop in a certain sequence. For instance if a mature community of forest is cleared, it is likely that a fresh forest community will eventually develop there but not straight away. It develops through succession of several types of seral communities which are different from forest communities. These seral communities are essential stages necessary to restore the environment for the development of the original type of community. Since succession is orderly, the stage could be predictable and could be deflected or arrested at a certain stage by manipulation of the environment by man or modification due to grazing, burning etc.

Odum (1969) has given three main attributes of succession viz., "(i) It is an orderly process of community development that is reasonably directional and therefore predictable. (ii) It results from modification of the physical environment by the community, that is succession is community controlled even though the physical environment determines the pattern, the rate of change often sets limits as to how far the development can go. (iii) It culminates in a stabilized ecosystem in which maximum biomass (or high information content) and symbiotic function between organisms are maintained per unit of available energy."

KINDS OF SUCCESSION

For the initiation of succession, it is essential that there should be some bare area. If an area is colonized by organisms for the first time, the succession is called *primary succession*. If the area under colonization has been cleared by whatsoever agency (like burning, grazing, clearing, felling of trees etc.) of the previous plants it is called *secondary succession*. Usually the rate of secondary succession is faster than that of primary succession because of better nutrient and other conditions in areas previously under plant cover. Further, depending upon the predominance of green plants or of heterotrophic organisms in the initial seral stages, successions are called autotrophic or heterotrophic. In *autotrophic successions*, initially green plants are much greater in quantity than animals and this takes place in a medium rich in inorganic substance. In *heterotrophic succession* the population of heterotrophic organisms like fungi, bacteria and animals are in greater quantity at the initial stage and such a succession begins in a medium rich in organic matter such as small areas of rivers, streams which are polluted heavily with sewage or in all small pools receiving leaf litter in larger quantities.

In most cases succession, or the replacement of one type of community by another is due to modification of the environment by the communities themselves. The effect of community on the environment becomes the cause of succession. Such a successional process is called *autogenic succession*. In some cases replacement of one community by another is largely due to forces other than the effects of communities on the environment. This is called *allogenic succession* such as might occur in a highly disturbed or eroded area or in ponds where nutrients and pollutants enter from outside and modify the environment and in turn the communities (Fig. 10.9).

Induced succession

Activities such as overgrazing, frequent scraping, shifting cultivation or industrial pollution may cause deterioration of an ecosystem. Agricultural practices are retrogression of a stable state to a young state by man's deliberate action. In a natural steady state the community respiration balances community production and there is little left which man can harvest. Therefore, man tries to control the succession in such a way that a managed steady state is maintained which is different from a natural steady state and wherefrom a good amount is harvested as removable product of the ecosystem. Diverse man made ecosystem, if properly managed on ecological principles, may therefore, reach a steady state not in relation to natural conditions but in relation to total environment including management practices.

Retrogressive succession

It means a return to simpler and less dense or even depauperate form of community from an advanced or climax community. In most cases the causes are *allogenic*, i.e. forces from outside the ecosystem become severe and demanding. For example most of our natural forest stands, are degrading into shrubs, savanna or even more depauperate desert-like stands by the severity of grazing animals brought from the surrounding villages. Excessive removal of wood, leaf and twig litter also leads to retrogressive succession.

Cyclic succession

It is of local occurrence within a large community. Cyclic refers to repeated occurrence of certain stages of succession, whenever there is an open condition created within a large community. The progression of succession on the overall community level is called *directional succession*. In a forest community, when certain old trees die, open space is created. This leads to invasion of new species in this sunny place. There is a rapid replacement of invaders by other seral species and soon the relatively stable stage is reached. Such a cyclic

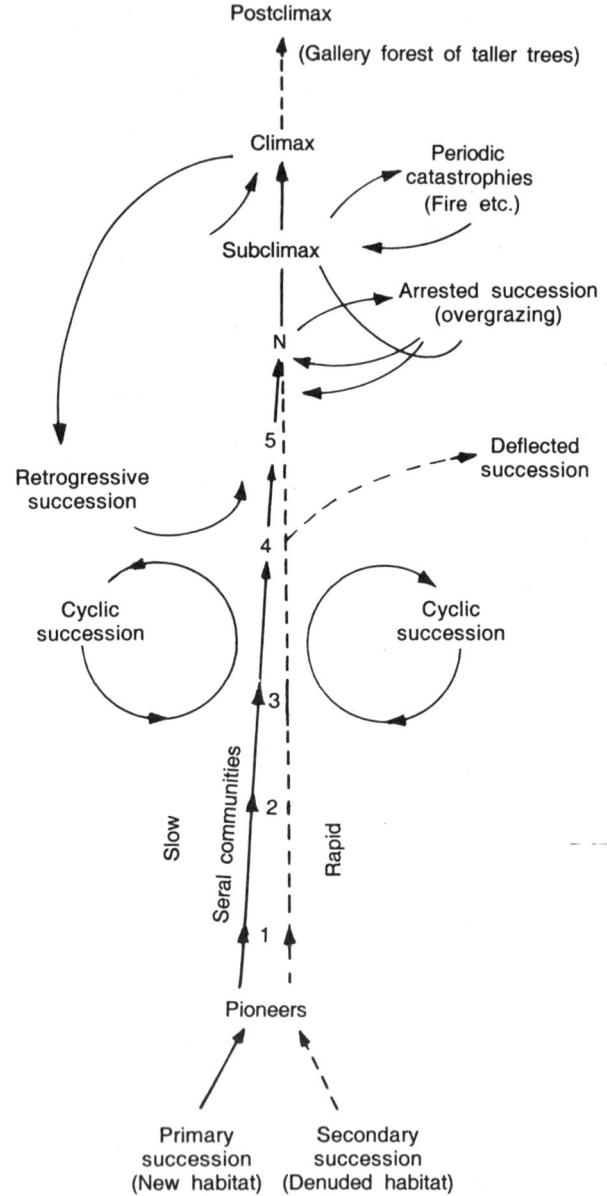

Fig. 10.9. Diagrammatic representation of successions of different kinds.

succession on isolated open patches periodically created by the death of old trees keeps taking place locally. In some adapted tree species, their own juvenile plants are found in the understorey at an arrested stage of growth. As soon as an old tree dies, the juvenile plants of the same species, on getting adequate light and space, suddenly show active growth and cover the open canopy space. In such adapted communities, cyclic succession is not met, as the open space freshly created is not ecised by successional stage plants.

Yeaton (1978) has described another kind of cyclic succession in Chihuahuan Desert of Texas, USA, where on bare desert *Larrea tridendata* invades and forms a shrubby cover. In its protection *Opuntia leptocaulis* succeeds to grow. Roots of *Opuntia* soon spread in soil sub-surface to wide areas and in competition. *Larrea* is totally eliminated. In absence of canopy cover of eliminated *Larrea*, soil erosion is accelerated. Surface soil is lost, the roots get exposed and *Cactus* dies. Again the bare stage is reached the cycle of same successional stages are repeated. Forcier (1975) has recorded a case of cyclic microsuccession in the climax forests in New Hampshire,

On the death of individuals of *Fagus* trees the space gets colonized by *Betula* and in turn it gets replaced by *Acer saccharum*. Finally *Fagus* again returns. The mechanism of this short cyclic succession is shown to be due to *positive association* between the seedling and saplings of *Fagus* sp. with the overstorey of *Acer* sp. Whereas the *Fagus* seedlings are negatively associated with its own overstorey.

Process of natural succession

Successions are variously designated according to the types of habitats in which they begin. The chief representative types are *Hydrosere* or *hydrach* (in water), and *xerosere* or *xerarch* (in dry situations).

Hydrosere or *hydrarch* succession begins in water (Fig. 10.10). In the initial stages the water is poor in nutrients and devoid of much life. The water is incapable of supporting large life forms. Generally, phytoplaknton consisting of microscopic algae, begin multiplying and quickly become the pioneer colonizers. With the death of phytoplankton and animals depending upon them the population of decomposing organisms like bacteria and fungi increases in the pond mud. Decomposition results in release of minerals and enrichment of aquatic

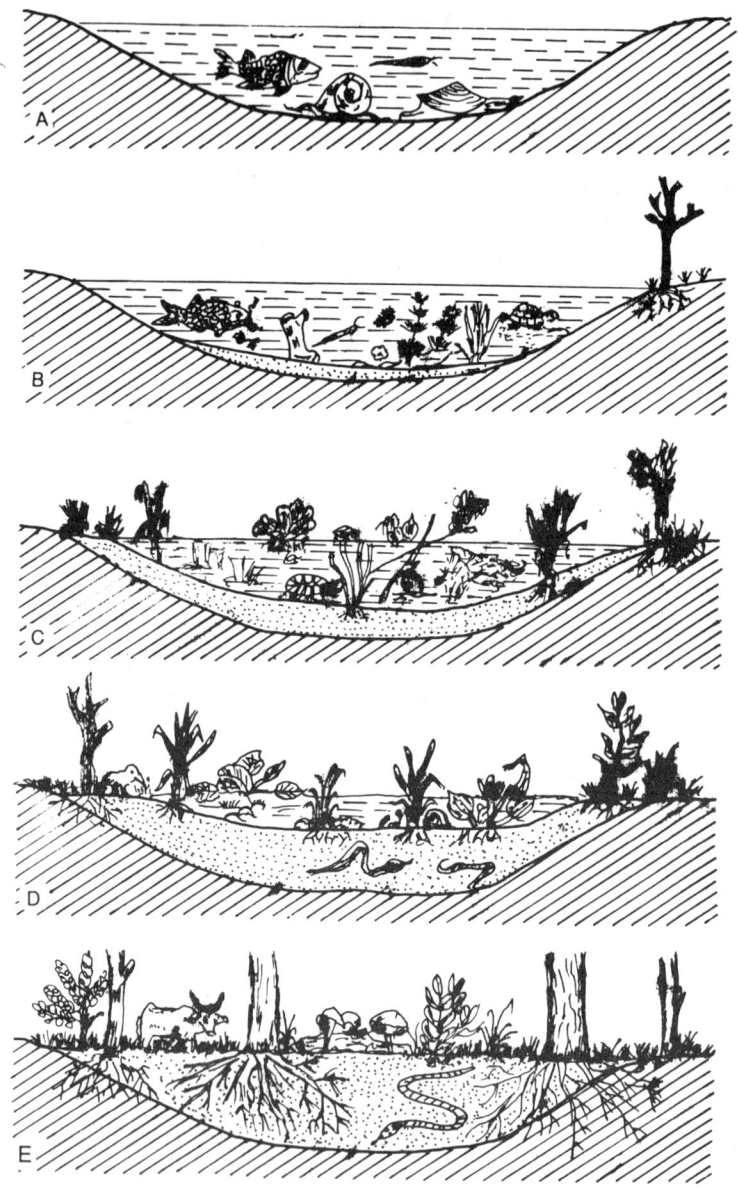

Fig. 10.10. Hydroseric succession showing through stages A to E, formation of different communities at the same place. Gradual silting of pond bed reduces the depth. Finally a terrestrial community is formed.

habitats. The rich mud now supports the growth of some rooted hydrophytes like *Vallisneria spiralis, Ceratophyllum, Potamogeton,* etc. on the shallow regions where light reaches the bottom in sufficient quantity. The death and decay of these plants further contribute to the enrichment of the medium. Deeper zones, are then occupied by such species which are rooted in mud but whose leaves reach the water surface and float. *Trapa bispinosa* (Singhara), *Nelumbo nucifera* (Kamal), *Nymphaea stellata* (Kumudani), *Monocharia* sp., *Aponogeton* sp. etc. are common examples of this type. Gradually, with evaporation of water the concentration of nutrients increases and becomes sufficient to support free floating species which are not rooted in mud. Free floating genera like *Lemna, Wolffia, Azolla, Pistia, Salvinia* etc. gradually cover the water surface. If *Eichhornia* happens to invade such ponds, it rapidly spreads all over the sheet of water due to its high rate of productivity, rapid vegetative reproduction and aggressiveness. The pond margin on the other hand, because of very good environmental condition of high moisture, enough light and aeration soon gets covered by emergent hydrophytes like *Eleocharis plantaginea, E. pallustris, Isoetes coromandelina, Typha angustata, Cyperus* spp., *Fimbristylis* sp., *Polygonum amphibium* etc.

Gradually with the passage of time the silt and dead organic matter deposit on bottom and raise its level. The activities of pioneers thus modify and enrich the medium which becomes more suited to other species. The floating leaved communities cast shade on their predecessor. Submerged communities get eliminated by the deficiency of light. The raised pond bed or shallow water in turn cause invasion by reed swamp species in the area. These in turn are eliminated and replaced by terrestrial and mesic communities as the water evaporates. Climax formation of trees like *Salix* is found in low lying lands in parts of Kashmir.

Xerosere or Xerarch succession begins on any kind of dry substrate like sand (*psammosere*) or rock (*lithosere*). On highly saline areas which are said to be physiologically dry the sere is called *halosere.*

On rock the successional trend is governed by the process of soil formation and accumulation. On rock, water is a scarce material and only such species can grow which can remain attached in the form of crusts on the dry rock and which remain in a dormant condition for most of the time. Lichens and blue green algae are common plants which colonise rocks. These grow and multiply in the rainy season. It is likely that the lichens roughen the rock surface due to corrosion. In the space, thus created, dust and dead organic material accumulate and provide space for foothold and growth conditions for higher life forms like foliose lichens and mosses. The pioneers are eliminated in competition with later arrivals. The cushion of mosses further catch dust and bigger mosses, some pteridophytes like *Selaginella*, and individuals of *Adiantum* also appear. Biological activities, action of carbonic acid on rocks, rapid decomposition of dead organic matter and more soil accumulation and better moisture status improve conditions for plants growth. Many aggressive and hardy grasses, weeds and tree seedlings grow and form characteristic communities. Finally, trees with their dominating influence form forests.

Within the same broad climatic belt, seral stages starting in water and on rock lead to the same type of forest climax community. Similarly in sand the pioneers which colonise initially are able to bind the sand and stabilize the soil. Once the soil disturbance is reduced other species also invade, competition sets in and some species are eliminated in the process.

Primary succession which begins on a new surface, hitherto not occupied by life earlier, takes very long time to reach a stable or climax stage as compared to the secondary succession whch starts on a newly cleared soil where life had existed earlier. There are several places where on account of volcanic lava flows the existing soil surface gets covered upto several metres killing all previous life beneath it. After cooling, on the newly deposited habitat, succession begins. New islands also emerge out of the sea bed and primary succession begins there. Mueller Dombois and Ellenberg (1974) have estimated that it takes around 400 years for succession to reach climax stage on lava flows in Hawaii region. Thornton (1984) has described the possible sequence of the thus far reached successional stage on the island of Krakatau which had emerged violently near Sumatra in Indonesia in 1883. After 3 years of eruption cyanobacteria and mosses appeared. Savanna greases dominated by *Saccharum spontaneum* appeared between 1897 to 1906. By 1908 mixed woodland species including *Ficus* first appeared. The tree species from among the pioneers were gradually replaced except one species of the primary tree by the year 1980 and from the seral savanna and woodland, the present vegetation of Krakatau island is of a rain forest type. It has yet to reach the climax level. But within 25 years of the emergence of the island, all major groups of soil organisms common in Indonesian soils including the nitrogen fixers had established.

Under the ecosystem concept, the succession of plants and animals together in their biotic communities is more usually considered. During seral stages the ratio between the photosynthesis and community respiration is either more than or less than one depending upon the autotrophic or heterotrophic nature of succession. The seral stages converge in either case towards a biotic community where the P/R ratio tends to be one and this is regarded as a stable or climax community. The species content in autotrophic succession in the initial stages has greater variety of green plants. The species composition changes rapidly in the beginning and number of species declines at advanced stages in succession. The animal populations and varieties increases until a balance is reached between the autotrophic and heterotrophic components. The total biomass, the dead organic material and chlorophyllous tissues increase in successive stage of ecological succession. The food webs also become complex and the gross production increases in certain conditions. The net community production which is initially higher at seral stages tends to become equal to community respiration in climax stages. Odum (1969) has arranged different types of ecosystems based on community metabolism (respiration and primary production) on a succession diagram. Deserts, poor lakes and oceans in general have low production whereas fertile estuaries and coral reefs have high production. Ponds, grasslands and rich forests occupy intermediate positions, but all these may be climax formation if their production rate is equally matched by community respiration (Fig. 10.11).

Croplands and some other managed ecosystems, where primary production is much higher than community respiration, are young seral stages and will tend to develop into self-sustaining communities like forests or grasslands. In order to utilize the community resources for the benefit of man, therefore, a good understanding of successional trend and means and methods to arrest the trend at a desired stage are very essential.

Odum (1969) has discussed the strategy of ecosystem development. He has taken twenty-four attributes of ecological

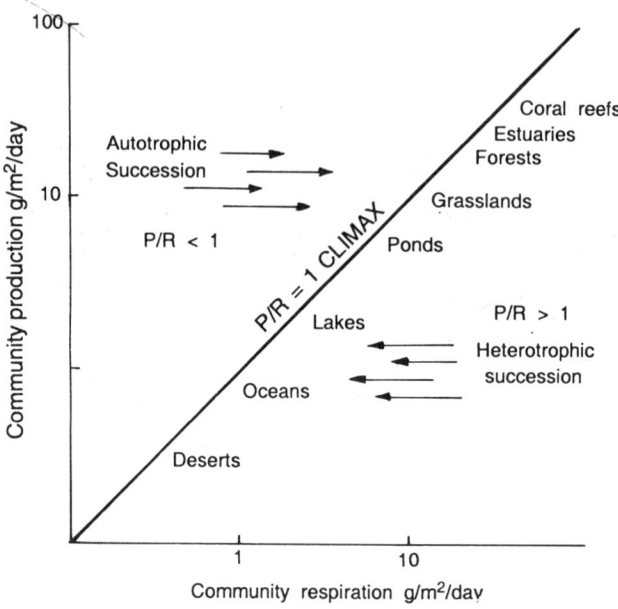

Fig. 10.11. Relationship of community photosynthesis and respiration ratio with successional stage. The communities with balanced (P/R = 1) relation are relatively stable. Ecosystems with increasing production values per unit area are arranged on the climax line (based on Odum, 1971).

systems grouped under six headings and shown the trends of changes from young to mature, i.e., in seral to climax stages. For instance, as indicated earlier, the gross production and respiration (P/R) ratio is greater or smaller than one in the successional stage to approach a value of one in climax stage. In most cases, the seral stage is predominantly autotrophic and *P* exceeds *R*, and therefore, the standing crop biomass increases. Thus P/B (gross production/standing crop biomass) will be high in the seral stages and will tend to decrease at climax stage. Therefore in climax formation the net community production in the annual cycle is very little or nil whereas in successional stages the yield or net production is more. The food web also becomes more and more complex as the community matures. In the young stages grazing food chain predominates, whereas at the climax, most energy flow is directed through the detritus food chain. Grazed material has a high production of indigestible parts which add organic matter to the soil in form of animal droppings. Organic matter in a mature or climax ecosystem increases and a greater part of the inorganic nutrients remain within the biotic components (intrabiotic) as against poor organic matter and extrabiotic nature of inorganic nutrients in the young or seral communities. The variety of species also increases as the community matures and greater stability of climax formation is achieved through better organisation, stratification and adjustment of biotic components as against poor diversity and adjustment in seral stages. As the diversity increases the ecological role of each species, i.e., its niche becomes more and more specialised and narrowed as against broad niches of only a few species in seral stage. The size of organism is also usually larger at the climax stage and they live longer and have more complex ecological life cycles. The climax communities entrap and hold nutrients within the biotic communities, i.e., the mineral cycle is closed and exchange of nutrients between the organisms and environment is slow compared to an open mineral cycle with fast exchanges between organisms and environment in the seral stages. The role of detritus also assumes a greater important in climax communities. There is in essence an overall better homeostasis with regard to mutual interdependency, conservation of nutrients and stability with higher information contents in mature ecological systems than in those of the seral stages.

Trends of changes in the main attributes of ecosystem development from a young to mature stages are as follows :

(i) The *biomass* per unit area *increases* as the system matures to climax.

(ii) The general appearance of the community or the *physiognomy* keeps on becoming more and more *complex* as the succession proceeds.

(iii) The rate of *gross primary production increases* but because of increase in biotic community respiration, the *P/R ratio decreases to about 1.*

(iv) The *canopy* cover gets distributed to *several layers* for a better exploitation of aerial space. *Roots* of different species also get adjusted to *different layers* for an overall *better niche specialization* as the succession proceeds towards the climax.

(v) *Nutrients* in the young stage are allocated mostly in the soil, but as the seral stages advance, nutrients get allocated more in the *vegetation* and less in soil. Further, the nutrient cycling becomes more *closed* or *intrabiotic* with an efficient cycling mechanism whereas in young stage the nutrients easily *leak out* from the system or the cycling is more of an *open* type.

(vi) The role of detritus becomes progressively more and more *important*.

(vii) The quality of the *habitat* gets progressively modified to a more *mesic* condition from *either too dry or too wet* condition in the early stage.

(viii) The *niche specialization increases* as also the *diversity increases*. This means that different functions are more effectively performed by specialist species in mature stage, whereas in early stage many functions are performed but less efficiently by a few species.

(ix) *The life cycle* of mature community species are *longer* and *more complex* and the *reproductive strategy* is more of *K-type* whereas there are more of annuals and other herbaceous species with short life cycle and the reproductive strategy is of *r-type* in young stage species with organisms of bigger size becomes more dominant in old stage while smaller sized organisms are dominant in young stage.

(x) The importance of *macroenvironment* is slowly dampened or it becomes less in later stages.

(xi) Relationship among component species becomes more and more *mutualistic* or helpful even though there exist enough competition and allelochemic activities to prevent invasion of outside elements.

(xii) The overall *resistance increases* while the *resilience decreases* with the progress of succession.

(xiii) *Dispersal* of seeds and propagules is by wind in young stage while by *animals* in mature stage.

(xiv) The *longevity* of seeds and other propagules are usually long in young and *short* in mature stages.

Connell and Slatyer (1977) have given an excellent discussion of succession and proposed three models to accommodate different possible pathways of succession. These are *facilitation* model, *tolerance* model and *inhibition* model.

1. *Facilitation model* is based on the Clements ideas of relay communities in which the seral community is supposed to modify the environmental complex in such a way as to facilitate its own replacement by another better suited community. This new community in course of time prepares suitable ground to facilitate its own replacement. So each community like a 'relay' process hands over the habitat to next or higher status community. This model is criticised for lack of proper evidence.

2. *Tolerance model* : There is a concept of *initial floristic composition* or *IFC* which suggests that arrival of new invaders of higher life form types necessarily does not eliminate the pioneers. Based on this concept the *tolerance model* has been proposed. Only such higher successional or climax species are able to join which can be tolerated by the early settlers. So this concept differs from the Clement's relay succession in the sense that as the succession proceeds, more and higher life form plants are tolerated to join and co-exist, than necessarily replacing the earlier component species. With the passage of time, species that mutually tolerate each other including the better competing long lived trees, gain control over the habitat to form the climax vegetation.

3. *Inhibition model* : This model suggests that the early arrivals on a new habitat may develop counter-mechanism to normal replacement process. Allelopathy may be the common counter-acting adaptation to thwart or *inhibit* the entry of later arrivals. This kind of highly adapted early stage communities may not be common on a wide range of habitats. Only a very restricted level, inhibition model might be operative such as on an intertidal habitat. After the death of such allelopathically armed plants, relay succession gains control. This model is also criticised for its lack of universality. McCook (1994) has discussed the succession models and stressed the need of further work on the above three succession pathways because of facultative and obligate stimulations by early colonisers to subsequent arrivals.

Tilman (1985) proposed '*Resource-Ratio*' Hypothesis of succession. He has agreed to Clement's ideas of replacement of communities and the resources of the habitat are regarded as the key factor. Limited resource leads to competition. For each kind of resource a resource ratio is created due to competition and depending upon the new ratio levels created, new species adapted to it succeed.

Thus, to sum up, succession is directional and progresses towards the climax. In some cases it may be non-directional i.e. it returns to a particular stage and it is *cyclic*. It takes hundreds and thousands of years to complete primary succession while the secondary succession is quicker (10-200 years). In some cases the initial stage community may inhibit altogether the process of succession by allelochemic inhibition of new arrivals. There could be many kinds of driving forces for succession. It may be *retrogressive* also. In some situations life forms of still higher types than the climax may be found (*post-climax*) in better microclimates.

Pandey and Singh (1984) while elucidating the mechanism of ecosystem recovery is degrading Kumaun Himalaya have shown that on land slide exposed land during succession, for the first six years herbaceous species mostly annuals colonize and flourish. For the next seven years, i.e., between 6 to 13 years after land slide, many perennial species gain control. There is a gradual appearance of trees and in about forty years the juvenile trees of the climax species of oaks come up.

CLIMAX

A climax community is the end product of the succession of several seral communities over a long period (hundreds or thousands of years) at one place. Once the climax community is formed, it may remain indefinitely, meaning thereby that organisms of the communities perpetuate there on the same site in more or less equal numbers and in similar associations. The community as a whole perpetuates and assuming no change in climate, the climax community retains its structure and composition permanently. It is regarded as a *steady state* which develops forces against perturbations and restores the community structure to its steady state and the mechanism is known as *homeostasis*. Odum (1963) states that "Recent studies on primary succession on each sites as sand dune or recent lava flows indicate that at least 1000 years may be required for the development of a climax". Secondary succession may take about 200 years to reach climax stage in cleared forests or abandoned agricultural land or only 50 years in certain grasslands.

The idea of climax originated more than a century ago, but it was only in 1916 with Clements' publication of '*Plant succession an analysis of development of vegetation*' that the concept acquired a very definite shape. Climax has been viewed as inseparable with the climate (not with all factors of environment) and very often climax is referred to as *climatic climax*. A whole range of the climatic spectrum is available in India from the driest conditions in Rajasthan, to the world's wettest in Assam, the world's highest mountain ranges of the Himalayas in the North with temperate and alpine condition, one of the most extensive mangrove forest in Bengal, tropical rain forests of tall evergreen broad-leaved trees on the Western coast and part of Assam, etc. Therfore, under the influence of such a wide variety of climatic conditions several relatively stable types of climax communities have been formed. In the last few centuries, however, man's increasing population and its ever-increasing demands have altered the soil, the natural communities, the hydrological cycle etc. to a great extent. The original community-environment equilibria are disturbed in different parts of the country and have attained new equilibria whereas instead of climate, other environmental aspects have dominating influence. Such equilibria or climaxes, where master factor is not the climate, are referred to with suitable prefixes by the Clementsian school of thought at a level subordinate to climatic climax. Many modern European phytosociologists do not consider these climaxes as subordinate to the climatic climax but equal in rank. The latter school of thought is, therefore referred to as the *Polyclimax* school and the former as the *Monoclimax* school.

Climax is regarded as an indicator of climatic conditions and in turn the climate indicates the type of climax community of an area. It is generally believed that once the climax is attained the community does not change at all, but this is not exactly so. Climax communities also change due to what may be said as '*ageing*'. In this process it is likely that the community and the environment in course of time get 'aged' and influences like disease and storms hasten the process. Odum (1963) suggests that with the aid of radioactive tracers the rate of mineral cycling and energy flow through climax formations could be studied over a period of time and if the rates are found to be getting slowed, it would suggest that climaxes, much like individuals, get old and die. Man is responsible to altering the community of most of the climaxes so that secondary successions are now taking place or that the seral communities are maintained at some, early developmental stage through repeated action of man and his domesticated animals. Occurrence of dominant species all over a climatic belt in various stages of growth and regeneration indicates that the area is under climax community.

Climax terminologies

Broadly speaking the climax is a well-understood term, but different interpretations of causative factors for the development of a mature community and their relative status have led to the coining of several terms to indicate various types of climaxes.

Pro-climax is almost a stable community, usually on the margins of climaxes where there are influence of another neighbouring climatic zone. These are relict associations of past climax and can easily undergo change in appearance on disturbances.

If the cause of disturbance is man or other animals as in the formation and maintenance of grasslands in a potentially forest type of climate, then this is called *dis-climax* (disturbed climax). If the climax formations are on margins or in localized regions in a dry zone where life-forms of plants are of lower order than actual climax association then such a strip of formation is called *preclimax* or *sub-climax*.

Post-climax on the other hand is one where edaphic conditions are more mesic or moist and due to this, strips of areas with life forms of higher orders than the extensive climax associations occur. Hill slopes facing the North have different moisture conditions from those facing South. In forests, strips of vegetation along streams bearing *post climax* formation can be recognised. In these climax terminologies the highest manifestation of succession is the climatic climax and all other relatively stable communities are only in arrested stages and therefore subordinate to the climatic climax. This approach supports *mono climax* (i.e. climatic climax) theory.

Another way of looking at the subject is that all factors of the environment are equally important and any relatively stable community in equilibrium with environment is a climax without any special relation to just one factor, i.e., climate. Therefore, under such a concept, climaxes may be of several types (*poly climax*) and the chief argument in favour of this theory is that the evolution of flora is largely dependent upon the genetic potential of the species available and the biotic potential of the environment. So a variety of stable communities may be expected in a climatic zone, whose stability is largely controlled by soil physiography, man, animal etc. (*edaphic climax, physiographic climax, biotic climax* etc.). The advocates of the monoclimax theory, however, argue that climaxes other than climatic are really not in the final stage of succession but they change rather slowly towards the climatic climax. There is no concrete evidence in support of this hypothesis.

CHAPTER
11

Biodiversity and
Major Biomes of the World

Until recently 'biodiversity' was a technical term used by ecologists, but in recent years, particularly during the United Nations Conference on Environment and Development (UNCED) in June 1992 held at Rio De Janeiro biodiversity became one of the most important issues and now it is discussed in non-technical sense and in a wide perspective. A great deal of effort was made by more than 100 countries for arriving at an international accord for the preservation and conservation of biodiversity. While other global environmental issues like of ozone hole or acid rain or global warming or even of desertification are easily observable and measurable, the problems related to loss of biodiversity is less measurable but not less alarming. The biodiversity essentially refers to the variability among the different life forms or the number of species or races within the species or 'gene pools'. It also includes the habitat diversity as well and within the present day concept of biodiversity. Biodiversity can be defined as the *genetic variability and diversity of life forms such as plants, animals and microbes.*

India has a very wide range of climate and habitat types. So there are very rich varieties of plants and animals. We have from very wet and warm regions in Western Ghats and Meghalaya to moist warm, to seasonally moist and seasonally dry warm and cold hilly plateau and extensive alluvial plains, to dry to very dry, hot and sandy deserts. We have also from extremely cold boreal or alpine in the Himalayan upper ranges of habitats for all kinds of forests, savannas, grasslands, deserts, wetlands, mangroves and marine biota including coral reefs. India is also a very ancient centre of culture where human races from different parts of the world have merged. Numerous ancient tribes are found in different parts of the country. India is indeed very rich in biodiversity. The biotic diversity is associated with habitat diversity. Therefore, with different kinds of stresses on natural habitats the diversity is also getting reduced. The rate of extinction of many species which were quite common in the past is fast increasing.

Biodiversity of a habitat is the *sum total of variety of life found there*. Mutation is the principal source for the creation of genetic variability. The process of creation of new variants and the natural extinction of some of the existing species or subspecies is always operative everywhere at a slow balanced state. Biodiversity loss is mainly due to rapid extinction rate generated in recent times by human activities, WRI, IUCN and UNEP (1992) in their volume Global Biodiversity Strategy have defined biodiversity as "*the totality of genes, species and ecosystems in a region*". These definitions look at the biodiversity at three different levels without much emphasis on their interrelationships. Younes and diCastri (1996) have formulated another definition to include interactions, as "*the ensemble and the interactions of genetic, the species and the ecological diversity in a given place at a given time*". To further integrate they have given another definition, "*the ensemble and hierarchical interactions of the genetic, taxonomic and ecological scales of organization and different levels of integration.*" UNESCO (1994) defined biodiversity as 'an umbrella term for variability among living organisms from all sources including terrestrial, marine and other aquatic ecosystems and ecological complexes of which they are part. Diversity can be within species, between species and ecosystems. Biodiversity is a function of time (evolution) and space (biogeographical).

Biodiversity is conceived at various levels, but the commonest are at gene species and ecosystem levels, i.e. genetic diversity, species diversity and ecosystem diversity. All the three are important but in ecology we shall consider here species and ecosystem diversity. No other science is more important for mankind than that of life. It is strange that while we have gone ahead in generating fantàstic quantity of precise information about other planets and stars, we do not know how many different species of plants and animals exist now or had existed in past. We are even not sure whether the so far scientifically recorded species would be three fourth, half, one fourth, or less of the possible total number of species. The guess given by different biogists is highly variable. There is absolutely no reliable guess work of the total number of individuals. Of course the census of humankind shows a tremendous increase in population during the past century, but there are many other species whose population have rapidly decreased to the status of rare, threatened or recently becoming extinct. WRI, IUCN and UNEP (1992) have brought out "Global Biodiversity Strategy", Solbrig, Edem and Oordt (1992) have edited a volume '*Biodiversity and Global Change*' and Perrings et al. (1995) have edited a volume '*Biodiversity Loss*'. Kormondy in his fourth edition of '*Concepts of Ecology*' (1996) has given an epilogue on biodiversity. Schulze and Mooney (1994) have edited the book '*Biodiversity and Ecosystem Functions*'.

Wilson and Peter (1988) in their edited book '*Biodiversity*' have given that the number of described species in blue green algae as 1700, green algae 7000, brown algae 1500, red algae 4000, chrysophytes 12,500, Ascomycetes 28,650, Basidiomycetes 16,000, Bryophytes 16,600, Pteridophytes 11,294, Gymnosperms 529, Dicotyledons 1,70,000 and Monocotyledons 50,000. UNEP (1995) has given the number of species of the main categories of plants e.g. Green Algae 10,000. Red Algae 5,000, Bryophytes 16,000, ferns 10,000, seed plants 24,000 and fungi 72,000. Very little is known about the diversity of microbes. For animals biodiversity, the estimated number is 13,20,000 of which Arthropoda accounts the maximum number of 10,85,000 and Chordata only 45,000. There is a wide gap between the number of species already described and possible existing estimated working figures. The gap is very much for viruses, bacteria, fungi, protozoa and algae. The number of already described species is not accurate as there are thousands of synonyms, i.e. more than one name for same species. Generally the level of biodiversity is lesser in cold climates than the warm and moist climate of the tropics. In the natural ecosystems the biodiversity, both in respect to the number of species and the number of individuals of each species, is controlled and stabilized as the ecosystem matures. But the overriding influence of human demands, unecological developmental and exploitive technology, there has been an increasing loss of biodiversity of species and more so of numbers. The tropical rainforests of the world are restricted to less than ten per cent of the land surface, and they harbour nearly 50% of the biodiversity. Both the density and diversity of plants is fantastic but the soil, on which such a rich plant biomass is sustained, is nutrient wise poor. Nutrients are largely intrabiotic and are released gradually to the needs of plants through leaf and litterfall and their rapid decomposition. So once the tropical biodiversity is lost, recovery becomes much more problematic because nutrient cycling on poor soil is disrupted. It can therefore be rightly said that the rich biodiversity is delicately balanced in a highly fragile state.

Economic importance

Tropical forests are the store house of many economically important species and therefore, liable to be overexploited. Improper deforestation is bound to lead to a process of no return for trees. Therefore, management of biodiversity and prevention of its loss are of great economic value also. Economic benefits derived from wild species especially for protein source from fishes and birds, timber, oil, gums etc. account for hundreds of billions of rupees annually. Share of wildlife directly and indirectly account for even more. Not only food, but wild biodiversity has been the main source of medicine. There is a rapid extinction of many of the traditional Indian medicinal plants. Even in the present day, modern medicines are largely derived from biotic sources including penicillin, tetracycline, cyclosporin, etc. Biodiversity is not only about the number of species or number of individuals of each species, but also about the structure and function. There are certain parts or regions on the globe which are not only exceptionally rich in biodiversity but are important for certain rare species, endemic nature, dominant trees and key stone species. Some such spots are extremely vulnerable to human disturbances. Most ecologists regard that there are at least 15 to 20 *hot spots of biodiversity* around the world. In India the main hot spots are in the rain forests of Western Ghats and in the North-East Himalaya. Some other important hot spots of the world are the Atlantic forests of Brazil, Tropical Andes, California, Madagascar and neighbouring islands, Central

America, forests of Kenya, Tanzania and West Africa, South West Australia and New Zealand.

Biodiversity plays a key role in the ecosystem stability. There is a certain range upto which the system is resilient and returns to its natural balance, but beyond this on further loss of biodiversity the system begins to collapse. Not all species have equally important role in respect of ecosystem stability. Some species are of high ratings while others are rated moderately or lowly. Therefore, in biodiversity conservation species with high ratings are preferred over others. Importance of biodiversity in ecosystem can be compared to a man-made machine like a bicycle. If some species of low rating are lost the system continues to work with lesser efficiency such a bicycle will continue to run with poor efficiency when some nuts or one or two spokes or part like the bell is lost, but when high rating parts like the joints of chain or ball bearings of the axil is broken, the bicycle stops functioning and in the same way the ecosystem collapses on loss of high rating or key stone species in the overall biodiversity complex.

Conservation International (CI), established in 1987, a private, non-profit organisation (headquarter in Washington D.C., USA) is now a frontline organisation for conservation of biodiversity and first published hot-spot map. Mittermeier et al. (2002) has described (1) *Megadiversity Countries*, (2) *Hotspots* and (3) *Wilderness Areas* in respect of setting priorities for saving life on Earth. They have given, "to qualify as a hotspot, an area must contain 0.5% of the global total of vascular plants (estimated at 300,000 species) or 1,500 species as endemics". When an area qualifies with about 15000 species of vascular plants and or 1500 endemic species the next point of importance is the *degree of threat*. If an area has lost in recent past about 70% or more of its species, it is in a threat condition. Based on major biodiversity constituents like plants, mammals, birds, reptiles, amphibians and selected group of insects, they report that, 17 countries qualified for biodiversity status. On terrestrial segment of earth 25 hotspots are identified. These cover about 11.8% of the terrestrial portion of earth (17,541,969 km^2) (Mittermeier, 2002). The real challenge to mankind is to conserve at least this 12% (hotspots) of the land area without further loss of time or extinction of species. If mankind fails in this leadership responsibility amongst fellow life forms, then the biodiversity losses would totally disrupt the biosphere's natural balance.

Mittermeier (2002) has reported that among all countries. Surinam (formerly Dutch Guiana) has the distinction of highest per cent cover, about 90% is still undisturbed in primary condition. A little more than 30% of Surinam's population is constituted by "Hindustanies of East Indian origin". Similar least disturbed country is Papua New Guinea. These countries must be protected from exploitation by rich countries for timber, wildlife and minerals to prevent the greatest ecological tragedy that may otherwise occur. CI and IUCN together have the biggest force of experts engaged in biodiversity conservation.

Genetic diversity in wild relatives of many economically useful plants is tremendous. There are hundreds of varieties of sweet potato, potato, rice with unexplored genetic abilities to adjust to new climates and resist various diseases. Most of the currently high yielding rice cultivars have been made disease resistant by the transfer of resistant genes obtained from an Indian wild rice *Oryza nivara*. Had just this one wild Indian rice species become extinct, before the discovery of its resistant gene, the loss of humanity would have run into billions of rupees per year that would have gone on pesticides to kill the vectors.

The unknown potentiality of genetic, species and ecosystem biodiversity is far more than hitherto known. Urgent and world wide efforts are needed to know more and more of earth's biodiversity and its genetic values. Ofcourse this would be an open ended effort as we would never be able to completely know the usefulness of biotic resources. Since we know so little and rate of biodiversity loss is so fast that we are on the point of losing much of the gene pools well before their potential for human welfare is established. Every nation in the interest of welfare of its citizen must do all that is possible to conserve biodiversity for a better food supply and economic prosperity. No other developmental programme can be more important than the biodiversity conservation. Human culture all over the world has always taught to respect and conserve other living beings, howsoever unimportant they might have appeared WRI, IUCN and UNEP (1992) have pleaded for a "shift from a defensive posture-protecting nature from the impacts of development to an offensive effort seeking to meet peoples needs from biological resources while ensuring long term sustainability of Earth's biotic wealth". Tolba and El-Kholy (1992) in the UNEP edited Book. 'The World Environment' have given estimates to economic gains by gene transfers from wild relatives to modern high yielding crops. Besides the well known increase in rice production by the transfer of disease resistant gene from an Indian wild rice, examples for other crops are also given. A wild wheat has increased wheat production worth US $ 50 million annually. A gene from Ethopian barley transferred to cultivated barley for resistance against yellow dwarf virus adds US $ 160 million annually in sales in USA. Similar gain is expected in *Zea mays* from an ancient Mexican variety with genes resistant to seven common corn diseases.

According to an estimate of WWF about the 1990 black market costs of some wildlife are US $ 25,000 per kg of *Rhino* horn, $ 300 per kg of elephant tusk and $ 2,500 for a necklace of grizzly bear claws (McKinney and Schoch, 1998). Some rare wild animals are also expensively priced as $ 30,000 for an Amazon Macaw and $ 15,000 for a mountain gorilla. Only a small fraction of this flourishing trade is confiscated and destroyed by government agencies.

Causes of biodiversity losses

As already given, the largest store house of biodiversity are the tropical forests. Man is deforesting these by about 17 million hectares every year (WRI, IUCN, UNEP 1992) which means that already much of the rich biodiversity habitat has been rendered devoid of its richness but even on realising this man has failed to check deforestation. This is the first and foremost cause. If we fail to arrest deforestation, we may lose thousands of plant and animal species for ever. Both the tropics and temperate belts are facing deforestation in the developed and developing countries alike.

The next most important ecosystem types facing loss of biodiversity are wetlands, rivers, estuaries and oceans. Aquatic pollution by organic wastes, synthetic chemicals and oil have killed thousands of species and the number of individuals of different species has been cut short by several billions.

Another cause is total neglect of traditional crop species at the cost of high yielding varieties. It is understandable that farmers have given up cultivation of most of the traditional rice varieties in Eastern and South Eastern Asia, but the importance of their *in situ* and *ex situ* conservation cannot be minimised or ignored. In absence of suitable resistant genes in high yielding varieties, there are numerous examples of outbreaks of diseases involving losses running into billions of dollars and we can expect similar disasters in near future especially where the genetic diversity has been ignored and only one or two genetic stocks have replaced all other traditional varieties.

Biological invasion and introduction of new aggressive weeds are very important reasons for the elimination of existing local species with poor competitive ability. These are several examples of spread of noxious weeds overtaking and even eliminating local species in different parts of the world. Tribals know much more about wild plants and animals, and have used this for their food, medicine and other uses. There is a global decrease in tribal human populations. Many of the tribes have disappeared and with them the cultural diversity and traditional knowledge have also disappeared. This cause of biodiversity loss is largely due to the loss of tribals natural habitats and to some extent their migration to villages and cities in search of livelihood. In India, tribals used to visit village bazars to sell traditional wild plant and animal products of potent medicinal and high energy values. Now they are mostly gone for ever and a few pseudotribals cheat ignorant public by selling fictitious specimens leading to the killings of already depleted wild life. Trade in wild life products for fur, bones, skin, tusks of elephants and horns of rhinoceros and so many other products has been officially regulated, but unofficially a roaring business exists. Most of the species producing the above are on the verge of extinction.

At the root of most of the biodiversity losses is the tremendous rise in human populations, complicated life style high consumerism, ignorance, poor implementation and enforcement of environmental laws at local, governmental and intergovernmental levels, poor environmental education, narrowing down of use of a few high yielding species, totally ignoring the wild traditional varieties. Human society is also divided into prosperity classes, some few rich consuming the most of the natural resources and in the process polluting the global environment. The whole human philosophy of prosperity is to be changed and the ideal of simple living and high thinking has to replace the present day air conditioned indoor and outdoor (cars and trains) life with wasteful habits.

To sum up the causes of biodiversity loss are (i) continued rise in unsustainably large human population, (ii) more deforestation than plantation, (iii) conversion of wetlands in terrestrial systems of landfills, (iv) pollution of water bodies, (v) global climatic changes including increasing tropospheric UV-B and ozone levels, (vi) discarding of hundreds of traditional varieties, (vii) narrowing spectrum of selected few cultivated varieties, (viii) overexploitation of medicinal species, (ix) destruction of representative habits, (x) illegal trading of wildlife and their products, (xi) lack of proper environmental education and required knowledge of biodiversity functions, (xii) invasion of aggressive alien weeds which eliminate many native species, (xiii) overexploitation of some wild plant and animal species for various economic consideration, (xiv) high level of consumerism and exacting demands of natural products by prosperity classes of human society, (xv) a 'waste producing' technology, including wasteful habits and above all, (xvi) a very poor enforcement of legislations.

Our purpose here is to propagate ecological knowledge on biodiversity conservation for a better world tomorrow.

How to conserve biodiversity

The biodiversity conservation has to do with the protection of genes, species and their numbers in population, and ecosystems or habitats. First step, therefore, is of an inventory of biodiversity wealth, their value and the causes of losses. Depending upon the dimensions of the problems, the method has to be applied. One point of cardinal importance is that conservation steps must have a component of 'use'. Without this, the concerned human society tends to ignore the governmental and scientists efforts. Common man must have a driving urge of participation. When the element of use comes, the need of regeneration. of resource is necessary, otherwise the resource will soon exhaust. The level of utilization and regeneration has to be balanced at a sustainable level. There is always a limit to growth. Conservation steps must keep an eye on this limit in concerned use and conservation balance.

Much of the cause of overuse or wastage is due to the decreasing concern for community life at the cost of personal life and self possession. For example each man wants to travel alone in his car rather than using public transports in day to day activity. This is responsible for the worst kind and level of atmospheric pollution which not only harms human kind but all other life forms. Excessive use of CFCs responsible for ozone depletion has led to enhanced UV-B radiations which not only kill smaller life forms but also affects the genes of plants, animals and man. These examples reveal that for the biodiversity conservation, a clean environment is necessary and pollution has to be controlled for air, water and soil. Action is required to be taken at local, village and farm levels of bioregions, national and international levels.

These days countries rich in biodiversity such as India must rise to the occasion and obtain necessary patterns to protect their flora and fauna from exploitation by others. American patents for Neem and Basmati rice would show the necessity of such a step. There are international conventions on biological diversity under the aegies of UNEP. The United Nations General Assembly in a resolution designated 1994-2003 as the decade of 'International Biodiversity'. This would provide opportunity to design and implement international efforts and creation of an Early Warming Network. The main items to be monitored by the Warming Network are to record the evidences of over exploitation of species, habitat losses, pollutant discharges and climatic changes that threaten biodiversity etc.

Besides general efforts in biodiversity conservation in all regions, there are some special efforts required especially for threatened genetic, species, population and ecosystem diversity.

The on site protection steps are called *in situ* and off site i.e. in zoos, botanical gardens, gene banks, aquaria, etc. is called *ex situ*.

Ex-situ conservation of wild varieties of domesticated crops provide opportunity to plant breeders and genetic engineers to transfer desired traits in high yielding varieties. *Ex-situ* raised wild offsprings can be released in their natural homes to restore dwindling population. Botanical gardens, zoos, arboreta can maintain populations of wild species, raise new generations, supply surplus plants and animals to other places or reintroduction in suitable wild habitats.

Biodiversity Conservation Centres or 'Biological Parks' are to be developed in all major biodiversity regions where a whole range of important plants and animals of the region can be perpetuated.

Ex-situ culture collection of microorganisms is done in big laboratories where algae, fungi, bacteria and virus could be maintained in culture conditions. Culture collection centres can be of any size from small housing local collections to large housing world collections. There is a risk of sudden destruction of biodiversity collection in *ex-situ* centres due to some local catastrophic events; hence world collections need to be maintained in different centres to safeguard against such casualties. 23 Microbiological Resource Centres located in 19 countries have been developed by UNESCO of which in Asia are in Bangkok, Osaka, Tokyo and Beijing.

A large number of tree species, medicinal plants, crop plants, ornamental species are being conserved both *in-situ* and *ex-situ* conditions. Over 12000 threatened plant species are being grown in over 1500 botanical gardens and arboreta all over the world (Tolba in WRI, IUCN, UNEP, 1992), the collections are maintained as trees, shrubs and herbs, as dried or refrigerated seed samples, as clonal collections and as tissue cultures. Aquaria are also used in the conservation of aquatic organisms.

In all such biodiversity centres, in element of research is must. Information flow on global network is being developed. Important agencies handling biodiversity conservation and information are World Conservation Monitoring Centre (UK), World Conservation Union (IUCN, Switzerland, World Wide Fund for Nature (WWF), FAO, Commission on Plant Genetic Resources (FAO, Rome), International Board for Plant Genetic Resources (IBPGR), United Nation's Environmental Programme (UNEP, Nairobi) etc.

In the normal process of evolution of plants and animals there is a very slow process of speciation or creation of new species. Naturally occurring extinction of species has also been equally slow except in case of major natural catastrophies like meteorite hits, sudden floods, glaciation, or other natural calamities on a major scale. These, however, occurred at intervals of thousands of years. In recent years there has been creation of high yielding, disease resistant strains of agricultural crops. Their overuse naturally resulted into disuse of hundreds of traditional species and strains leading to their extinction. Many of the wild varieties, which have been rich store house of gene pools, have also been lost or getting lost. Future generation of scientists will not have access to pick up genes of positive advantage in their biotechnological efforts to incorporate desired traits in popular cultivars when such natural gene pools are lost i.e. become extinct. Further, each species in an ecosystem has some specific function to perform. It has a niche specialization. When a species disappears from the ecosystem, its specific function may be partly taken up by some others but full replacement is often not possible and the natural balance of the ecosystem gets disturbed. The scale of loss of biodiversity has tremendously increased, thus leading to widespread ecological imbalances. Ecosystem is like a network of components with very definite interlinkages and interdependencies. These interlocking mechanisms are badly affected upon environmental degradation, habitat degradation and loss of biodiversity. While man can take corrective measures to withdraw pollution sources and rehabilitate the habitat, but very little can be done once a gene pool or species is lost. It is lost for ever and to reconstruct the species as a whole is just impossible.

There are two methods available for maintenance of biodiversity. A large number of species are listed by the IUCN in their Red Data Book as '*threatened*' i.e., the number of individuals of those species have become less and less and any time they may become altogether extinct in near future. These can be preserved and prevented from getting extinct by preserving their natural habitats in form of bio-sphere reserves. This is essentially an *in situ* method. Another method is to rehabilitate such species to new homes in sanctuaries, national parks, botanical gardens, etc. This is an *ex situ* method. In the same way threatened wild life animals can be *in situ* or *ex situ* preserved. In some rare cases the desired traits can be permanently preserved for future biotechnological use by storing them at very low temperature. This is possible for reproductive material and tissue cultures and the method is called *Cryogenic* technique.

Red Data Book of IUCN has listed disappearing species under (1) *Endangered* for those on verge of extinction unless saved by man, (2) *Vulnerable* for those species likely to become endangered soon unless not saved immediately, and (3) *Rare* in which the number of individuals has considerably reduced but not yet vulnerable, it is on decline. Heywood and Watson (1995) have added another 4th category of '*Indeterminate*' type where the status of the taxa with respect to extinction is uncertain. They have given the present number of plant species under these categories as 3632 endangered, 56867 vulnerable, 11485 as rare, and 5302 as indeterminate. The current rate of extinction is about 2000 plant species per year (Reid, 1992a).

For a proper assessment of biodiversity and its speed of extinction, it is necessary to have periodic data collection on the number of species and a general census of each on some representative sites. We do not have such initial data as there is little unanimity on the number of plant and animal species found on earth. There are scientific descriptions of thousands of plants and animals but a very large number is yet to be discovered. Many may become extinct even before they are discovered. It is estimated that there may be in all about 10 million of plant and animal species of which only 1.4 million (14 lakhs) are scientifically described. Further, it is estimated that more than 80 to 90 million species must have become extinct during the millions of years of the evolution of life on our earth. Among the animals, the largest share of biodiversity is of insects (about 750 thousand) while vertebrates are much less (41 thousands). Plants account for 250 thousand (Tolba and Elkholy, 1992). Many species are confined to small or restricted area and have failed to migrate to new places. These are called *endemic*. Man has moved in search of new places for exploitation of natural resources or for gainful employment. He has destroyed the natural habitats and converted them for his own settlement, activities and exploitations. In this process many endemic flora and fauna are completely lost. A large percentage of the island dwelling birds and mammals have become extinct during the past 400 years.

As already indicated India is very rich in plant and animal varieties as well as in cultural diversity. Hence it is regarded as a 'mega-diversity' region. The variety richness in the tropics is far greater than in the temperature region. Tropical forests have as many as 500 tree species per hectare as against about 40 tree species per hectare in the temperate forests. In tropical oceans also there are 5 to 6 times more planktonic varieties than in the temperate belt. So, the biodiversity problems and preservations must be viewed much more strongly in the tropical region than in other climatic belts.

For complete survey, taxonomic identification of the flora of Indian and to enlist endangered species, undertake effective conservation measures and to bring out National, State and Regional 'Floras', the Botanical Survey of India (BSI) was established on 13th February, 1890 at Calcutta. Important Floras published are *Flora of Bombay Presidency* by Cooke (1901-08), *Flora of Presidency* of Madras by Gamble (1915-36), *Bengal Plants* by Prain (1903), *Flora of Upper Gangetic Plains* by Duthie (1903-29) and *Botany of Bihar* by Hains (1921-25). *Flora of British India* (1872-1897) was however prepared before the establishment of BSI. The BSI is now organized into the Headquarters (Kolkata) and Regional Circles at Coimbatore, Pune, Jodhpur, Dehradun, Allahabad, Gangtok, Itanagar and Port Blair.

BIODIVERSITY IN INDIA

There are about 45,000 plant species in India (Tewari, 1993) which constitute about 12% of the global plant biodiversity. Gadgil and Meher-Homji (1983) have analysed India's biological diversity into 43 vegetation types. Angiosperms account for about 15,000 species. Tewari (1993) mentions that there are 372 species of mammals, 693 species of fishes and 60,000 species of insects in India. Among the forty three (plant biodiversity) vegetation types of Gadgil and Meher-Homji 3 types are in (1) **thorny** vegetation like *Salvadora, Zizyphus, Acacias,,* 16 types in (2) **deciduous** category such as dominated by *Anogeissus, Terminalia, Tectona, Shorea, Buchanania*; 2 types in (3) **semi-evergreen** characterized by *Toona-Garuga, Bridelia-Ficus-Syzigium* types : six vegetation types in the (4) **evergreens** including the sholas of *Dipterocarpus, Mesua*, etc., 3 types in (5) **Southern Indian** vegetation characterised by *Acacia, Anogeissus*, etc; and 11 vegetation types in the (6) **Himalayan** category. The last or seventh category has **mangroves**. Chauhan (1993) has regarded Western Ghat as the richest biodiversity area and from this region possibly many cultivated plants have originated, as for example turmeric, ginger, pepper, cardamum, jackfruit and mango. Besides the above terrestrial biodiversity areas, the wetlands are also important habitats of aquatic flora and fauna. Disturbance, pollution, landfills, dam containing nutrients (causing eutrophication) and toxic pesticides, are responsible for destruction of aquatic biodiversity. Many fishes which breed in a particular microenvironmental conditions fail to complete their reproductive cycle in the event of degradation of the specific habitat conditions. Biswas and Trisal (1993) have given a list of important large lakes and reservoirs of India, which includes 2164 lakes of natural origin and 65250 of man-made types which respectively cover 14,50,861 ha and 25,89.266 ha area. Some of the very important wetlands of the country are Dal and Wular in Kashmir, Harike and Sukhna in Punjab; Sambhar, Bharatpur and Pichola in Rajasthan, Gujar and Surha in UP, Kabar in Bihar, Chilka and Bhitarkanika in Orissa, Loktak in Manipur and Koleru in A.P. etc.

Conservation of endangered genetic resources is described later in the chapter on conservation. Threatened plant species are listed there. Some of the important animal species of endangered category are black buck, blue whale, cheetah, chinkara, clouded leopard, dolphin, flying squirrel, Gir lion, Kashmir stag, musk deer, one horned rhinoceros, tiger, Ridley turtle, pythons and birds like crane, monal, baaz, Great Indian bustard etc. For biodiversity conservation important National Parks in India are Dachigam, Corbett, Kaziranga, Namadapha, Gir, Ghana, Periyar and Bandipur. Among biosphere reserves are : Nanda Devi, Sunderbans, Nilgiri, Manas and among sanctuaries are : Abhor, Hazaribagh, Jagdalpur.

Ambasht, Srivastava and Ambasht (1994) have briefly reviewed the conservation aspects of biodiversity in India and highlighted the basic ecological processes involved in their loss as well as in their conservation. They have listed seventeen points in this context which need to be taken into account while handling biodiversity problems.

MAJOR BIOMES

Biomes or major communities are the extensive formations of vegetation. Details about the kind of plants occurring in different geographical belts are described later in the chapters on phytogeography and vegetation of India. The main biomes are : (1) Tundra biome, (2) Boreal coniferous forest biomes, (3) Temperate forest biomes, (4) Temperate grassland biomes, (5) Temperate and tropical desert biomes, (6) Tropical rain forest biomes, (7) Tropical deciduous forest biomes, (8) Tropical savanna biomes, (9) Wetland biomes and (10) Marine biomes.

1. Tundra biomes

These lie north of 60°N latitude or beyond the timberline characterised by an extremely cold condition not favouring the growth of trees. Usually, the climate is moist and the growing season is short. Plants are of low height and even perennial herbs and less than 10 cm tall. Shrubs are also dwarfed. Mosses and lichens are common forming the important base of food chain of primary producers, reindeers and Eskimo (man). Because of extreme cold, the frozen soil moisture of only the top few centimetres gets melted and moisture is available from this top layer. Lower soil layer remains permanently frozen and is called *permafrost*. This solid layer impedes the movement of water, nutrients and root penetration. Tundra is Russian means marshy and unforested land. Tundra is of two main types, (i) the *Arctic tundra* in the extreme northern latitudes and (ii) the Alpine tundra on mountain tops even at lesser latitudes but at altitudes above the *timberline* i.e. where the trees do not grow. In the chapter on phytogeography the vegetation of this zone is described under *boreal* and *arctic* zones. Totally snow covered vegetationless icy expanse and mountain tops are called as *nival* zone.

Growth phase is of two to three months in a year when it is some-what warm and the average maximum is only around 10°C but air temperature may at times reach warmer levels of 20°C or so. Temperature fluctuation is more on alpine tundra than the arctic tundra. Photoperiod shows the widest fluctuation as the dry length in summer may reach 22-23 hours or more and in winter to only less than 1 to 2 hours. Winter temperature falls to minus forty or fifty degrees celcius. Wind speed is normally high in the range of 50-90 km per hour. Rainfall is widely different in different parts, but because of very little evaporation the effective rainfall is high. Soils are rich in organic matter because of very slow decomposition rate. *Histosol* with more than 20% organic matter is characteristic of bogs while *entisols* are coarse textured with undecomposed organic matter with the top *talik* layer full of roots and lower *permafrost* or frozen layer. The surface topography plays an important role on the availability of water during the brief summer when ice melts. The depression appears as wet *meadows* while ridges or slopes look like stony wasteland or desert.

Plant phenology is strongly influenced by the temperature and day length features. Dormant buds get activated as soon as the snow melts. Leaves unfold, photosynthetic activity is soon maximised and flowers are usually brightly coloured since insect pollination is common. Seeds are usually long lived and seed output is high. Net primary production is low, 1-2 tons per hectare per year.

Respiratory utilization of gross primary production is about 30% which is much higher than in tropical plants. Dominant families are Caryophyllaceae, Compositae, Cruciferae Cyperaceae, Gramineae and Rosaceae.

2. Boreal coniferous forest biome

The ecotone region between the tundra in the north and the boreal coniferous forest biomes in the south, has characteristic trees of dwarfed, rather twisted nature.

3. Temperate forest biome

Temperate belt extends from 30° to 55° latitudes and floristic details are given in the chapter on phytogeography. On mountains they are on higher reaches upto the timberline, and divided into subalpine and montane zones. Details about Himalayan temperate vegetation are given in the chapter on Flora and Vegetation of India. In the American temperate forests tall trees forming an overstorey of about 75 metres are *Sequoia, Pinus, Thuja, Abies, Larix, Picea, Tsuga,* etc. which accumulate huge biomass of about 2000 tons ha^{-1} and the life span of trees is well above 500 to 1000 years.

Sequoia sempervirens or redwood tree, the tallest among all living things, crosses 100 metres height and it lives for 1000 to 2000 years. Of course there are reports of some living *Pinus* plants much older than the red woods. Pines are the commonest trees found round the world in this kind of biome and their different species often form pure stands in distinct bioclimatic and edaphic zones. Role of mycorrhizae and actinorhizae are very important in nutrient dynamics in temperate belt. The alders (*Alnus* spp.) are most effective actinorhizal trees that fix atmospheric nitrogen. Forest fire is a very important recurrent phenomenon that regulates secondary succession, vegetation composition and land use patterns. Jhum cultivation in North-East India is fire regulated (see chapter on fire factor).

In the temperate forests, there are (i) *Evergreen biomes* in which leaf fall is continuous and the canopy is never naked and (ii) *Deciduous biomes* which all the leaf fall is more or less synchronous usually in March and April, rendering the trees naked for brief periods between the fall of old leaves and emergence of new ones.

Soil in this biome is mostly of *podzolic* type of ash, grey brown or ash coloured with acidic pH around 5 or 6. As we move from the Tundra to temperate to tropical biomes, the pH keeps slowly increasing from highly acid, to acidic, to neutral, to alkaline types. Mycorrhizae help the trees to compete effectively with grasses and in the absence often tree stands are replaced by grasslands. In the North-American temperate forest, the commonest trees are *Pinus, Tsuga, Quercus, Fagus, Liriodendron, Tilia, Castaenia, Carya, Acer,* etc. and the climbers are *Smilax* and *Vitis*.

4. Temperate grassland biome

Grasslands are also among the most extensive formations or vegetation types found all over the world and in all ranges of climates from mesic to xeric and from cold to warm conditions. The temperate grasslands are extensive in the North America and are called as *prairies*. They are *tall grass prairie, mid grass prairie* and *short grass prairie* depending upon the height of the herbage portion. The natural grassland plains have largely been converted into croplands or managed grazing lands. In temperate savanna of Australia, trees are being burnt in order to develop extensive grasslands for maintaining huge herd of cattle for meat, milk, butter and leather production. Grasslands are dominated by *graminoids* i.e. grasses and sedges and forbs or the non-graminoids like dicot weeds.

Perennial grasses usually propagate and perennate by runners and rhizomes, while the annuals have a high seed output. Savanna are also regarded as a grassland since the ground cover is largely by graminoids, but there occur trees or shrubs at regular intervals.

Temperate grasslands experience a wide range of temperature from about –40°C to +40°C in extreme winter and warm summer. Rainfall is highly variable and evapotranspiration is high. Common grass genera are *Panicum, Poa, Bouteloua, Stipa, Sporobolus, Agropyron, Andropogon, Buchloe, Aristida, Festuca, Bromus* etc. Soils are near neutral of *Chernozem* type containing high organic matter in dark upper zone followed by a clay loamy mineral matter. In tall grasslands, the canopy is so dense as to give a leaf area index of 5 to 8. The standing crop biomass and LAI decrease in mixed mid prairie and short grass prairie. There are a few C4 grasses in temperate belt but they are much less than in the tropics. In the mixed grasslands the commonest forbs are *Polygonum, Helianthus, Aster* and *Rosa* in the North Dakota, but grasses are dominant.

5. Temperate and tropical desert biomes

Desert refers to a general loss of plant life. They are usually dry and show isolated growth of xerophytes. Deserts are divisible into the (1) cold deserts and the (ii) warm and hot deserts. Scrublands have rather open canopy shrubs, usually of spiny thickets. Rainfall is scanty and grazing by goat and sheep is common. Frost and snow are common in cold deserts. *Artemisia tridentata* is commonest species in the American cold deserts while *Larrea tridentata* or *creosote bush* in warm deserts. Due to poor canopy and sparse distribution of plants, the leaf area index is less than one. Productivity is also low. Soil topography in deserts is of different but characteristic types. In the rocky areas the slopes or valleys have alluvial soil. *Bajada* a Spanish name is given to gentle slope rocky deserts with *coarse alluvium the plant diversity is high*. In Arizona bajada there are many perennial species wherever the texture and moisture are favourable. Bajada soil is called *aridisol* (Barbour, Burk and Pitts, 1980) with poor profile development and pH at 7-8.5, Calcium carbonate layer is called *Caliche* in USA. It is much like the *Kankar* in north India and it impedes root growth and affects plant composition. In poorly drained and highly saline bajada the pH rises to 9-11 much as we have the *usar* and *reh* soil causing desertification in North India. Only highly specialized species thrive on such deserts where soil moisture is very poor. Temperate desert plants may shed their leaves either in the drought phase or in cold winter season. Succulents are Crassullacean Acid Metabolism or *CAM plants*. Stomata open during night in such plants which greatly help in a better photosynthetic efficiency (see chapter on primary production).

Opuntia, Pachycereus and *Cereus* are commonest in American deserts. There are extremely slow growing plants. Details of species composition are given in the chapter on phytogeography and in the section 'West Indian Deserts' in the chapter on flora and vegetation of India. CAM and C4 plants are commonly adapted plants of desert climates. Common desert genera are *Acacia, Prosopis, Tamarix, Ephedra, Capparis, Zizyphus, Salvadora, Calotropis, Tribulus, Suaeda* in the Asian tropical deserts, while in the American warm deserts common genera are *Ambrosia, Yucca, Encelia,* large variety of cacti (like *Opuntia, Cereus* and *Pachycereus*) *Parthenium, Jatropha, Agave,* etc.

6. Tropical rain forest biome

This biome represents the most majestic and complex formation or vegetation found in warm and wet climates of the tropics. The luxuriance of vegetation is at its peak. Diversity of vegetation is also maximum. Richness of life forms such as sciophyte herbs, shrubs, climbers, liana small, medium, tall and very tall trees and epiphytes are most abundantly represented. The diversity is so high that often it is difficult to find two individuals of same species in close vicinity. Animal life is also very rich with many kinds of insects, snakes, birds and mammals. The tree trunks are full of epiphytic growth of mosses, ferns and orchids. In the multistoreyed stands some very tall or 'flag' trees emerge distinctly well above the main canopy layer. Their huge size is best supported on the plank buttresses of the basal part of main trunk. Since the annual rainfall is not only high but spread over 9-10 months in the year and the relative humidity is very high, the epiphytes grow most luxuriantly. There being no prolonged dry or adverse season, all leaves do not fall together, but do so slowly and canopy is never naked. Hence they remain evergreen all the year round. Leaves are strong, firm often glossy. Evapotranspiration is high. Decomposition of litter is also rapid, but the nutrients so released are most rapidly and efficiently absorbed by higher plants, leaving very little nutrients in soil. As such the tropical rain forests represent a kind of climax vegetation which is supported on its own decomposition products, and of course on the abundant rainfall and bright sun shine, whereas the soil is nutrient poor. On the ground layer, there are numerous plants of Scitaminceae, Piperaceae, Urticaceae, Araceae, Rubiaceae, small palms, large varieties of ferns, tree ferns, and mosses. Saprophytes are also common. Water drops keep on falling from the leaf surface drip even while it is not raining.

In India, the rainforests are confined to North-East in Assam, Meghalaya, Arunachal etc., and in South-Western region in Karnataka, Tamil Nadu and Kerala. Vegetation of these regions are described in the chapter on Vegetation of India. *Hopea parviflora, Dipterocarpus indicus, Mesua ferrea, Calophyllum elatum* and *Artocarpus* spp. are the common dominant trees.

Tropical rainforests are common in equatorial belt in Indonesia, Malaysia, Singapore, Hawaii, Amazonia, and in Central Africa particularly in Zaire basin region. One important point that must be noted about these forests is that once destroyed, it is difficult to recover the wonderful vegetation since the soils are poor. There are numerous examples of failure of cultivation or other exploitation attempts once the native trees were cut.

7. Tropical deciduous forest biome

Deciduous refers to the nature of trees in which all leaves fall together in one season or month so that the plant becomes leafless or naked for some period until the new leaves emerge. This period is largely controlled by either the stress condition or the photoperiod. In nature we find all grades of evergreen to deciduous characters in different climatic geographical belts. In cold climates, winter season with snow is the cause of leaf fall. In the tropics, in March and April i.e. just before the onset of summer season, when the day length begins to increase, the leaf fall starts. But contrary to old belief that this helps to reduce transpirational water losses, the new thin leaves emerging in April-May transpire more water than old thick leaves. Deciduous forests are also called as 'seasonal' forests, because during summer the ground vegetation dries, the shrubs and trees lose their leaves and the normal forest like appearance is lost, while the same becomes a thick, impenetrable forest during the rainy and winter season. These forests are characterized by a rich diversity of flora of two to three layers of canopy in the wet deciduous forests and 1-2 layers in dry deciduous forests. Wet deciduous are of course more productive, dense and full of tall and medium height trees, climbers and epiphytes while in the dry deciduous belts the epiphytes and climbers are less and the canopy cover is rather open, and some light reaches the ground particularly in dry months. Degraded forests and in dry habitats scrublands or thornwoods are found which are intermediate between forests and savannas. Despite the loss of tree life forms, the savanna and tropical grasslands show almost an equal or sometimes more net primary production because of rapid seasonal growths particularly by the C_4 grasses. *Acacia arabica, A. senegal, A. leucophloea* are most common in India particularly in biotically disturbed areas and many a times the regeneration of non-spiny tree seedling is possible only under the protection of spiny plants. This type of thorn forest is called as "*bush*" in Australia and *Caatinga* in Brazil.

In tropical deciduous forests the common trees in India are *Michelia* sp., *Emblica officinalis, Syzigium cumini, Odina wodier, Dillenia pentagyna, Artocarpus* spp., *Dipterocarpus indicus, Hopea odorata, Salmalia malabaricum, Lannea grandis* and *Terminalia* spp.

The common timber tree plants are *Shorea robusta* and *Tectona grandis* in moist as well as not so moist region. In dry deciduous forests, *Anogeissus latifolia, Boswellia serrata, Buchanania lanzan, Diospyros melanoxylon, Ougenia dalbergioides* are more common.

8. Tropical savanna and grassland biomes

Wangeri and Sanford (1985) have given that grass covered biomes constitute about 42-57% in Africa, 6-12% in Asia, over 50% in Australia and about 80% in South America. Grass dominated lands are called as (i) *Savanna* when there are some interspersed shrubs or trees as is extensively found in tropics and (2) *grasslands* when graminoid dominated stands are free from shrubs or trees. However, in modern literature, savanna and grassland terms are often used interchangeably to mean all stands dominated by grasses, sedges and forbs. Savanna is derived from the Caribbean Indian language in which *sabana* means forest clearings. In Venezuela (South America) the grass dominated savannas are called (i) *Llanos* and in Brazil they are called (ii) *Cerrado*. The scrub thorn savanna in tropical America is called (iii) *Catinga* while in East Africa the woody savanna is known as (iv) *Miambo* and grass formation in South Africa is (v) *Kuroo*. Extensive grassy plains in some South American regions are called (vi) *Pampas.*

Savannas are usually characterized by a prolonged dry winter and summer seasons and brief wet rainy season. So the species are capable of survival in prolonged dry condition. Grasses support a large variety of grazing animals like deer, antelope, wild and domesticated goats and sheep and wild asses. In African savanna, zebra and giraffe are common while in Australia, kangaroos are characteristic. While in most of the grasslands and savannas, predator animals like leopards, hyenas, lions are common, there are no such ferocious carnivorous animals in Australia, and because of this, the once few rabbits introduced there, have, now multiplied to uncontrolled large size populations.

The most comprehensive account of Indian rangelands is the voluminous 213 pages Presidential address of the III International Rangeland Congress by Prof. S.C. Pandeya (1988).

Dabadghao and Shankernarayan (1973) have distinguished four main types of grass cover in the country. These are :

(i) *Sehima-Dichanthium* type of peninsular south India dominated by these two grass species in association with many more grasses. The shrub elements are spiny thickets of *Mimosa rubicaulis, Acacia catechu*, and the fleshy and spiny *Euphorbia* sp.

(ii) *Dichanthium-Cenchrus-Lasiurus* type in parts of Gujarat, Rajasthan, Delhi, Punjab and West Uttar Pradesh in which several other perennial and annual grasses and legumes are found. The shrubs are *Acacia senegal, Calotropis gigantea, Prosopis cineraria* and *Salvadora olioides*.

(iii) *Phragmites-Saccharum-Imperata* type : A characteristic of sub-humid or humid condition of Gangetic plains when the other important associated grasses are *Bothriochloa pertusa, Dichanthium annulatum* and *Cynodon dactylon, Zizyphus nummularia, Acacia arabica* and *Butea monosperma* are common shrubs.

(iv) *Themeda-Arundinella* type in humid hilly tracts of Assam, Manipur, West Bengal, Uttar Pradesh, Himachal Pradesh and Jammu and Kashmir. These are principally derived from the forests of account of excessive grazing and shifting cultivation.

9. Wetland biomes

Wetlands are characteristic transitional conditions between terrestrial lands and deep water bodies and represented by paddy fields, riverine flood plains, lakes and *lacustrine* marshy lands, *palustrine* ecosystems and *marine* backwaters and coastal wetlands. Wetlands are often seasonal, i.e. with free overlying water in the rainy season, marshy and often dry in summer. Wetlands support specialized vegetation and fauna and serve as the breeding grounds of many migratory birds. Food plains and marshes are sometimes not easily accessible and hence serve as a secure place for many kinds of wild life. Sunderban is one such habitat in the coastal belt of the Bay of Bengal. Wetlands, on account of supraoptimal water, are quite productive. In tropics, because of associated bright sunshine and warm condition their primary productivity is very high. Rice, which is the major food crop is best grown in wetlands. Biomass is well distributed between the above ground foliage and underground rhizomatous and cormous parts. *Typha* and *Phragmites* have very high production efficiency. *Euryale ferox* a prized dry fruit crop is most abundantly grown in wetlands of North Bihar. Wetlands are fast disappearing in many parts of the world because of dumping of wastes and reclamation for agriculture, housing, aforestation, etc.

The main kinds of wetlands are :

1. *Shallow* and *seasonal rivers* and the extensive shallow *embankments* and *flood plains* of major rivers, springs and streams.
2. Marshes, peats fens, bogs and coastal lagoons.
3. Low-lying lands under agriculture, particularly rice.
4. Fishery ponds, shallow lakes, *Trapa* and *Euryale* cultivation ponds and other aquaculture habitats.

The flowing of lotic wetlands of the first category are most easily polluted by city sewage, industrial effluents and runoff pesticides and other agro-chemicals applied in the drained upland crop fields. This aspect has been described in the chapter on Environmental pollution. The second category of wetlands are fast shrinking in their total area and depth due to their conversion into otherwise usable lands. The third category of cultivated wetlands are subject to intensive application of agrochemicals and weeding practices, as a result of which most of the characteristic and rare gene pools restricted to low-lying wild conditions are fast becoming extinct or endangered. Fishery and aquaculture water bodies are also losing their natural identity, diversity of flora and fauna and only a few selectively cultivated ones are thriving.

10. Mangroves and coral reefs

(a) Mangroves

Most of the tropical sea shores with shallow water are occupied by halophytic shrub and tree forests called as *Mangroves*. They are rich in biodiversity and are of high economic and ecological values. They occur in intertidal belt and get partially submerged during *high* or *spring tides* at new and full moon periods and get exposed during low or neap tide between the new and full moon periods. Mangroves and sea grasses produce enough dead organic matter to add enough fertility to ecosystem necessary for other life forms. Mangrove vegetation shows distinct zonations, some species occupying the shores, some others intermediate and still others the inner zone. Some genera are of worldwide occurrence in mangroves, such as *Acanthus, Nipa* and *Acrosticum*. In India, mangroves are extensive in Sundarbans in Bengal, and eastern coasts of Orissa, Andhra and Tamil Nadu and in Western Coasts of Kerala, Maharashtra and Goa. Some of the common plants are given in the chapter 4 on Water. *Rhizophora mucronata, R. conjugata, Avicennia officinalis, Acanthus ilicifolius, Ceriops* sp., *Bruguiera gymnorrhiza, Excoecaria agallocha, Sonneratia acida* are common in Indian mangroves. Some of the important adaptive features are occurrence of stilt roots, buttress roots, knee roots, pneumatophores and anchor roots. Some species are adapted as salt excluders, some salt tolerant and some salt excretors. Pollen grains may be stiky and get cross pollinated by butterflies and birds. In some plants, the seeds germinate while still attached to the mother plant (*vivipary*) and escape high salinity conditions of mud at germination stage. In mangrove forests on marshy habitat, Royal Bengal Tiger is common. Crocodiles, sea snakes and pythons are also common. Highly expensive crabs, prawns and fishes are raised in estuaries. Quite a good part of leaf litter is washed to coastal land by tides and wave-action and thus add to the fertility of coastal land as well. Mangroves are efficient primary producers and are rich source for fuel wood, charcoal, timber and house thatching and construction material. Tannins, dyes, medicines and food are also obtained.

(b) Coral reefs

Coral reefs found in sea are amongst the high biodiversity and high primary productivity ecosystems. Corals are sea animals producing polyps which produce calcium carbonate skeletons. Specialised symbiotic algae - *Zooxanthellae* are the source of high primary production. Corals form reef of calcium carbonate. In the coral reefs a vast variety of animals live from top carnivores like sharks to herbivores and algae at the base of the pyramid. Sea urchins, crustaceons and herbivore population, marine algae like *Padina* and *Dictyota* increase. Maximum diversity is at about 20-30 metres depth and 5 to 6 thousand marine arthrozoan animal varieties are found in coral reefs in the tropical belt. Diversity rapidly decreases as we move to temperate belt. Coral reefs are more abundant in Indian Ocean and the Tropical Pacific and much less in the Atlantic ocean. Coral reefs are divided into three types : (1) fringing reef, (2) barrier reef, and (3) attols (Ozborne, 2000). As the name suggests fringing reefs are on the short fringes. Barrier reefs are further away from shore. The Great Barrier Reef extends for about 2000 km in the North East of Australian rocky coast. It is a good example of fringe type. Attol is ring shaped reef formed at post volcanic eruption places after cooling as coral island in lagoons.

11. Marine biome

This is the largest and most extensive of all biomes in area and volume, covering about three fourth of the earths surface. It is the store house of about 97% of the total water. High salt content of an average 3.5% or 35 g per litre render it unfit for biological needs of man, and terrestrial plants and animals. Only such organisms that can withstand the osmotic pressure associated with high dissolved salt content live in marine biomes. The extensive sea regions are called *oceans* and *seas* like the Pacific, Atlantic, Indian, Arctic and the Antarctic oceans and Arabian, Baltic and Mediterranean seas.

The coastal shallow marine biomes are highly productive and divisible into *tidal* and *neritic* zones and the continental shelfs. Away from the coast are the *oceanic belts* divisible into the upper surface *euphotic* zone (epipelagic and mesopelagic zone) and lower *bathyal* (200 m to 2000 m deep) and *abyssal* dark deep zones. Life is abundant in euphotic zones. Planktonic life forms are abundant. Large marine algae, dominated by green, brown and red algae are abundant. The overall marine primary productivity is much lower than the terrestrial biomes, but certain marine biomes such as the coral reef and estuaries and highly productive and very rich in biological diversity and activity. Marine biome is the principal source of sea foods like fishes, shrimps and certain edible algae.

Besides food, marine biomes abound in them huge deposit of petroleum and gases and they are being increasingly obtained from high seas. Sea beds are rich in many minerals. Marine biomes regulate the global rainfall and atmospheric CO_2 balance. Sea is being polluted most severely, particularly by oil and highly toxic chemicals. A large number of international and intergovernmental agencies are now busy in controlling marine pollution. A number of highly hazardous pollutants are '*black listed*' and their discharge in sea is totally prohibited. These are halogenated organic compounds, tin, mercury and cadmium compounds, radioactive isotopes and carcinogenic material. Somewhat less toxic material are "*grey listed*" like compounds of arsenic, zinc and antimony, cyanides, organosilicon compounds, crude oils, foam forming detergents and surfactants. Their discharge into marine biomes is permitted in small quantities only.

12

Ecological Energetics

Energy is the basic force responsible for running the machine of life. In fact energy is the capacity to do work and all living things must work. A virus, a bacterium or a tall tree, or any other organism, needs energy for metabolic activities. Green plants capture solar energy through photosynthesis and convert it into a chemical form. In this stored form the energy is passed on from organism to organism as food. Thus the sun is the source of all energy for organisms on the earth. Even the coal and petroleum which we use to provide energy for many purpose is the solar energy which was converted and stored by organisms in the geological past and which became fossilised.

In the sun exceedingly high temperature acts a nuclear furnace where nuclear rearrangement like continuous transmutation of hydrogen atoms into helium is taking place. This process released fantastically high amounts of energy which radiates out in all directions in the form of electromagnetic waves called solar radiation. A very minute portion of this radiation, i.e. about one fifty millionth of the sun's energy output reaches the earth's atmosphere. The green plants have evolved a mechanism of trapping and converting it into the chemical form of energy.

The chief forms of energy are : (1) radiant energy, (2) heat energy, (3) chemical energy and (4) mechanical energy.

Solar energy travelling in form of radiation is called *radiant energy*. Part of this radiant energy in wavelengths between 400 mμ to 760 mμ is the visible range or light energy. Light energy is readily converted into heat energy as can be seen in the open sun when light waves striking the ground convert partly into heat which warms the soil or other objects. The light energy entering the chloroplast machinery of the plant body is taken up by electrons of the chlorophyll molecule and passed on through series of stages into chemical energy in the bonds of organic compounds.

Mechanical energy may be of two forms : *potential energy* or the stored energy which remains in reserve and which can be used only on conversion into *'free'* form or *kinetic energy* or *useful energy*.

Different forms of energy interchange their forms under certain set rules. These are defined in two laws of thermodynamics.

The first law may be stated : "Energy cannot be created or destroyed but may be converted from one form to another."

The second law, for the purpose of ecology, may be stated : "processes involving energy transformations will not occur spontaneously unless there is a degradation of energy from a non-random to random form" (Phillipson, 1966).

The unit of energy measurement is *erg*, the work done in lifting 1 gram of weight to a height of 1 cm against the force of gravity is equal to 981 ergs. One crore ergs (10^7 ergs) is equal to one *joule*. All forms of energy can be completely converted into heat energy. For a better and uniform expression, in ecology, therefore, energy is measured not in terms of ergs but in joules or units of heat measurement. Heat is measured in calories. Once calorie is equal to the heat energy required to raise the temperature of 1 gram of water from 14.5°C to 15.5°C, and one calorie is equal to 4.2 joules or 4.2×10^7 ergs. One thousand calories (10^3) make one kilo calorie or kilogram calorie (kcal or Cal). There is now a trend of expressing energy in ecological literature in terms of kilojoules.

In ecological energetics we are interested mainly in the (i) quantity of energy reaching an ecosystem per unit area (say a square metre) per unit time (say one hour, day or year) and (ii) quantity of energy trapped by green plants and converted to a chemical form and (iii) the quantity and path of energy flow from green plants to organisms of different trophic levels over a period of time in a known area. As already described in Chapter 2 there is a progressive loss of energy among the biotic components of the ecosystem and hence at successive stages from producers to consumers of different orders the quantity of energy available decreases.

In the earth's atmosphere about 15×10^8 calories/metre^2/year of solar energy is received (Phillipson, 1966). A large fraction of it does not reach the ground surface and is scattered by suspended particles in the atmosphere. The constant rate at which solar energy reaches the earth's outer atmosphere is called *solar constant*. Increase in the concentration of greenhouse gases and decrease in ozone have changed the level of solar constant. About 33% of the incoming solar radiation is reflected by clouds and another 9% by suspended dust particles. Nine per cent is further held ozone, water vapour and other atmospheric gases, thus leaving about 47% to reach the surface of the earth (Fig. 12.1). The quantity of solar radiation received at any place, therefore, depends upon the clarity of the atmosphere. Latitude is another important factor and the equatorial region receives maximum solar radiation followed by other regions of the tropics. The quantity of energy goes on decreasing with increase in latitude both in the Northern and Southern hemispheres. This figure is approximately 2.5×10^8 cal/m^2/yr in Britain (Phillipson, 1966), 4.7×10^8 cal/m^2/yr in Michigan (Golley, 1960) and 6.0×10^8 cal/m^2/yr at Varanasi. Only about one to five per cent of the energy reaching the ground is converted by the green plants into chemical energy which is used in building up the plant body and in metabolic activities. The rest of the energy is lost in various forms especially as heat dissipated and heat used to evaporate water. Part of the chemical energy in plant tissues is passed on from organism to organism as they are successively eaten. Ultimately the entire energy trapped by green plants at one time is lost from the ecosystem in several stages. One important concept of ecological energetics is that *energy always flows* in one *direction* in the ecosystem while materials like carbon, nitrogen, hydrogen, oxygen etc. are repeatedly used in the ecosystem cycling from inorganic to organic and back to inorganic forms. Energy does not move in a free condition, but rather through organic materials from one trophic level to another. In assimilation of energy the green plants build up organic material from inorganic substance and gain in their organic biomass. This gain in weight (dry matter weight) as a result of conversion of solar energy into chemical energy per unit area of space for unit time is called *Primary Productivity*. Out of this increase in organic production some

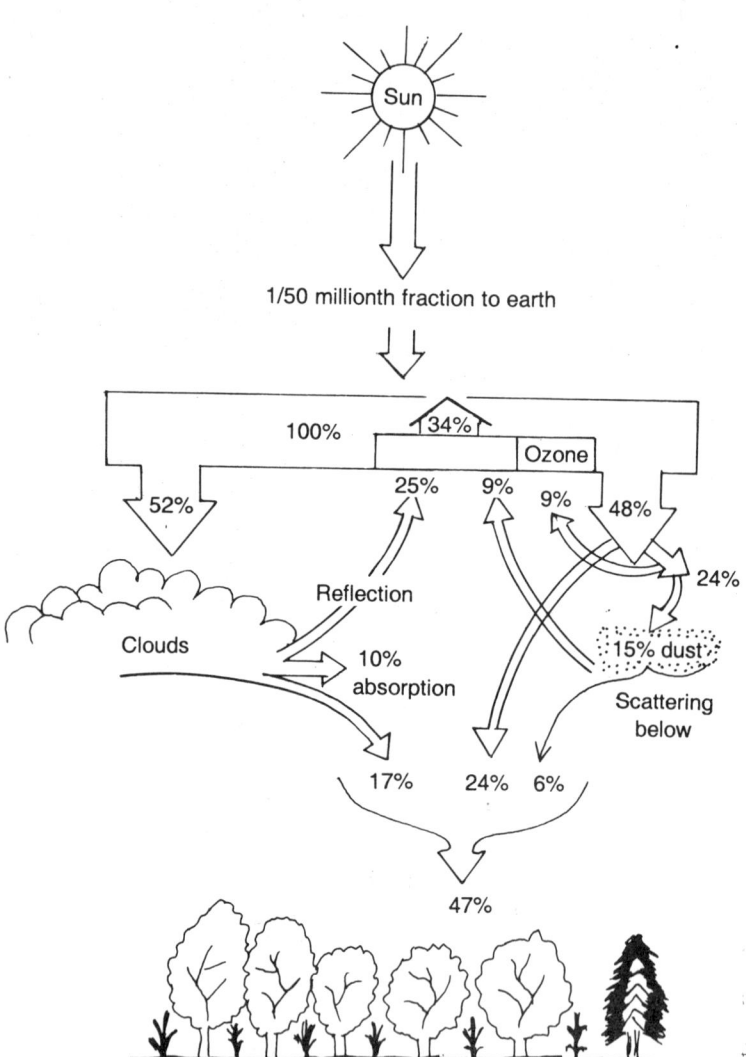

Fig. 12.1. Fate of solar radiation upto its incidence on earth's surface.

energy is lost by way of respiration of the primary producers and the net gain in weight is only the difference between the gross production and the quantity lost due to respiration. Animals derive their food from the net primary production. Only a very small quantity of the organic matter eaten by herbivores is actually used in their weight and the rest is lost as heat or is lost unassimilated through excretion. The production of organic matter by the animals or other heterotrophic organisms is called *secondary production*. An ecosystem remains balanced as long as the energy utilization by animals is matched by the energy trapped by the plants. Otherwise food becomes short, a famine sets in and the consumers of an ecosystem perish (unless food is made available from outside the ecosystem). Ecological energetics is mostly concerned with energy transformations occurring between the time it enters an ecosystem and the time it leaves.

About 20% of the energy primarily fixed by green plants is lost as heat of respiration and the rest goes to net production. Among herbivores only about 10% of the food eaten is used in gross secondary production and much of it is lost in respiration. Carnivores assimilate somewhat more of diet of animal tissues.

It is of great ecological significance to understand in any ecosystem (1) its efficiency of primary production of organic material, and hence, at successive stage, of secondary production of herbivores, carnivores of first order, second order, and so on to the top carnivore, (2) the total input of energy in the form of food and its efficiency of assimilation, (3) loss through excretion, (4) loss by way of respiration, (5) gross and net production, etc. At each stage environmental factors play a leading role and a thorough understanding of ecosystem structure and function especially from the energetics and production view point, leads to a better management of resources for human welfare. In fact, every citizen should be educated at least in non-technical terms about the constant cycling of materials and flow of energy in any type of ecosystem of which he is a part or of those he is connected with for his usual requirement. This will result in proper regulation of his action. In the cultural evolution of man there are numerous examples of his actions which have upset ecosystem balance and, in the course of time, have caused disadvantage to man. For example indiscriminate destruction of certain animals in greater quantity than their regeneration rate results in a population increase of those plant species which are selectively eaten by the animal. The ecosystem structure of cycling of materials and energy flow also gets disturbed. Similarly, indiscriminate felling of forest vegetation upsets the entire balance of water and other mineral cycles as well as plant and animal populations. Rajasthan, a few centuries back, was under a forest vegetation. But with little idea of ecological interrelationships and ecological balances, man removed plants at a faster rate than they could be regenerated. In the absence of adequate primary production, the native fauna also perished and now we find desert there.

A few ecologists have studied energy budgets for the western countries for the amount of energy received and the proportion of it used for different purposes. For a Wisconsin lake in the United States it has been found by Juday (1940) that out of the total solar energy entering the lake 49.5% is reflected or otherwise lost, 25% is absorbed in the evaporation of water, 21.7% is used in raising the temperature of the lake, 3% in melting ice in the spring and only 0.8% is used directly by the organisms. Thus a very small fraction is used in the lake for organic production. In many tropical conditions 2.5% of the energy is used by green plants. Sugarcane crops even utilize up to 10-12%. The energy of solar radiation has many further important roles in the maintenance of environments such as temperature which in turn governs the metabolic activities of almost all organisms. Based partly on actual data, and partly on assumption, a few ecologists have prepared models of energy flow in community dynamics. Lindeman (1942) has designated the stages of energy input, flow and loss out of the ecosystem by different symbols. The Greek letter little lambda (λ) represents the energy contribution of one system to another next (trophic levels and capital lambda (Λ) represents the actual energy content of a trophic level. R represents the energy lost through respiration and λ represents the sum of energy contributed to next trophic level ($\lambda n + 1$) and the energy lost through respiration (R).

Based on these notations a simplified line diagram has been prepared to illustrate flow of energy in one direction (Fig. 12.2).

Any quantitative estimation of energy must be recorded in a known area and known period of time. Following Lindeman's nomenclature Phillipson (1966) has given a generalized equation for expressing the rate of change of energy content of any trophic (Λn) level.

$$\frac{\delta \Lambda n}{\delta t} = \lambda n + n$$

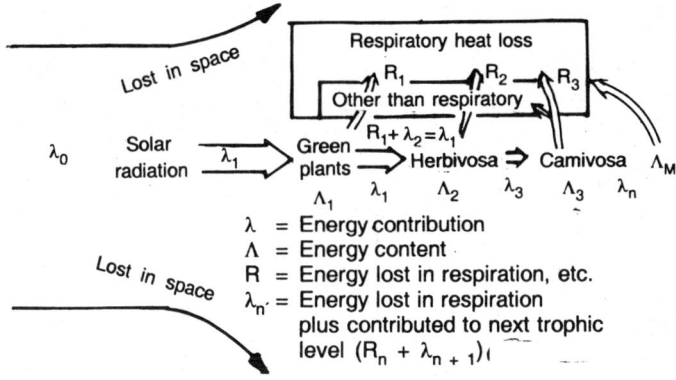

Fig. 12.2. Simplified diagram of flow of energy among organisms of different trophic levels together with respiratory losses. The notations are based on Lindeman's paper (1942).

This means the rate of change of energy content of a standing crop is equal to the rate at which that standing crop absorbs energy minus the rate of loss of energy from it.

Lindeman (1942) was first to give an energy diagram through different trophic levels in Cedar Bog Lake of Minnesota. Of the total 1,18,8720 kcal/m²/yr incident solar radiations the primary producers could capture only 1110 kcal/m²/yr i.e. less than 0.1%. This energy capturing efficiency is rather much less than many other ecosystems worked out in recent years. In corresponding area and time, energy passed on to herbivores was only 150 kilocalories and to carnivores only 30 kilo calories (Fig. 12.3). It is unfortunate that this pioneer paper on trophic dynamic aspects published in Ecology had to be the last one of the brilliant ecologist Raymond L. Lindeman as he died at the young age of 26 years.

Golley (1969) has given the clorific value of wet tropical forest vegetation from Panama and reports that the energy content per gram of organic matter is lower in tropical vegetation compared to temperate or alpine vegetation.

Teal (1957) has prepared an energy flow diagram for Root Spring in USA and has given the quantity of energy that flows from one trophic level to another. Similarly Odum (1957) has also given figures and diagrams for energy flow in Silver springs, Florida, USA. The

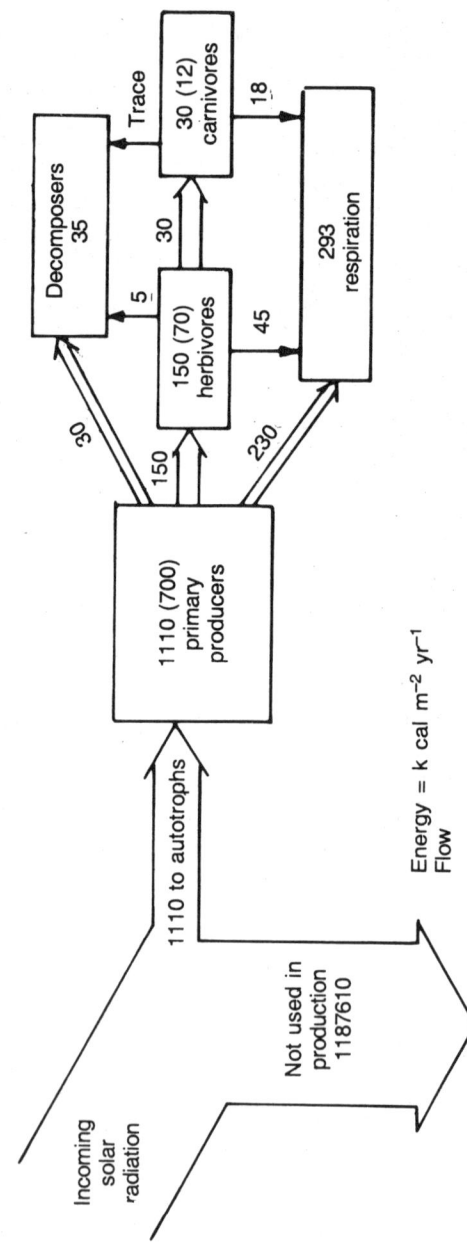

Fig. 12.3. Energy-flow diagram of Cedar Bog in terms of kcal m^{-1} yr^{-1} on the basis of data of Lindeman (1942).

incoming solar radiation is 1,700,000 Kcal/m^2/yr of which the effective solar radiation for the purpose of primary production is only 410,000 kilocalories per square metre per year and of this, 389,190 kilocalories of energy is lost as heat and only 20,810 kilocalories/m^2/yr is used in gross production through photosynthesis. Again out of this, 11,977 kilocalories of energy is lost by way of respiration, etc. and only 1,833 kcal, remains in net production. At each trophic stage energy may be passed on in three main directions; first to the next trophic level organisms as food; second through death to decomposers (many saprophytes etc.) and third as heat of respiration in metabolism. Ambasht and Srivastava (1988) and Srivastava and Ambasht (1989) have worked out the seasonal variation in the energy concentration and standing crop energy contents in the macrophytes found in River Ganga at Varanasi. They have found that the peak energy content per unit area is in transition between the winter and summer (March).

U.N. Singh (1972) has found that out of 750,000 kcal/m^2/yr of incident solar energy in *Heteropogon* grassland near Varanasi 23,909 kcal/m^2/yr or about 3% goes into net primary production. Thus there is a progressive decrease in total energy content (Λ n) at each subsequent trophic level and hence the pyramid of energy is always upright. In all studies of ecological energetics the factors of space and time are always considered. Ambasht and Singh (1975) have compared the biomass and energy structures of *Vetiveria* and *Heteropogon* grass stands on the basis of monthly variations in identical area and general climate but different pressures of grazing. Ambasht, Singh and Misra (1982) have compared a gradient of grassland, savanna and tree plantation forest stands under similar climatic and edaphic conditions on Vindhyan Hills for their energy conserving efficiencies and rate of primary production. Herbaceous vegetation is found to be more efficient solar energy converter than the tree components in this degraded forest ecosystem.

This has been illustrated in Fig. 12.4. In a range of communities growing nearby on the same edaphic and climatic environment in Varanasi forest division, the quantity of incident solar radiation, its fixation in net production and ecological energy conserving efficiency

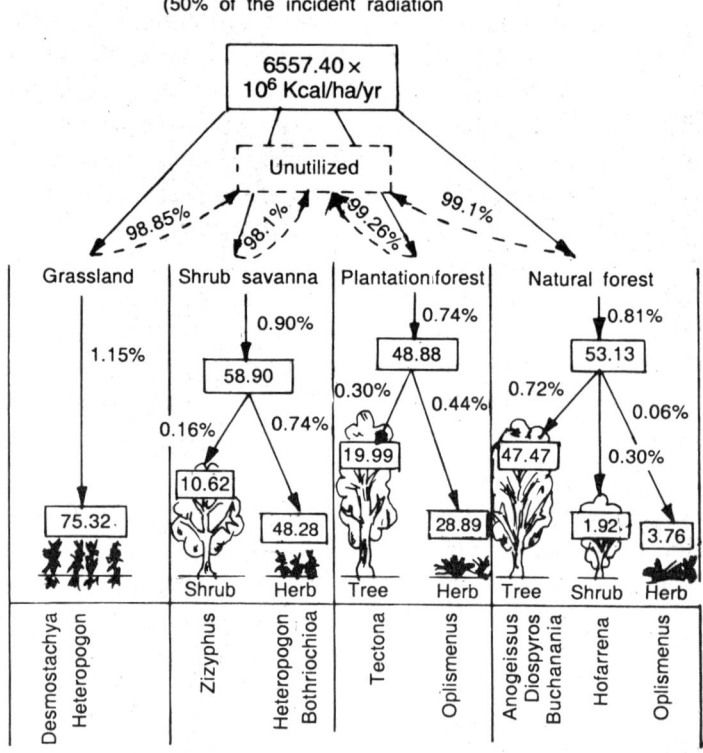

Fig. 12.4. Incident solar radiation and net energy fixation rates in grassland, shrub savanna, teak tree plantation and natural mixed forest stands in Chandraprabha sanctuary in Varanasi forest division (Data generated under the supervision of R.S. Ambasht).

(E.C.E.) are given. It is somewhat surprising that the low statured protected grass stands dominated by *Desmostachya-Heteropogon* grasses with 1.15% E.C.E. are more efficient producers than shrub savanna with 0.9%, tree plantation with 0.74% and natural mixed forest stand with 0.81% E.C.E. Even in some multistoreyed stands, the herbaceous communities account for a greater share than the shrub or tree stands. The fifty per cent of the incident solar radiation on this site in 1976-77 period was 6557 × 10⁶ kcal/ha/yr and out of this the grassland showed net energy fixation of 75.32×10^6 kcal/ha/ yr. In savanna the shrubs of *Zizyphus jujuba* account for 10.62×10^6 kcal and *Heteropogon-Bothriochloa* grass layer for 48.28×10^6 kcal/ ha/yr. In teak tree plantation stands of 16 yr age stand only 19.99×10^6 kcal/ha/yr goes into the net storage and the ground layer plants dominated by *Oplismenus burmannii* account for as much as 28.99×10^6 kcal/ha/yr. But in near natural mixed forest stand protected against cattle grazing 47.4×10^6 kcal of energy goes into tree layer dominated by *Anogeissus-Diospyros-Buchanania* species followed by only 3.76×10^6 kcal in grass layer of *Oplismenus* and 1.92×10^6 kcal/ha/ yr in the shrub layer dominated by *Holarrhena antidysentrica*.

The energy crisis of the present day is largely due to the fact that all life maintained on this earth is dependent on finite amount of solar radiation that the green plants fix through photosynthesis. Man is diverting most of this fixed solar energy for the use of just one species, i.e., *Homo sapiens* which is responsible for upsetting natural balance and ecosystem functions. Man is also destroying the forests, i.e. reducing the net energy fixation capacity of the earth.

In a fully balanced climax ecosystem the total energy fixed by autotrophs over a period of time is consumed in the respiration by the heterotrophs, herbivores, carnivores and saprotrophs. The energy fixation rate and energy utilization rate becomes equal, as a result of which there is no net gain and no new net growth of the system, i.e., the system maintains itself in its existing state perpetually. The ecosystem is stabilized or attains climax state. But in young or successional stages the ecosystem shows a greater quantity of gross energy fixation by the autotrophs than the combined respiratory needs of energy by all the heterotrophs of the ecosystem. This leaves enough of produce in form of yield, upon which the system grows in biomass and energy storage capacity. The process continues until climax stage is achieved.

Oceans are poorer energy fixing systems than the land. Their area accounts for 361×10^6 km^2 or about two-third of the total earth's surface of 510×10^6 km^2, but their net biomass production is estimated to be 5.5×10^{10} tons per year at an average of 155 g/m^2/yr. Terrestrial vegetation on the other hand occupying the land surface of

149×10^6 km^2 area accounts for 10.9×10^{10} tons/yr at an average rate of 730 g/m^2/yr (Whittaker, 1970). The present level of harvesting the energy from the ocean in form of fish output is quite high when we realise that it is a poor productive system. Further oceans are being dumped with pollutants of diverse kinds. Therefore, man has to look upon the terrestrial systems as the principal source of future food, fuel and other energy source for the increasing human needs.

Production Ecology

Great importance is being laid on researches concerning the rate of energy storage by green plants in diverse ecosystems. It is a well known fact that the ultimate source of all energy needed by the organisms is the sun and that only the photosynthetic organisms are capable of trapping solar energy. The demand of ever increasing human populations on food resources is tremendous. In the process of meeting this demand most of the ecosystems are at present influenced or maintained by man. Production ecology deals with production processes and productivity by green plants, herbivores and carnivores in different ecosystems. This type of study is of fundamental importance in the management of resources. The aim of the International Biological programme "*Organic Productivity* and *Human Welfare*" was towards an extensive investigation of the productivity of terrestrial, freshwater and marine ecosystems and their conservation on global basis.

More and more ecologists are now concentrating on production ecology and data are being discussed and published through a number of seminars and symposia organised by various ecological societies, national and international committees and United Nation's agencies.

The rate at which energy is stored by green plants is called *primary production*, and by the heterotrophs, *secondary production*. The total energy trapped in a given area by plants is called *gross* primary productivity. Some of the energy is lost by the respiratory process of the primary producer system. In *net* primary productivity the organic matter used up in respiration is excluded from the gross organic matter produced during photosynthesis. Several research papers on this subject from studies on forests, grasslands and other herbaceous systems and freshwaters in tropical regions were discussed at an International Symposium on *Tropical Ecology Emphasizing Organic Production'*, at New Delhi (January, 1971) and published by Golley and Golley (1972).

TURNOVER

Turnover concept has been used in somewhat different ways by different persons. But essentially it is the rate of rebuilding of organic matter with respect to the total biomass. Sometimes turnover rate is also used for certain elements like phosphorus returning from plankton to water (Whittaker, 1970) in fractions per hour.

   Dahlman and Kucera (1965), from periodic estimations of root biomass over one year period, have calculated *turnover per year* in grasses. Lewis (1970) using this method has given the formula for calculation of turnover as :

$$T = \frac{B \max - B \min}{\text{Total B}}$$

where T = turnover, B is biomass at minimum and maximum levels during the course of the year. Using the same method Ambasht, Maurya and Singh (1972) have estimated and compared turnover per year (in terms of fraction of total biomass) in alluvial and hill grasslands in protected condition. In *Dichanthium* grasslands a fraction of 0.66 is reproduced in a period of one year and in *Heteropogon* grassland 0.88 per year. But if the same values, instead of being considered as fraction reproduced in one year, are considered in terms of *time required* to produce a quantity equal to the maximum biomass then the turnover becomes 1.51 years in *Dichanthium* and 1.13 years in *Heteropogon* respectively.

PRIMARY PRODUCTION PROCESSES

Primary production is brought about by the process of photosynthesis and chemosynthesis. These processes are subjects of extensive laboratory studies, chiefly by plant physiologists and biochemists, and a vast quantity of literature has been fast developing on the subject.

The net result of photosynthetic reactions is the formation of carbohydrates and oxygen from CO_2 and H_2O in the presence of light and chlorophyll. For the fixation of each gram atomic weight of carbon from CO_2 to organic compound during photosynthesis, 477 kJ or 144 kCal of solar energy is trapped. But a number of steps, each governed by ecological factors, take place in between. Briefly the process can be summarised as follows :

The light energy is absorbed by chlorophyll *a* and other accessory pigments like chlorophyll *b* and carotenoids. Chlorophyll concentration is an important factor that determines the rate of primary production. Chlorophyll deficiency or *chlorosis* is as a result of ageing, mineral deficiency, pollution effects, drought and mutation. The chlorophyll molecule on receiving *photons* or light energy gets *excited*, i.e., assumes a higher energy level in which an electron is lifted into a new orbit. This energy of excitation is used quickly in two types of photochemical processes, firstly in splitting two molecules of water into one molecule of oxygen and four electrons and four protons which reduce a hydrogen acceptor $NADP^+$ or $NADPH.H^+$ and secondly in the formation of ATP from ADP and inorganic phosphates. Thus these reactions viz. excitation of the chlorophyll molecule, splitting of the water molecule, release of oxygen, reduction of a hydrogen acceptor and formation of ATP (photophosphorylation) are dependent upon light factor. Two pigment systems, viz., photosystem I or pigment 700 nm and photosystem II or pigment 680 nm are involved in light driven reaction in which besides chlorophyll *a* and *b*, other accessory pigments are also involved. In sun leaves there is greater concentration of chloroplasts in each mesophyll cell and the ratio of chlorophyll *a* is much higher than *b* in shade leaves. The ratio of photochemical activity say in terms of mol of oxygen liberated upon the amount of light quanta absorbed in terms of photons or *einsteins* as called as quantum or $\varnothing$.

Another set of reactions are strongly dependent upon temperature and can continue in the dark. Carbon dioxide entry into the green tissues such as leaves is dependent upon stomatal pore and resistance of wall layers. Carbon dioxide forms H_2CO_3 in combination with water and then it ionizes into the carbonate ion which becomes readily available for photosynthetic reactions. Carbon dioxide reacts with a five carbon compound, ribulose 1,5-bisphosphate (RuBP) and makes an unstable six carbon compound, which readily breaks into two molecules of the three carbon compound 3-phosphoglyceric acid (3-PGA). The products of photoreactions, namely ATP and $NADPH_2$ also enter the chain of reactions through triose phosphate and then fructose are formed and in the process ribulose diphosphate is reconstituted for another cycle of reaction. Detailed descriptions are available in most of the modern books on plant physiology and handbooks on photosynthesis. The photosynthetic products are quickly removed from their place of formation to other regions through phloem and this keeps the pace of forward reactions at a high level. Oxygen is liberated into the atmosphere.

Production process in C-4 plants

It has been known that the CO_2 is taken up in green plants by a 5 carbon compound called RuBP (Ribulose 1,5-bisphosphate) and then broken in two molecules of 3-carbon compound 3-PGA (3-Phosphoglyceric acid). This pathway of fixation is known as C-3 cycle or after the discoverers' names, Calvin-Benson cycle. Besides this, in recent years, it has been discovered that in the mesophyll cells of certain grasses, corn, sugarcane and certain weeds, the CO_2 first gets bound to PEP (Phosphoenol pyruvate) with help of the enzyme PEP carboxylase and a four carbon compound OAA (oxaloacetic acid) is formed. This pathway of carbon fixation involving a four carbon atoms containing compound OAA is called C-4 or Hatch and Slack of Hatch-Slack-Kortschak pathways (after the discoverers' names) and plants showing such a primary production process are called C-4 plants. In many of these plants OAA gets reduced into malic acid. It must be noted that in C-4 plants capture of CO_2 by PEP and formation of OAA and then aspartates and malates take places in the mesophyll cells. After these stages in these plants the remaining stages take place in the bundle sheath cell where the usual C-3 cycle reactions take over. Aspartates and malates formed in mesophyll are transported to the well developed bundle sheath cells full of chloroplasts, where they are broken into CO_2 and pyruvate. The CO_2 released is captured by RuBP and through Calvin cycle, carbohydrate is formed. The main adaptational advantage of C-4 mechanism is the fact that the CO_2 is drawn from air with far greater efficiency even when the CO_2 concentration falls to extremely low levels. PEP has the ability to draw CO_2 from air even if present in traces. After releasing the CO_2 in bundle sheath, the pyruvate returns to mesophyll tissue where it forms new PEP molecules for recapture of fresh CO_2.

Production process in CAM plants

A somewhat similar mechanism of capturing CO_2, formation of malic acid, its decarboxylation, release of CO_2 for subsequent efficient utilization by RuBP through normal C-3 cycle is found in Crassulaceae family succulent plants. This process is called Crassulacean acid metabolism or CAM. Besides Crassulaceae, this mechanism is met in many other families as Euphorbiaceae, Lilaceae, Agavaceae, Cactaceae, Portulacaceae, Asclepiadaceae and in the leathery leaves of the rare gymnosperm *Welwitschia*. In these plants, the stomata remain open even in night when large amount of CO_2 is captured by PEP, present in the cytoplasm. From PEP, oxaloacetic acid (OAA) is formed which in turn gets converted into malic acid and gets stored into vacuoles. During day time the acid gets transferred to chloroplasts where through decarboxylation (as in C-4 plants) it releases CO_2 to be taken up by RuBP. Thus, there is a decrease in pH during night due to accumulation of acids and during day pH rises as the acids are decarboxylated.

Ecological adaptations in C-4 plants for an efficient primary production

The main features of C-4 and CAM plants are an efficient intake of CO_2, its conversion into four carbon compounds, transfer to photosynthesis site, decarboxylation and making available the released CO_2 to RuBP for a rapid photosynthesis. There are a large variety of ecological implications involved towards a better adaptation for rapid primary production in C-4 plants.

From structural view point, as against the normal mesophyll cells around the parenchymatous bundle sheaths of C-3 plants, in C-4 plants the highly chlorenchymatous well developed bundle sheath is surrounded by densely radiating mesophyll cells called as '*Kranz*' type of anatomy. '*Kranz*' term was given by Volkens (1887) in his book

'Flora of Egypt'. In CAM leaves, there are very large vacuoles present which serve as storage tanks for malic acid during night time. In C-4 plants the ratio of chlorophyll a : b is somewhat higher (4 : 1) than in C-3 plants (3 : 1). It is understood that Kranz anatomy does not permit CO_2 molecules produced in photorespiration in bundle sheaths of C-4 plants to escape out of leaves but acts as a trap for CO_2 to be refixed by PEP present in the mesophyll cells. Further in C-4 plants photorespiration is limited to bundle sheat cells, whereas in C-3 plants all chloroplast containing cells show photorespiraton. *Photorespiration* is a metabolic process taking place in green tissues only during day (i.e. in presence of light) in which O_2 is taken up and CO_2 is released and in high intensity of light this process becomes very fast. This process neutralises to a good deal the primary production process, thus reducing the overall rate of net photosynthesis. RuBP, the key compound in fixation of CO_2 also acts as the key compound of photorespiration by taking up O_2 and releasing CO_2 through a complex enzymatic mechanism of RuBP carboxylase/oxygenase system. Since C-4 plants have the restricted photorespiration (in bundle sheaths only) the CO_2 so liberated is trapped in mesophyll, the adverse effect of increasing light intensity via photorespiratory losses is not felt by C-4 plants. Hence, whereas the C-3 plants show a light saturation upto 40,000 to 50,000 lux beyond which photosynthesis rate does not increase, in C-4 plants photosynthesis rate rises upto the full sunlight in tropics upto 90,000 to 1,00,000 lux. Ecologically, C-4 plants are more efficient to exploit high light intensity, a higher range of temperature regimes up to 45 to 50°C (as against upto about 30°C in C-3 plants) and much lower carbon dioxide compensation concentration levels.

A high concentration of oxygen in the air has a depressing effect on the rate of photosynthesis. O_2 concentration when falls down from the normal 21% content in the air this depressing effect is reduced and plants photosynthesise more efficiently, but C-4 plants do not show depressing effect of high oxygen content. They, thus have a higher net production rate under normal oxygen concentration.

C-4 plants are found more in xeric tropical climates particularly in members of Gramineae family. So far about one thousand angiospermic plants belonging to 18 families (and in some blue green algae) have been recorded to follow C-4 cycle (Larcher, 2003). There are many genera with some C-3 and some C-4 species and even in a single species of grass *Alloteropsis semialata* there are a C-3 and a C-4 ecotypes.

Ecological efficiencies related to productivity

Ratios or percentages of input versus output in biomass from one trophic level to the next, or in terms of solar energy input versus the energy captured by primary producers or between the energy intake by herbivores in form of fodder to energy stored in their biomass, etc. are described as ecological efficiencies by different workers in some different ways. Some of the more common ones are as follows:

(i) Energy conserving efficiency (%)

This is calculated by the following formula

$$ECE\ (\%) = \frac{\text{Solar energy captured by (kcal ha}^{-1}\ \text{yr}^{-1}\text{) vegetation}}{\text{Usable solar radiation falling on the area (kcal ha}^{-1}\text{yr}^{-1}\text{)}}$$

The usable solar radiation is the usable range of wavelengths i.e. 'light' waves which constituted about 45 to 50% of the total radiation. So most workers use 1/2 solar radiation values. In most of the grasslands, savanna and forests it varies between 0.75 to 1.75%. In mesic or wet grasslands and young tree plantation stands like *Alnus nepalensis* (7 years old trees) it may be 3 to 4% and in highly productive sugar cane crops or in coral reefs it may be more than 5%.

(ii) *Trophic level energy efficiency (Tr E.E.)* is the ratio of total gross production (P_G) upon the total light (L) or total absorbed light (LA):

$$\text{Tr E.E.} = \frac{PG}{LA}$$

(iii) Exploitation efficiency % as given by Barbour, Burk and Pitts (1980) as modified from Ricklef (1973) is :

$$\text{E.E.} = \frac{\text{Gross primary production}}{\text{Solar radiation}} \times 100$$

where the production and radiation have to be expressed in common unit and for same land area and time period.

(iv) Net production efficiency (%) as also given by the above authors is :

$$NPE = \frac{\text{Net primary productivity}}{\text{Gross primary productivity}} \times 100$$

Efficiency between successive trophic levels is usually about 10%. This means that if a herbivore eats 100 g of grass, about 10 g of it is converted into secondary production or out of 100 g of flesh eaten by a carnivore, 10 g goes in the production of the carnivore.

The rate of primary production can be measured by a number of direct and indirect methods.

MEASUREMENT OF PRIMARY PRODUCTIVITY

In most ecosystems the standing crop biomass of green plants is over 98% of the total living biomass. Therefore, determination of primary productivity gives approximately a total picture of the ecosystem productivity. Primary productivity is actually the *rate of energy conversion* from the radiant form to the chemical bonds of organic substances. This results in *increase in weight.* The process of energy trapping is called photosynthesis and it involves *exchange of gases*—intake of carbon dioxide and release of oxygen. The machinery to do this job is the *chlorophyll* molecule. Therefore, productivity can be measured by measuring any of the above items. Estimations of caloric value energy content of plant parts at intervals of time are made to record the net increase in energy. The instrument used to find out the energy content of organic substance is called a *bomb calorimeter.* Periodic estimations of biomass of plant materials is known or can be done by harvesting the plants (all plant parts) and estimating their oven dried weights. This method is usually called *harvest technique.* The method involving measurement of rate of photosynthesis and respiration by collection and analysis of carbon dioxide and oxygen gases entering and leaving sample of an ecosystem is called the *gas exchange* method. This method when applied in aquatic communities with the use of bottles is called the *light and dark bottle* technique. Chlorophyll estimations also give an idea about the rate of production since this is proportional to the chlorophyll content.

1. Bomb calorimetry (Calorific value)

Several varieties of bomb calorimeters are now in use for determining the caloric (or calorific) values of biological materials. Lieth (1968 and 1975) has described methods of energy estimations with the aid of an oxygen bomb calorimeter and its importance.

A bomb calorimeter (Fig. 13.1) consists of a thick walled stainless steel cylindrical pot known as the 'bomb'. It has an air tight cap fitted with the help of screws and has four main connections. On the two sides of the cap are outlet and inlet valve devices for filling oxygen gas and its release. Through the other two points electricity passes.

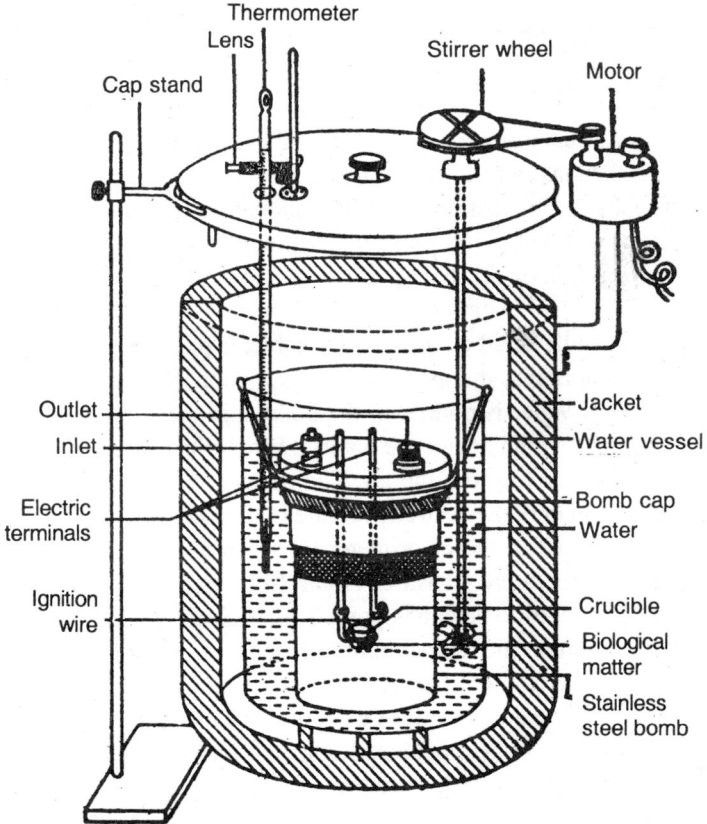

Fig. 13.1. Bomb calorimeter and its various components in a vertical section view.

Inside the bomb a small crucible is placed in which dried plant material is kept. An ignition wire passes through the material and its ends are connected to the two electric ignition terminals. The bomb is kept in a thin-walled pot called kettle which is removable. It is filled with clear water sufficient to cover the bomb. This again is put in a bigger double walled plastic container. The container cap has holes through which a Beckmann thermometer passes below in the kettle water. An electric operated stirrer is also provided to stir the water.

In the bomb calorimeter a known weight of dry biological material is burnt within the stainless *bomb* in an atmosphere of pure oxygen at about 30 atm pressure. The bomb is kept immersed in a bath of water all the time. The bath is kept in a container adequately insulated against loss of heat from the water.

The plant material is first dried in an oven usually at 80°C for 48 hours or more. The dried material is powdered in an electric or a mechanical grinder or even in a mortar. A small quantity of (1 g) the powder is then taken and pressed in the form of small pellet or tablet in a special pressing instrument. While preparing the tablet the ignition wire is inserted in the powder. The pellet is carefully weighed, preferably in chainomatic balance, and should be about half to 1 gram in case of the plant materials. There are special types of bomb calorimeters to handle small quantities. The tablet is placed in an ignition crucible in the bomb, the ignition wires are connected to terminals. About 5 ml of water is poured in the bomb and the bomb is closed tight. Through an inlet device the bomb is filled with oxygen gas from a gas cylinder, driving out the air. The outlet is closed and further oxygen is fed in until the desired pressure is attained. Known quantity of water is placed in the calorimeter in which an electric stirrer is fitted. The temperature of water is carefully recorded. Through a press button electric current is passed through the tablet and is it completely ignited. The temperature of bomb and, therefore of water surrounding it, rises. This is recorded with the aid of a Beckman thermometer. The calorific value (V) of the biological material is then calculated. In this connection, besides the dry weight (G) of material and the temperature rise (Δt) two more things should be known. They are the water equivalent value of the system (W) and the temperature rise value due to ignition by the wire (Σc). With the aid of these the calorific value can be calculated in terms of calories per gram with the aid of the following formula (Lieth, 1968) :

$$V = \frac{W \cdot \Delta t - \Sigma c}{G}$$

In determining ecological efficiencies the rate of input of radiant energy is measured first for area and time. The calorific value of the plant community is then calculated at regular intervals from identical areas and the ratio between energy receipt and energy converted into the biological materials gives the *ecological efficiency* value. Calorific values of different types of materials differ and in most cases of plant tissues it varies between 3000 to 5000 calories per gram of dry matter. In fact the recent trend is that instead of giving productivity figures in terms of weight it is given in terms of energy. The weight measurements are not exactly comparable due to differences in energy content per gram of different plant materials. In animal tissues, especially, energy content is much higher. Lieth (1975) has given approximate energy value of different chemical compounds. Starch and cellulose have about 6300 cal/g and fat about 9300 cal/g. In plant and animal materials after complete burning some ash is left. The energy values are sometimes expressed on ash free dry weight basis also by subtracting the weight of ash from the weight of plant or animal material taken for ignition.

2. Harvest method (biomass estimation at periodic intervals)

In this method the primary productivity is measured by harvesting the crop of plants of known ages from known areas at periodic intervals. This gives the value of net increase in weight. The chief difficulty concerns protection of the primary produce from grazing animals or insects eating the leaves. It is easier to apply this technique on cultivated crop plants where date of sowing is taken as the starting point. At periodic intervals plants from known area are harvested, dried in an oven and weighed.

The biomass change (ΔB) between time t_1 and t_2 is obtained by deducting biomass value at t_1 from the biomass value at t_2 i.e. :

$$\Delta B = B_2 - B_1$$

If the losses due to shedding of leaves, twigs, etc. (litter or L) and due to consumption by herbivores, parasites, etc. (grazed or G) between time t_1 and t_2 are added to ΔB, then net production (Pn) can be calculated as :

$$Pn = \Delta B + L + G$$

The net primary productivity in such cases is expressed in terms of grams/square metre/day or sometimes kg/hectare per year. In the case of forest trees, the productivity values are estimated usually on the basis of fruit, leaf, twig, trunk and root biomass. On account of the difficulty and labour involved in unearthing the roots their biomass is very often not estimated directly, but approximate corrections are made on that account. Instead of harvesting big trees, there are indirect methods also to predict the biomass by measuring other correlative characters. One common method is to measure the diameter of trees at breast height (DBH) and determine its correlation with biomass. Another method is to measure the increase in weight of leaf punches of known area and number in *light* and the decrease in weight of an equal number of punched leaf in the *dark*. This gives the rate of net and gross production.

3. Gas exchange method

This method is based on the concept that rate of primary production is directly related to the quantity of oxygen evolved during photosynthesis. In natural conditions, however, oxygen is being used by animals, fungi and bacteria. With certain modifications, the method is being used in crop and grass communities. Small areas of sample plot are covered by a glass chamber in which air enters through an inlet by the aid of an aspirator from one end and leaves the chamber by an outlet. Three such sets are prepared two enclosing plants of almost identical age and spread, and the third enclosing same area without plants. One of the two plant covering jars is opaque to make the interior dark. The air leaving out of the systems is passed through a solution of barium hydroxide. The rate of flow of air (that is suction applied) is uniformly the same. Carbon dioxide is absorbed by the barium hydroxide and the content of carbon dioxide so absorbed is found out by titrating it against 0.1N Hydrochloric acid. The one without plants gives total carbon dioxide passing through the area in unit time, the dark chamber gives carbon dioxide passing through air plus that released in respiration and the third gives CO_2 passing in air plus released in respiration and minus carbon dioxide used in primary production process. From these values the actual rate of carbon dioxide utilized in production process per square metre can be calculated. Dwivedi (1969) has described this method after several field trials. Based on a similar principle, Odum (1971) has described the light and dark bottle method for measuring primary production by phytoplankton in water. Pond or lake water containing phytoplankton are taken in two glass bottles. One bottle is darkened to preclude photosynthesis. Both the light and dark bottles are suspended in water. In the light bottle oxygen is liberated due to photosynthesis and some of it is used in respiration whereas in the dark bottle only oxygen is used in respiration. Thus if the oxygen value of the light bottle is added to the quantity of oxygen used in respiration the total oxygen evolved can be calculated. This method is commonly used in primary production studies in freshwaters.

4. Radioisotope method

Use of radioisotopes especially C^{14} in the form of sodium bicarbonate $(NaHC^{14}O_3)$, has been made to measure the rate of the primary production in aquatic autotrophs. The C^{14} gets assimilated alongwith stable carbon to form carbohydrates. After a few hours of introduction of radiocarbon in light and dark bottles, the samples are taken out. With the aid of a Geiger counter the quantity of radiocarbon synthesized is recorded and this is proportional to the synthesis of stable carbon.

5. Leaf Area Index (LAI)

Since the leaf is the principal organ in which primary production takes place, its area in relation to ground area is of great significance. Leaf Area Index is the total area of leaf (of one surface only) per unit area of ground.

$$LAI = \frac{TLA}{TGA}$$

TLA is total leaf area and TGA is total ground area which is under the canopy of sampled leaves.

The rate of production increases with increase in leaf area index up to a certain extent provided other factors, especially moisture, are not limiting. Generally beyond 4 or 5 LAI value, the increase in the production value stops or falls depending upon the type of species and prevailing conditions. In vertical leaves a higher leaf area index is needed to intercept the same amount of light and thus a higher photosynthetic rate is achieved.

The area of leaf can be measured by (1) *planimeter* or by (2) use of graph paper or by (3) the weight method. At first the leaf is placed on a paper and its margin is mapped on it. On arm of the plantimeter is then retraced on the leaf map from the beginning to the end and the area is read on the instrument. If the map is drawn on graph paper the area is directly calculated by counting the squares covered by the leaf on the graph paper. Alternatively a paper is superimposed oin the leaf and it is cut out along the leaf margin. This paper of shape and size equal to the leaf, is weighed. Another piece of the same paper of known area say 25 × 25 cm is cut out and weighed. Thus from the weight values the area of leaf can be easily calculated.

It is also useful actually to measure the area of a few leaves and also to determine the product values of their length and breadth.

These two values are not identical because the shape of a leaf is not rectangular. Another value K (constant) has to be determined in order to find out the actual area directly from the length (L) × breadth (B); and it can be found out when actual area is already measured in a standardization process.

$$\text{Actual area} = L \times B \times K$$

This K is usually 0.9 for narrow leaves as in most monocots and 0.6 for broad leaves. Thus once K is standardized, leaf area can be readily measured simply by measuring the length and breadth.

6. Chlorophyll estimation method

Chlorophyll content has certainly a relationship with primary productivity. In an ecosystem like a desert or an oligotrophic lake the quantity of chlorophyll is low per square metre or per hectare and hence the primary productivity will also be low. Similarly in multistoreyed forests in the tropics the quantity of green tissues or the quantity of chlorophyll is very high and the rate of production is also high. Many ecologists have found that the assimilation ratio between the quantity of chlorophyll and the quantity of biological materials produced is variable. Photosynthesis depends upon a number of factors. But even the chlorophyll content gives a good idea about the productivity status of an ecosystem. The biological green materials are sampled periodically and chlorophyll is extracted in soluble organic solvents and the chlorophyll concentration is then determined by spectrophotometer or spectrocolourimeter.

PRODUCTIVITY IN DIFFERENT ECOSYSTEMS

Rates of production have been estimated by different methods in freshwaters, grasslands and forests, in different parts of the world. There is a fairly wide range in the percentage of primary produce that is lost through respiration. Odum (1957) and Riley (1957) record 57% and 53% losses by way of respiration for Silver Springs and Sargasso sea ecosystem respectively. Under the auspices of the International Biological Programme symposia have been held on primary productivity in freshwaters and terrestrial communities. Proceedings of the I.B.P. symposium on Primary Productivity in Freshwater (Goldman, 1969) and Copenhagen Symposium on 'Functioning of Terrestrial Ecosystems at primary level' (Eckardt, 1968) gave a fairly comprehensive range of papers on the subject.

Lieth (1964) has given an account of primary production in the biosphere and has prepared a world map of primary production both on land and sea (Fig. 13.2). The highest production on land is found throughout the tropical rain forest and in the humid tropics. But in marine regions production is highest in temperate belt between 45° to 60° latitudes both in the Northern and Southern Hemispheres.

Whittaker (1970) has compiled a table showing net primary production of major ecosystem. He has given the total area of earth as 510×10^6 km^2 of which land accounts for 149×10^6 km^2 and total oceans 361×10^6 km^2. The net primary production on land is much higher, the mean value being 730 g/m^2/year about 10.9×10^{10} tons/year on total land as compared to mean of 155 g/m^2/year in oceans or 5.5×10^{10} tons per year in the entire ocean. Thus the total annual primary production of earth is estimated to be 16.4×10^{10} tons (dry). Of course the standing biomass at any one time is much higher being about 185.5×10^{10} tons for the entire earth. With the new discoveries of occurrence of subsurface ultramicroscopic algae the oceanic productivity estimates may have to be revised.

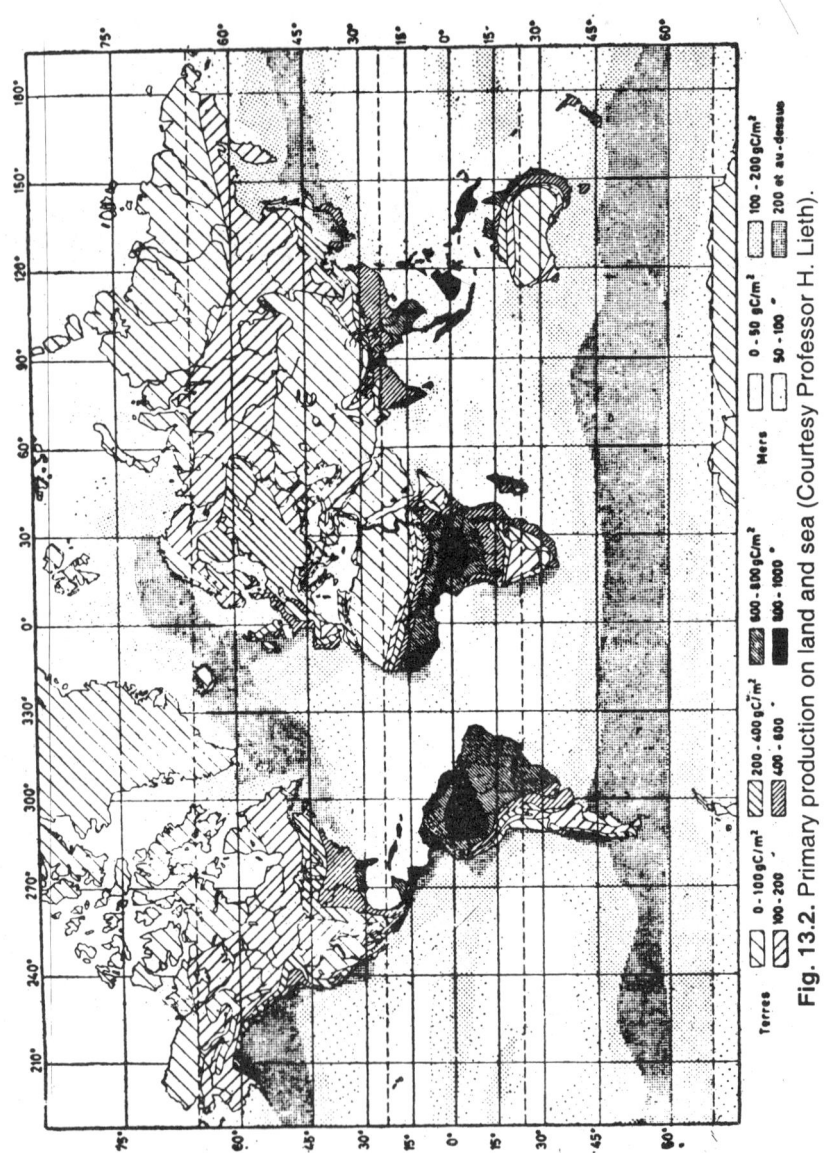

Fig. 13.2. Primary production on land and sea (Courtesy Professor H. Lieth).

The tropical forests occupy about 13% of the land surface but because the production rate here is much higher ranging between 1 kg to 5 kg per m^2 per year with a long life span of trees, their total world biomass is 90×10^{10} dry tons. The average world production rate in tropical forests is about 4×10^{10} tons per year.

Temperate forests occupy about 12% land. Their average production is 0.6 kg to 2.5 kg/m^2/year and for the entire area about 2.34 $\times 10^{10}$ tons/year and the world biomass for temperate forests is 54×10^{10} tons. In the Central Himalayan Pine forests, Chaturvedi and Singh (1987a, b) have reported production rate of 10 to 20 t/ha/year with a standing crop biomass of 115-286 t/ha. They have worked out the nutrient dynamics also in these forests and reported turnover time of 1.5 to 2.5 years.

Sharma and Ambasht (1990) have measured biomass, net primary production and energy fixation in age sequence of *Alnus nepalensis* tree plantation in Kalimpong Himalaya. Biomass ranged from 106 t ha^{-1} in 56 year old stand but the rate of primary production was between 25 t ha^{-1} year^{-1} in 7 year old and 13 t ha^{-1} year^{-1} in 56 year old stands.

Swamps and marsh lands have the higher rate of production of about 1 kg to 4 kg/m^2/yr but because they are restricted in area their overall world biomass is as low as 2.4×10^{10} tons. Their productivity value is 0.4×10^{10} tons/yr (Whittaker, 1970).

Ambasht (1971) has compared the standing crop biomass of tropical pond macrophytes with that of temperate ponds. Ambasht and K. Ram (1976) have made a stratified productive structure study of a large Indian lake and found that on the basis of green biomass distribution there were three distinct profile structures; some showing heavy concentration of biomass in the basal region, some in the middle of the water column and some on the top layer.

Golley and Lieth (1972) have also given a tabulated account of net primary productivity estimates. The tropical grasslands which are spread over 15×10^6 km^2 and account for 0.45×10^{10} tons per year.

Cultivated lands occupy 14.0×10^6 km^2 and their approximate total production is about 1×10^{10} tons.

There is a wide difference between the production rate in temperate and tropical grasslands. Whittaker's (1970) compilation from diverse sources indicates an average productivity of 0.5 kg/m^2/yr in temperate regions. In the tropics estimations are very few. At Chakia and Varanasi grasslands in the Indian tropics, Ambasht, Maurya and Singh (1971) found that the production rate was three to five times greater than the temperate average rate. Agricultural lands also have a high production rate reaching in some cases upto 4 kg/m^2/yr but because of several factors including the fact that fields often remain fallow for a few months, the overall production rate is about 0.65 kg/m^2/yr and the world production is slightly above 0.9 × 10^{10} dry tons/year. Ambasht and Pandey (1975a, b and 1976) and Pandey and Ambasht (1979) have given a comprehensive account of *Aristida cyanantha* dominated grasslands for phytosociology and productivity and Ambasht (1974) and Ambasht and Singh (1980) have given a compilation of uptodate data on Varanasi grasslands.

Lieth (1979) has again compiled the biomass and energy fixation values of the different kinds of biomes on the 149 × 10^6 km^2 land area of the earth. Tropical rain forests top with 139.4 × 10^{18} cal/yr energy fixation followed by tropical grasslands with 42.0 × 10^{18} cal/yr. Cultivated lands account for 37.1 × 10^{18} cal/yr. The total energy fixation of all continents is approximately 426.1 × 10^{18} cal/yr.

Odum (1983) has computed the gross primary productivity in terms of kcal/m^2/yr and the total GPP for different major biomes in terms of 10^{16} kcal/yr by doubling the available NPP values in low production and trebling the NPP values of high production regions given by Lieth and Whittaker (because of higher respiration rate in high production belt) for terrestrial biomes. For marine biomes the GPP values were obtained by multiplying Ryther's (1969) net carbon production figures by ten to get in terms of kcal and then doubling it to get approximate GPP. The values thus obtained show that in the total marine ecosystem spreading over 362.4 × 10^6 km^2 the GPP is 43.6 × 10^{16} kcal/yr. Open oceans account for the largest share of marine area of 326 × 10^6 km^2 and show an average GPP of 1000 kcal/m^2/yr or for the entire area (or volume) 32.6 × 10^6 kcal/yr. But estuaries and reefs, with only 2.0 × 10^6 km^2 area, have a high average GPP of 20,000 kcal/m^2/yr. On land this GPP rate is matched by wet tropical and subtropical broad leaved forests spread over 14.7 ×

10^6 km^2 accounting for 29.0×10^6 kcal/yr. Compared to these highly natural systems with 20,000 kcal/yr GPP, the fuel subsidized agriculture spread over 4.0×10^6 km^2 area of earth fixes in GPP 4.8×10^{10} kcal/yr at an average rate of only 12,000 kcal/m^2/yr and very little or no energy subsidized agriculture practiced on about 10×10^6 km^2 produces approximately (at the rate of 3000 kcal/m^2/yr) 3.0×10^6 yr.

Whittakar (1975) has given the idea of *biomass accumulation ratio* or BAR. This is the ratio between the standing crop biomass to net annual primary productivity. It gives a measure of how much of the primary produce usually goes into the woody biomass. Usually in the annual grass and forb dominated vegetation it is 1, in perennial grasslands 1.5 to 4, in shrublands 3 to 12, in deciduous open forests 10-15 and in dense forests from 20 to 50. Leaf area index is one of the most important factor determining the productivity. BAR value starting from 1 in pioneer stage rises slowly with the successional process and may reach upto 30-40 or 50 in climax forest community. Productivity also rises with the rise in temperature, say upto 30°C in C3 and upto 45°C in C4 plants. Increase in the mean annual precipitation also increases production upto about 3000 mm.

The human needs of bioproductivity in terms of food energy for an estimated world population of 4.3 billion (in 1980) has been computed to be 4.3×10^{15} kcal/yr by Odum (1983) at the rate of 1 million kcal per person per year. The world food (not GPP of organic production) for the same period was much higher being 6.7×10^{15} kcal/yr but due to uneven distribution, wastage in storage, pests and other reasons, there has been food shortages in many regions.

Modelling of primary productivity of the world has been done by Lieth on the basis of a number correlation factors. Temperature and precipitation have been found to be closely influencing productivity. On the basis of these and other environmental conditions of a chain of selected sites on a global basis, computer model of world primary production has been prepared. Thus environmental map could be converted into productivity map. Lieth and Box (1972) have infact prepared a predictive productivity map on the basis of actual *evapotranspiration* values only. The factor of time, i.e., the vegetation period of a biome has also been incorporated in these modelling.

LITTER

Production

A good part of primary produce is returned to soil in the form of litter through the death of fine roots of perennials, whole plants of annuals, leaf and twig fall, exfoliation of bark and through fruits. Depending upon the nature of litter, temperature and available moisture, the litters decompose at different speeds and return the nutrients to soil for fresh growth. Thus the rate of litter fall and its decomposition are of great importance in production ecology. Singh and Ambasht (1980) and Singh, Ambasht and Misra (1980) have shown that in tropical forests there is rapid turnover of organic matter and about ninety per cent of annual litter production is decomposed within one year and only very hard and fibrous tissues take more time in their complete humification and mineralisation. Sandhu, Sinha and Ambasht (1990) in a study of litter decomposition of *Leaucaena leucocephala* plantation floor have found that out of 256 kg ha^{-1} nitrogen in the litter of which 208 kg is released to soil within one year period (81%).

The average litterfall in tropical rainforest is about 1600 g m^2 yr^2 and in subtropical forests it is about 1200 g m^2 yr^2 (Rodin and Basilevic, 1968). In temperate forests and other communities litter production is 300-500 g m^{-2} yr^{-1}. But the decomposition rate is very different in different climatic belts. It may be much less than one year to complete total decomposition in the moist tropics, 3-5 years in moist subtropics and 30 years in cool and dry regions. Moisture and temperature are regarded as the key factors for high microbial activity and decomposition. Actual evapotranspiration (AET) as a measure of energy and water in environment, and lignin content as the measure of litter quality have been used by Meentmeyer (1978) to predict decomposition rates.

Decomposition

Litter reach the soil surface and sub-surface from leaf and twig fall, exfoliation of bark, death of fine roots and death of animals, plants, of soil fauna and flora, etc. The breakdown of complex organic components on decomposition produce humus which is further acted upon by the microbes to produce in inorganic form the soil nutrients, CO_2 etc. through the process of mineralisation. The process is extremely slow in boreal or very cold conditions and it is fastest in warm and moist tropics. So there is a progressive accumulation of litter in cold climates as only a fraction of the litter produced in one year is decomposed in the same period. Cellulose contributes maximum weight to the litter mass and are used by decomposer microbes for their energy and resynthesis of their body. Microbial polysaccharides so formed bind soil clay and silt particle due to their slimy nature. Lignin is another important litter constituent which is decomposed slowly due to their high molecular weight, chemical structure and nature. Laccase-forming fungi are better decomposer of lignin and produce humic substances. Decomposition of waxes and lipids (long chain hydrocarbons) are more difficult.

14

Conservation & Management

The subject of applied ecology deals with the special aspect of study which is connected with the welfare of human society. There are numerous ways in which the subject of ecology is directly and indirectly concerned with the immediate and long range necessity of man. With the cultural evolution of man, he began to use more and more of the natural resources for his immediate use. Primitive man hunted for food all the time and lived like other organisms as very interdependent component of the ecosystem. Many of his attributes, especially the capacity to pass on accumulated experiences to new generations through his faculty of speech, memory and the art of writing, and reading led to an altogether new type of evolution which was never witnessed before in over 2.5 billion years of earth's history since life arose. Man soon became a powerful force in upsetting ecosystem balances and has become the most dominant organism of the earth. In a naturally balanced ecosystem the biological attributes of all species are such that they produce offsprings in greater number than the system could actually sustain if all of them become adult. This is because of heavy mortality due to immunerable factors but only a few better fitted individuals survive. Competition for food and space, disease, predation, natural calamities like foods, famine, landslides, fires, etc. are some of the many forces that account for heavy mortality. Man with his gradual dominance in ecosystem, has developed several methods to combat these forces leading to reduced early mortality of his offspring. However, at the same time, he continues to overproduce. The rate of increase in human population of the world has thus kept on increasing. The problem of space and food were met by clearing land of undesirable organisms like trees, wild carnivores, etc. in favour of production of crop plants, domestication of cattle and poultry and so on. The organism-environment balances are repeatedly being upset and many resources like soil, forests, lakes, etc. are getting depleted and impoverished at rates faster than the rebuilding process can replenish. Thus present day man finds many valuable natural assets and resources which were available yesterday are not there today. Many cleared forests brought under plough or under industrialization may not remain useful land in future, if the ecology of the system is not understood or ecological principles are disregarded. The soil may get eroded, industrial fumes may kill the trees and expose the soil to forces of erosion. The copper hills of a technically advanced country like the USA is one such example where not a blade of grass grows now. In the past the fumes from copper smelting factories have killed the natural vegetation.

There are numerous ways in which ecological knowledge can be applied for human welfare. The major aspects are *conservation of different types of ecosystem* for yielding economically useful material over a long period at a fairly high rate.

CONSERVATION ECOLOGY

Conservation is derived from the Latin words *con* = 'together' and *servare* = 'guard'. The term conservation as understood in ecology has been derived from the title 'conservares' given to the British officers who were at one time stationed in India to manage natural resources.

Conservation ecology deals with practices and customs through which man tries to ensure a continuous yield of biological material and protection of non-renewable resources from wastage (Ambasht, 1966, 1980). It deals with biogeochemical cycles so that the process of or removal or dispersion of energy and materials are only as fast as they are synthesized. A good conservationist always aims at continuous yield of renewable resources through the inbuilt resources of the ecosystem, the external application of material and methods for increasing the output of desired products, and the restriction of undesirable ones. Many dwindling resources and attractive animals need protection in sanctuaries where the ecosystem is specially maintained to suit selectively the protected species. Preservation of rare species is certainly not all that conservation means. Conservation is in fact never a 'hoarding' or even 'rationed use' of a limited resource but a constant effort to increase the resources and rapid resynthesis of the materials. Lack of understanding in any ecosystem dynamics, succession and ecological balances has resulted in widespread damage to vegetation, fauna and the land. Despite this fact, quite a few persons in power like administrators and politicians or even persons with specialisation in a narrow field, think that any piece of land or river as a matter of right may be overnight turned for immediate use for raising crops or building dams or constructing factories etc. The results of such actions in both the remote and recent past have resulted in the widespread formation of deserts, unplaughable ups and downs, rugged gullies, *usarlands,* rocky outcrops etc. Therefore activities of men with regard to biotic and abiotic resources such as croplands, grasslands, forests, freshwater biota, soil water and minerals should be so managed as to cause least damage to the rebuilding part of the ecosystem. Natural resources have been classified usually into : (1) *renewable* such as plants and animals whose continued harvest is possible upon proper planning and management and (2) *non-renewable* which, once gone, have very little chance of recovery or resynthesis. Oliver S. Owen (1971) has made a very broad classification of natural resources into : (a) *Immutable* which are less likely to be changed adversely by man's activities like *atomic energy, wind power, precipitation or the water power of the tides,* or (b) *Misusable* where although chances of complete exhaustion are remote but if improperly used the quantity of the resource may be degraded. For instance *solar power* received by an organism may be impaired in its quality by severe air pollution due to dust or atmospheric gases. The quality of *atmosphere* and *water* in oceans, lakes, streams or rivers and the scenic beauty of landscape may be degraded badly. The exhaustible resources, such as forests, grasslands, agricultural and commercial crops and animals of land or water, etc. are easily *renewable* and *dwindling wild life if* lost among *non-renewable* resources and (b) *Non-maintainable,* whose supply is limited like metals and fossil fuel. The use of some can be *non-consumptive* and therefore they could be recycled as in the case of precious metals, or they could be used up consumptively like petroleum and coal which once used are lost for ever.

For the conservation and management of any region, it is essential to make an assessment of the state of affairs of the prevailing environmental stresses and an inventory of renewable and non-renewable resources. Ambasht (1981) has given such an account of Varanasi and some of the measures that are needed to manage diverse ecosystems. Ambasht, Srivastava and Ambasht (2000) have regarded the conservation and restoration of tropical ecosystem as the key of human survival on the planet earth.

Conservation and management of some of the important resources are discussed under the following heading :

1. Agriculture

Crop plants, like other organisms, react to their environment and grow best under a certain range of climatic and edaphic conditions. Through trial and error farmers have found the season, quantity and frequency of irrigation, the type of fertilizers and many other practices that suit best for a high yield of a desired crop. Thus, agriculture is one of the best examples of applied ecology. The science of agriculture can progress best if an overall picture of the cropland ecosystem is understood. Cultivation of land over a period of time leads to impoverishment of soil due to removal of minerals by crop plants. In primitive agriculture, a piece of land was kept under cultivation for some years. Depending upon the fertility of land, sloping soils, cleared forest (especially, those burnt for cultivation) yielded bumper crops initially, but in the absence of protection by vegetal cover the soil soon used to get eroded exposing the underlying rocks. Space was abundantly available and people moved to newer areas after turning fertile tracts into degraded wastelands.

Gradually with increase in human population on the one hand and destruction of fertile and on the other, the necessity of repeated use of the same land over longer periods of time for raising crops was felt and the practice of fertilizing the soil started. Different crop plants have different requirements and ecological amplitudes. Therefore, good experimental and field studies are required on the autecology of individual crop species like wheat, paddy, sugarcane, etc. Within each species already a large number of varieties and races have been developed to suit different climates, soil and biotic conditions as well as to resist various fungal diseases.

Besides autecology another ecological aspect is an understanding of nutrient cycling and turnover of different species so that a crop rotation on the same site may be practised with advantage. This insures a high production of crop and at the same time maintenance of soil at high level of fertility.

Species interaction, competition and other ecological behaviour of associated weeds are other important aspects. Detailed autecological studies of weeds on the Ganga river bank have shown that *Alhagi camelorum* in field of wheat and mustard does not compete for food and water as its absorptive zone of roots is over one metre deep (Ambasht, 1958, 1963). Instead of eliminating the weed, ordinary practice of weeding by pulling the aerial parts and throwing them always, in fact increased their populations. Damage to aerial part activates several vegetative buds present on the rhizomes. The cut away parts are also capable of striking new roots and producing new plantlets. Another weed *Xanthium strumarium* is found to be a short-day plant (Kaul, 1959) and hence exposure of the species to flashes of light at night prevents flowering. Through this *Xanthium* could be eliminated in a year by checking flowering and therefore, seed formation. Weed ecology has assumed a very important position in agroecosystem management since weeds reduce the crop yield very considerably. Soni and Ambasht (1977) have found a remarkable reduction in the level of nitrogen and phosphorus uptake in wheat crops if the weed *Anagallis arvensis* is allowed to grow in the field. Ambasht (1982) has reviewed Indian works on the phytosociology and autecology of agricultural weeds in the book '*Biology and Ecology of Weeds*' (Ed. Holzner and Numata, 1982).

River banks or riparian lands are fertile alluvial sloping habitats where utmost care and adjustment are needed in raising agricultural crops and at the same time preventing erosional losses and river meandering. Ambasht, Singh and Sharma (1983) have studied the fractionated biomass, chlorophyll and density of crops and weeds on R. Gomati banks. They have also given a working model for the ecodevelopment of the region (Singh, Ambasht and Sharma, 1983).

A thorough study of soil conditions, its microflora and microfauna, topography and contours and many other aspects have also their importance in the cultivation of crops of cereals or cash crops.

Biotic problems to crops due to rats, insects like caterpillers, grasshoppers and locusts, grazing animals, fungi, bacteria, viruses etc. are also important ecological aspects which need thorough study by devising correct remedial measures.

2. Range management

Another important necessity of man is to provide grazing ground for cows, buffaloes, sheep and goats. For this purpose pasture or range is maintained in areas which are specially suited to biotic climaxes of grasslands. Grasslands are naturally dominated by a large number of individuals of the family Gramineae. In fact cultural evolution is intimately related to cultivation of grass family crops of rice in Indian and Chinese culture, wheat in the Mediterranean region culture and maize in the New World. Through a long evolutionary history grasses have been subjected to intensive pressure of grazing and other adverse situations. Hence generally they have become quite hardy. Lands which are therefore, less stable under crop cultivation, are usually suited for maintenance of ranges. Range management is essentially an important branch of ecology in which the effect of grazing is studied experimentally by maintaining sets of extensive plots under native or better suited palatable grass and forb (other weeds) species under different stresses of grazing intensity. Perennial grasses uaually recover to a certain extent from physical damage and removal of vegetative parts in grazing. The intensity and frequency of grazing have, therefore, to be regulated after careful research to maintain the forage production at a high rate. In fact in management of all ecological systems the rate of removal of resource should be regulated to a level at which the system can rebuild itself. Therefore, the productivity of grassland species has to be studied in exclosed plots free from grazing and under different degree of grazing to ascertain the extent to which grazing may ultimately be permitted in pastures. Researches on reproductive capacity and other behaviour of plant species in grazing grounds indicate that in some species the average seed weight is reduced under grazed conditions leading to lower reproductive capacity through seeds but often to higher vegetative reproductive capacity as in *Dichanthium annulatum* (Ambasht and Maurya, 1970b). The population of some species increases with greater degree of grazing intensity due to the fact that the competition from those species which are palatable and therefore, selectively grazed is reduced. The grazing effects on seed output, reproductive capacity, establishment, vegetative growth and flowering in relation to climate, soil and biotic pressures of grazing and scraping are some of the more important ecological aspects. Distribution, geographical races, palatability, nutrient value, etc. may lead to introduction of better 'foreign' elements in ranges. Gupta and Ambasht (1979) have discussed tropical works on use and management of grasslands in the I.B.P. synthesis book *Grassland Ecosystems of the World* (Cambridge Univ. Press, UK).

Heavy grazing is almost always detrimental to the grassland structure and productivity. But moderate grazing, in a grassland dominated by *Aristida cyanantha* on Vindhyan hills, is found to increase the diversity of flora as well as the net primary productivity (Pandey and Ambasht, 1981).

Since rangeland is a source of nutritious forage to cattle and other livestock, its management on ecological principles is essential for a perpetual production and availability of forage. A *range condition* refers to the general health of grazing land i.e. their present condition with respect to their potential productivity. Ranges are qualitatively classified into (1) *excellent*, (2) *good*, (3) *fair* and (4) *poor* classes. Level of biotic disturbance particularly the grazing pressure, scraping and fire alter the plant composition of a rangeland. Grazing reduces such species which are more palatable and nutritious. These species are called *decreasers*. Less palatable ones, in absence of decreasers get a chance for competition free growth and increase in their density and hence such less palatable species are called as *increasers*. Very heavy grazing causes creation of bare lands on which spiny plants resistant to cattle grazing invade and establish. These are called as *invaders*. In any region, there are characteristic species which by their presence indicate the level and kind of biotic pressure. These species have a narrow range of tolerance to particular types of stresses. These are called as *indicators*.

Some of the important methods of range management are as follows :

1. Stock level policy

Every rangeland has an ideal carrying capacity, i.e., the average number of grazing animal classes like buffaloes, cows, goats or sheep that can be properly maintained over a given area. But in nature, all years are not equally productive due to fluctuations in the climate. In some years there are droughts and plant productivity falls down considerably and a number of animals die of starvation. Therefore, ranchers evolve a method of manipulating the number of cattle in different ways so that the problem of overgrazing and starvation deaths are reduced to a minimum level. This is achieved either by maintaining a high level of stock in good years and selling away a portion in drought years or permanently keeping a lower population at about 60 to 70% of the carrying capacity.

2. Deferred grazing

Under this system the rangeland is divided usually into three segments. The cattle are kept in the first segment for two years, leaving the other two segments ungrazed. This affords opportunity to the ungrazed segments to attain a high level of plant biomass. In the third and fourth years the cattle are moved to the second segment leaving the first and third ungrazed. Finally the cattle are moved to the third segment in the fifth and sixth years. The process is recycled in the same way. This helps in proper recovery of the rangeland from getting permanently degraded, and it remains full of nutritious grasses all the year round.

3. Fire

Sometimes rangelands are burnt at regular intervals of annual or longer cycles in order to destroy less palatable and hardy species which otherwise overtake range lands on account of grazing in course of time. Fire is a good method of range management provided there is little possibility of rain immediately after the fire as that causes erosional losses of the much fertile ash of the burnt up vegetation.

4. Reseeding

Rangelands also sometimes need grasses and legumes for their successful growth against naturally invading less desirable species. Aeroplanes are used in broadcasting seeds on a large scale.

Various kinds of techniques of killing plant pests are also applied, but enough precaution is needed to avoid harm to non-target organisms and cumulative pollution effects.

Sometimes animals gather in one region and overgraze there leaving other parts ungrazed. This is particularly around water bodies or under the shade of a few trees or near salt blocks which they like to lick. Therefore in large rangelands, waterholes and placement of large blocks of common salts are evenly distributed so that grazing also becomes even all over the rangeland.

3. Forest management

For most part of the terrestrial habitats the natural vegetation was forest for millions of years. All the organisms sharing the habitat had a balanced but quite competitive and cooperative adjustment among themselves. The dominance of trees was perpetual and the trees formed the broad base of ecological pyramids, supporting and influencing other life-forms. For several thousands of years primitive man lived in forests as hunter and food gatherer component of forest ecosystem. Human population was small and it was maintained at low levels due to high mortality especially of young ones, predation by ferocious wild animals, non-existence of medical or health care and a general lack of adequate food all the year round. Gradually, with the cultural evolution primitive agriculture was evolved for which land had to be cleared of trees. That is how the trees were first cut down to clear the ground for crop fields and housing, and the wood was used for cooking food and constructing crude agricultural equipments and carts. Domestication of some selected animals also began. The balancing effect of forests was somewhat disturbed as in absence of adequate vegetal cover, the cleared lands often got denuded due to accelerated soil erosion and nutrient losses. Even as early as over two thousand five hundred years back, Lord Buddha in his teachings has emphasised the necessity of preserving the trees from wanton cutting and had enumerated the advantages of forest trees for mankind and the ill consequences of their destruction. But the situation was not the least alarming as it appeared that forests were limitless and any amount of new encroachments could be made when the old cleared lands become depauperate after obtaining a few years of bumper crops.

Jhum cultivation or slashing the forest trees and burning them and then raising agricultural crops on ash rich soil for few years became a common practice. With the development of extensive navigation in high seas, warfares and then industrial growth, large scale timber and other forest produce were used. The most massive onslaught on forest trees began about 150 years ago, when the railways came into existence. Throughout USA, Canada, UK, Europe, India etc. forests were cut to obtain thick slabs of railway sleepers to cushion the railway lines. Indian forests suffered the most as the resource was harvested and exported to UK and other European countries. Many of the tall and dense stands of forests particularly in the rich tropics were destroyed. They could not return to their original climax condition even on leaving them to recover. Nutrient cycles largely controlled by intrabiotic nutrient storage in plant biomass and their slow release through litter fall and rapid decomposition, get adversely disrupted due to removal of tree biomass and accelerated soil erosion. In the post independence, Indian forests were taken from private owners and rulers of different princely states by the State Governments and extensive forest management policies had to be devised to retard degradation of forests into open woodlands, savanna, grasslands and deserts. However, the onslaught of rising human population, developmental activities like construction of numerous dams (submerging much of the forests), industrialization and mining operations directly cutting the forests and through pollution destroying surrounding forests etc. have alarmingly decreased the forest cover of India.

Forests (i) supply us a large variety of natural resources, (ii) provide habitat for wild animals and plants many of which are getting extinct, (iii) conserve the soil, water and nutrients, thus regulating conversion of rainfall into stream flow and prevention of floods in rivers and regulate the climate. They are also useful in many other ways. Thus, there is a real and urgent need to manage our forests in such a way that there is a sustained supply of useful material from the forests and there is a continuous regeneration of trees either naturally or through afforestation methods. The extent of forest product use can be better appreciated from an estimate given by Owen (1973) for USA where a small but, most affluent segment of world human population lives. Even 25 years back they used 37.3 billion board feet of wood in 1962 (1 board foot = 1 foot square of one inch thickness wood volume). This would have been sufficient to build 3.5 million six room wooden houses or enough to bridge the distance from earth to moon 70 times with four feet broad planks. Every individual in all countries uses forest produce from a small tooth pick, or match box or a pencil to large carts, carriages, railway coaches, boat, ship, telephone poles and numerous other forest products like sealing wax, turpentine oil (in paints) and paper for writing, books, newspapers, napkins and packing materials. In order to maintain the supply line of the above and many more things in ever increasing quantities, there is no other way than to cultivate trees or manage the forests in a state of sustained regeneration and harvests.

Indian forests yield valuable commercial materials like *timber* (*Tectona grandis*—teak, *Shorea robusta*—sal, *Dalbergia sissoo*—shishum, *Pinus* sp—chir, *Cedrus deodara*—deodar, *Boswellia serrata*—salai, *Santalum*—chandan, *Diospyros*—ebony, etc.), fibre of bamboos, *Eucalyptus, Boswellia*, etc. for paper manufacture; lac or sealing wax from insects living on forest trees, leaves for wrapping biri (the commonest smoke or poormen's cigarette) from *Diospyros melanoxylon*; fruits like *Anacardium occidentale* or cashew nut, *Buchanania lanzan (chiraunji)*, medicinal plants like *Terminalia arjuna* or arjuna, etc. and a large variety of game animals.

Silviculture is the term applied for cultivation of trees. It also concerns to methods of forest regeneration. The basic feature of all forest management methods are based on *sustained yield concept*. This means that forests have to be used for taking out necessary and desired products at a maximum possible level of harvest but at the same time ensuring the continuous afforestation or regeneration so that always an adequate forest cover is maintained. In silviculture a knowledge of climate, drainage pattern, species content, successional trend, methods of arresting or diverting the stages that yield economically best products, the ecological life cycle of more important species, biotic interrelationships, etc. have to be gathered for a successful management of forests.

There are a few main important kinds of management practices such as (i) selective harvesting from mixed stands, (ii) *block cutting*, (iii) *taungya* system of block cutting and reforestation through seedlings, raised in nursery, (iv) coppice regeneration, (v) forest extension and social forestry, etc.

(i) Selective harvesting

In natural mixed stands of forests there are diverse kinds of trees, shrubs and herbs of different ages adjusted in several storeys. In closed canopy forests the seedlings of many dominant trees either remain dormant at stunted growth stages or die due to paucity of adequate light and aerial space for canopy to develop. But in the forest openings created due to death of old trees or otherwise, these stunted seedlings suddenly become active and grow rapidly to cover the openings. In managing such forest stands a working plan is developed, the trees are marked according to the time sequence they are to be harvested and suitable steps are taken to fill up the gap. In case the species can coppice (i.e., send out new shoots from the base of stem or stump of wood left after cutting the shoot) then from among the several new shoot branches, the most robust one is retained and the rest are cut away. The new coppice branch then takes up the place of main trunk and rapidly grows to fill up the gap. In other species, the seedlings raised in forest nursery (from seeds collected from a robust and healthy tree) are planted in preprepared pits, filled with humus and fertilizer.

(ii) Block cutting

This practice is usually followed in plantation forests where all trees in a particular block are of same age. The forest range is divided into 30-40 or more blocks. All trees on maturity in a particular block are harvested and the cleared land is planted with the seedlings of same species and they rapidly grow in absence of other competing species. Next year the next plot is harvested and the process is continued year after year endlessly as always new blocks become ready with mature trees. In many species, the harvested trees coppice very well as in *Tectona, Eucalyptus* or *Leucaena*, where planting new seedlings are not necessary. The old *stumps* or *stools* as they are also called successfully regenerate.

(iii) Taungya system

In order to reduce the expenses over plantation of trees—a system of forest culture plus agriculture has been evolved. It is called the *Taungya system* in India. Under this system villagers or poor people without any fields of their own are allotted cleared forest plots. There, they cultivate agricultural crops in the first year and take all the harvest; next year they prepare rows of beds for raising seedlings of trees and in between these rows they cultivate crop species. Crops are raised for 2-3 years during which period root system of tree seedlings grow deep down due to competition with crop plants in the upper soil horizons. Once the seedlings have grown to a metre or more, the labourers are moved on to the next cleared block. Thus, the forest department gets free labourers to keep planting trees on the one hand and cut the mature trees for the forest department on the other whereas the labourers get free land to live and raise agricultural crops for themselves.

In between forest openings grassland species for grazing purposes are raised. Heavy grazing is very often the cause of damage to tree seedlings and, therefore, it leads to the development of scrubs and spiny thickets. Seedlings of more dominant forms are usually able to escape the grazers in protection of spiny bushes. Therefore, regeneration of non-spiny or bushy plants in pockets protected by spiny bushes indicates a heavy degree of grazing.

(iv) Forest extension and social forestry

Forests in India are under the control of state governments. In many other countries the forests may be owned by individuals as well as governments. Even though there may be about 20 to 30% of the land area under the forest departments in different Indian states, yet it is found through satellite imagery, that not all this is covered with trees. About half of the forest lands are in different stages of degradation into woodlands, thickets or savanna. Unauthorised damages and excessive removal of timber and fuel wood and overgrazing have resulted into over exploitation and regeneration failures. Realising the importance of keeping a certain percentage of total land under forests, it has become necessary to extend the area of existing forest cover and also to raise trees in and around every village to meet basic wood needs of rising human population.

Waste lands, road sides, along railway line tracks, village panchayat lands, uncultivable hilly slopes, riparian lands, wide buffer belts around dams and industrial premises and all other unused lands are included for social forestry and forest extension purposes. All human habitation such as villages, towns, cities, industries must have a social forestry plan of continuous plantation of tree seedlings, their protection and harvests as a social plan for better economy, stability and cleaner environment. People's participation in social forestry as partners with forest department in protecting and harvesting, i.e., for input and sharing the profit is a must for the success of social forestry. A sense of social cooperative ownership and benefits has to be built in order to succeed social forestry programme. Every village needs fast growing trees for fuel wood, leaf fodder, fruits, leaves for preparing leaf plates (*pattal*) and leaf cups (*dona*), thatching material, bamboos and canes for basket, timber for making plough, cots, carts, small shops (*gumties*), etc. An ideal village tree plantation, therefore should be of mixed type with fruit trees, timber trees, fuel wood trees, bamboos, etc. at regular and suitable intervals. Large sized trees may be planted at distances 10 metres apart which may yield fruits year after year such as mangoes, jackfruit, *Eugenia jambolana*, etc. and they need to be replaced at intervals of 50-60 years. The timber trees like *Dalbergia sissoo, Acacia nilotica, Madhuca indica*, etc. may be planted in between the rows to be harvested after 20-30 years. Besides timber, *Madhuca* provides leaves used in preparing leaf cups, flowers are eaten and alcohol is prepared, seeds provide cooking oil. Smaller fast growing trees providing fuel wood and green fodder like *Leucaena, Sesbania, Diospyros,Prosopis*, may be planted at 2 metres apart and harvested after 4-6 years. Many species for specific purposes such as *Terminalia arjuna* and mulberry can be grown to raise tassar and silk cocoons. Cultivation of tall trees along railway tracks are to be avoided lest they may cause damage to tracks if they fall down during big storms. Social forestry provides several direct and numerous indirect advantages.

In silviculture, knowledge of the ecology of the habitat and of succession is a *must* and without it many silvicultural attempts have yielded poor results. Sometimes the timber is of poor quality. Sometimes regeneration does not proceed properly and sometimes operational costs make the enterprise uneconomical. Usually forest management takes into consideration the multiple use of the ecosystem. These are principally for (a) timber and other economically important tree products like fibre, fruits etc., (b) a proper maintenance of the water cycle, (c) use forest openings for grazing and forage, (d) preservation and multiplication of wild life, (e) habitat for migratory birds in low-lying areas and (f) recreational spots for city dwellers.

It is necessary that all sloping lands particularly the steep hill slopes and alongside rivers must be left under the cover of forest trees in order to maximize conservation of soil, water and nutrients of the habitat. Selective harvesting of fuel, wood, fodder, leaf, etc., has to be kept at a minimal level on such fragile habitats. For rural fuel wood and timber needs, social forestry is being introduced for every village. In many plantation stands of herbs or shrubs like cardamom and coffee, additional canopy layers of fast growing nitrogen fixing trees are recommended which provide necessary protection and fertility to cash crop and timber, fuel wood and twig fodder for rural needs. Sharma (1988) and Sharma and Ambasht (1988) have measured nitrogen fixing rates in *Alnus nepalansis* or alder trees on the hill slopes of Kalimpong in the Eastern Himalaya. Alder regenerating between 1000-2500 m altitude range showed best performance in the middle zone (1500-2000 m). Nitrogenase activities were between 5 and 19 µmol ethylene per gram nodule dry weight per hour in a peak activity month (July). The activity was largely dependent on mean soil temperature and root nodule moisture. Nitrogenase activity was negligible at night and increased during the day with a midday maximum. Alder regeneration, seedling growth, root nodule biomass and nitrogenase activity were temperature dependent along the gradient of altitude and fell off below 1500 m and above 2000 m (Sharma, 1988).

The per cent contribution of biological nitrogen fixation to annual total nitrogen uptake in alder plantations of Eastern Himalayas were 33%, 19%, 15%, 13% and 11% in 7–, 17–, 30–, 46– and 56 year old stands, respectively. Nitrogen accretion through fixation reduced with the decrease in nitrogen demand as trees aged. Low nitrogen fixation efficiency and low net energy allocation in root nodules with increase in tree age caused sharp reduction in nitrogen accretion through fixation in older stands, while nitrogen uptake from soil remained nearly the same in young to old plantations. Nitrogen build up strategy is based on a sound energy conservation dynamics and the alder replenishes nitrogen in erosion prone hill slopes of the Himalayas (Sharma and Ambasht, 1988).

WATER AND WETLAND MANAGEMENT

Ambasht (2003(a), (b) and 2005) and Ambasht and Ambasht (2006) have described the water and wetlands in India and given updated new information on global perspectives. Wetlands are inundated or water saturated areas where abundance of water is maintained for sufficient duration in the year to support hydrophytes. They are man-made as well as natural. In fact the most comprehensive definition is given by the Ramsar Bureau 'Switzerland'. It includes marshlands, peats, fens, permanent and temporary flowing and lentic fresh or brackish and saline waters upto six meter depth (at low tides). Wetlands are *lotic* (flowing) and lentic (stationary) the former are riverine while the latter are lacustrine (lakes) and palustrine (saturated) marshes. Amongst very large wetlands are the European and West Siberian (780,000 km^2) plus bogs over 220,000 km^2, the Amazonian flood plains spread out 800,000 km^2 while the Pantanals (wetlands) of S. America cover 140,000 km^2. Nearly 11$\2$ countries are signatories of Ramsar convention and this body keeps watch on numerous world wetlands designated as Ramsar sites. In India, Ramsar wetlands are Chilika lake in Orissa (1,14,000 ha), Keoladeo Ghana (3000 ha), Sambhar (7000 ha) in Rajasthan, Loktak (26000 ha) in Manipur, Harike (4,100 ha) in Punjab and Wular (18,900 ha) in Jammu & Kashmir states.

Mitsch et al. (1994) estimated that about 6% of the global land surface (7 to 8 million km^2) is covered by wetlands. On a worldwide scale, there has been a drastic cut in the overall wetland cover due to land fills, encroachments, siltation, expanding urbanisation, etc. Even the village, town and city ponds have been filled up for house con-structions. Some of the common names of different kinds of wet-lands are *wet meadows* and *prairies* for saturated pockets in grass-lands in America, *Vernal pools* for shallow water covered meadows in the Mediterranean climate, *swamps* with trees or tall reeds in Eu-rope. Deep water deltaic coasts in Europe are known as *lagoons* while similar wetlands in Australia are called *Billabong*. Wetlands full of partially decomposed plant material (due to cold climate) are called *peat* and *moor* which have acid loving mosses like *Sphagnum*. If these wetlands have poor inflow-outflow drainage, they are called *Bog* and if good drainage then *Fen*. Forested wetlands with alder and willow trees in Europe are called *Carr*. In Canada peatlands are called *Muskeg*. On sea coasts, partially water covered forests are called *Mangrove*. These are the habitat for halophytic trees and shrubs, Royal Bengal tigers, crocodiles, turtles, fishes and shrimps. Mangroves are of much economic and ecological importance for producing fuel wood and charcoal, medicines, fishes and prawns and in maintaining a rich biodiversity and they provide stability to sea coast against erosion and hurricanes.

Wetlands perform a number of services in nature (1) They are hotspots of biodiversity and speciation. (2) Rivers are perennial source of water to towns and villages abundantly located on river banks. They also serve as river course for boat transports for men and merchandise. Rivers easily drain out excess water and prevent waterlogging and flooding of low-lands. (3) Wetlands are the habitats for cultivation of rice, the staple food of man. Wetlands act in regulating recharge of ground water and as sinks of many harmful material. (4) They are rich gene-pool habitats and home of wild relatives of some important cultivated crops, particularly rice. (5) Wetland embankment vegetation conserve soil and water and act as filters (kidney) of toxic substances. (6) Coastal shores (wetlands) are very big source of petroleum and gases. (7) Common salt which we put in food is largely from coastal salt-pans. Besides large wetlands, the value of small wetlands so extensively found naturally and constructed tanks is very high. These provide day-to-day needs of fishes, water for human beings and domestic animals etc. and help in maintaining a high water levels in wells.

Pollution aspects are described in the next chapter on Environmental Pollution.

Virtual water

It is a new aspect in the ecology of water resource. Hoekstra and Hung (2002) have defined it, *"The water that is used in production process of an agricultural or industrial products, is called the 'virtual water' contained in the product"*. So while food grains and industrial products are imported or exported, there is a huge involvement of virtual water. For each ton of cereals and pulses there may be 500 to 1500 tons of virtual water involvement. Chapagain and Hoekstra have given virtual water values for each ton of production of potato, maize, wheat, soybean and rice are 160, 450, 1200, 2300 and 2700 tons respectively. For one ton of milk, poultry, eggs, cheese, pork and beef the respective virtual water is 900, 2800, 4700, 5300, 5900 and 16000 tons.

Thus virtual water estimates have great bearing in a working out the *'Water Foot Print'* of any country. WFP = WU + NVWI i.e. domestic water use and net virtual water import.

Extensively exporter of commodity countries are generally net virtual water exporters also.

Management of freshwater bodies

Inland water bodies like rivers, lakes, ponds, pools and puddles are of common occurrences all over the world. They are of great economic importance to man for the water and organisms that abound in them, and man has been using these with little consideration for their ecology. The demand of a growing human population for space and food is ever increasing. Freshwater bodies are being more and more used for these purposes. Management of freshwater bodies should aim to maintain them in useful form at a high productive level with provision for a high rate of harvest of plants and animals for human use. Thus a thorough knowledge of drainage, total water and runoff, mineral receipts, the biotic communities, the ecological life cycle of economically important species, energy flow pattern, quantitative study of mineral cycling from abiotic to chains of organism, turnover, should be understood. Then only the management with regard to its economic use such as quantity and periodicity of water removal for irrigation, maintenance of certain fish and plant crops, the extent of carrying capacity of the water body towards pollutants like town refuse etc. should be undertaken. A well managed river or lake is one which serves in a multipurpose way the human society without itself degenerating whereas in unmanaged or mismanaged freshwaters the utility is short term, less efficient and in most cases the system degenerates due to excessive silting, pollution, or rapid drying or due to preponderance of noxious weeds, etc.

Ambasht and Shardendu (1988a and 1988b),Ambasht (1989a, 1989b), have shown that the role of weedy species of Ganga riparian slopes on conserving soil is tremendous. Ambasht and Singh (1988) and Singh and Ambasht (1989) have quantified such roles of weeds on the wastelands of Gomati river banks. Nelson, Ambasht and others (1989) in a global paper on wetlands and floodplain management have focussed the attention on worldwide wetland problem areas and have formed a World-Wetland Partnership.

Freshwaters cover a relatively very small area of the earth's surface, but due to their utility the human cultures and settlements have developed mostly along rivers. Freshwater bodies like rivers, lakes, ponds and puddles are used variously. Rivers are used for : (a) irrigation, (b) transport, (c) supply of drinking water, (d) hydroelectric power, (e) aquatic sports and recreation, (f) bathing, (g) for draining out the municipal and industrial refuse, (h) fishing and so on. There is a range up to which any of these items could be put to use and they are interrelated. For instance excess removal of water for irrigation would affect all other aspects; the pollutants will not be sufficiently diluted due to reduced quantity of water, and may become toxic and injurious for drinking and bathing. It is essential, therefore, that aquatic ecosystems should be properly studied and management should be done considering the interrelation among multiple items of utility.

Shardendu and Ambasht (1988a, 1988b, 1989, 1990 and 1991) have measured the impact of urban environmental stresses on physiochemical and biological qualities of ponds and on primary productivity.

Management of water resource is perhaps the most pressing need of humanity, much more than the development of nuclear technology or space satellites. Abuse of water uses and wasteful technologies are going to create famine for potable water and abundance of disease causing foul water. From primarily anthropocentric, we must change to sustainable and ecocentric objectives.

SOIL CONSERVATION

The nature, properties, process of formation of soil and its role as an ecological factor have been described and discussed earlier. The weathered rock intermixed with decomposed organic matter is the soil. There is about an equal volume of space left between the packs of soil particles. In these pores moisture containing dissolved nutrients and air remain interlocked and root hairs absorbs them. It is a common experience that the soil is liable to removal from one place to another whenever there is physical force like storms or running water, usually after rains. The dust blows along with wind or flows in runoff water. This is called erosion. *Erosion* is derived from the Latin word *Erodere* which means to 'gnaw out' or tear away. It refers to physically detaching soil particles from their original place and transporting to some other place. While it takes a very long time to build the soil, its erosion, in absence of good plant cover, by the forces of rain and runoff water, wind action, etc. is a rapid process. Volume and intensity of precipitation, slope conditions, vegetation cover and wind speed are important factors. Erosion is one of the greatest problems in soil conservation. Like conservation of any other resource, in soil also conservation means protection, improvement and sustained renewal of the soil at any place. Soil erosion control is only one aspect of soil conservation and protection.

Soil erosion and its control : The principal causes of removal of soil from one place to another are water and wind.

Water erodes or cuts and removes soil chiefly in four ways : (1) *sheet* erosion, (2) *rill* erosion, (3) *gully* erosion and (4) *riparian* erosion. In almost all cases man is involved in some way or another in accelerating the rate of erosion. In the normal course of development of communities the edaphic conditions and other environmental conditions attain some sort of dynamic equilibrium with the vegetation and other biota. As a result of this the soil is held in position by vegetation cover. Killing of lions, tigers and other carnivores by man results in increased population of herbivores. The natural balance gets upset; herbivores in greater numbers damage the natural vegetation in greater quantities. Natural vegetation is also damaged by scraping and grazing. All these lead to the removal of protective plant cover from the soil surface. The beating of rain drops raises soil particles selectively; finer particles of clay and humus being lighter are raised to greater heights than sand particles. Thus, the finer components, which are incidently more fertile brought upon the surface and these seal and clog the pores. The accumulated water fails to percolate down and therefore, runs down the slope. The fertile top soil also gets washed away.

1. Sheet erosion

The soil is removed in small but uniform amounts from all over and therefore, does not leave a mark behind. The evidence of such a type of erosion can be found in the heavy quantity of silt that deposit elsewhere.

2. Rill erosion

The run off water moves rapidly and cults small stream-like structures. The movement of water is very rapid and hence the cutting effect is quite evident.

3. Gully erosion

Several rills converge towards the steep slopes and join to form broad channels of water called *gullies*. Whenever rills and gullies form in abundance, ploughing becomes impossible due to uneven topography and cultivation of crops is not advisable in these areas either.

4. Riparian erosion

It takes place on the banks of fast running rivers. Here the surface current cuts the margin of the bank laterally. Once much of the soil from beneath is cut away the top soil of the river banks in chunks tumble down into the water with big splash.

Erosion due to wind is very common. Everyone has experienced dust storms where huge quantity of dust is raised high and transported to great distances. The wind lifts finer particles high up whereas the coarser and heavier particles roll along the surface. The rolling particles rub the ground and due to abrasive action help in loosening the soil. The process continues and more dust particles gather as the storm advances.

The problem of checking erosion is mainly concerned with methods of keeping the soil in place, of keeping the nutrient level in soil at a certain desirable level and with educating about the science of the ecosystem and conservation to persons concerned with the eroding lands.

The principal methods aim at reducing the physical and biological forces that cause erosion and protecting the soil by plant cover against erosive forces.

Plant cover is the best method to check erosion of soil. Conservation, however, not only aims, at preservation of resources, but also its judicious and long range use for the maximum advantage of the mankind. Therefore, the plant cover should be such as to yield food, fodder or other resources. For proper utilization and management, land is divided into capability classes. Class I and II are level lands with very little scope for erosion and are recommended for intensive agriculture. Progressively poorer soils with greater scope for erosion on destruction of natural vegetation are placed in class III to VIII and on them restricted or no agriculture is permitted. They could be kept as grazing lands or national parks or natural forests. On lands with gentle slopes, *contour farming* is recommended in which ploughing is done at right angles to the direction of the slope. This reduces water runoff as each furrow acts as a reservoir and crop productivity is increased. On somewhat greater slopes, *strip cropping* is done where strips of land are ploughed at right angles of the direction of the slope and different crops are raised in adjacent strips. In different years the one crop of the strip is rotated with another crop. On hilly terrains where there is some good soil profile, terrace cropping is in practice since time immemorial to reduce erosion and obtain a reasonable crop productivity.

In preparing terrace, one block of flat land abruptly ends and a new terrace begins at a lower level. The soil is cut and laid to make broad steps. On the margins between two steps the slope is vertical. This zone is planted with some perennial soil binding grass whereas the rest of the space is used for cultivation. The runoff of rain water is considerably reduced and the soil becomes stable. This type of practice is called *terrace cropping*. In this flat strips or plots are prepared and ploughed according to the contour of the land.

Slopes are roughly divided into gentle, moderate and steep slopes. On such habitats a careful study of native vegetation, climate, rainfall, soil properties, forces of erosion or forces helping in erosion are to be made thoroughly. Then experimental studies to determine the effectiveness of different native or otherwise suited species in checking erosion are made. The best species may not necessarily be very useful economically. However, the growth of these species if allowed at least in strips will considerably stabilize the soil and save the land. Ambasht (1963, 1970b) has found out that on the banks of the rivers Ganga and Varuna at Varanasi, there is considerable erosion due to the sloping topography. Wherever the slope is under the cover of plants the rate of erosion is very slow. Through culture experiments it has been found that conservation values of native grasses *Saccharum munja* and *Cynodon dactylon* are 92-96% and 90-97% respectively. Weeds like *Euphorbia hirta* conserve only 10-12%. By conservation value is meant that the species by its presence in a given habitat condition conserves or checks that much percentage of soil which otherwise would get eroded if the land was bare. This may be calculated by experimentally finding out the quantity of soil eroded from identical plots under identical erosive forces, one under the cover of a plant species and the other bare. With these data, application of the following formula (Ambasht, 1963a), the conservation value is obtained :

$$C_v = 100 - \left(\frac{S_{wp}}{S_{wo}} \right) \times 100$$

where C_v = conservation value of species; S_{wp} = weight of soil (dry) washed from a plot under cover of the species; S_{wo} = weight of soil washed from an identical but bare plot under identical erosive factor. Ambasht, Singh and Sharma (1983) have recently assessed the soil, water and nutrient conserving efficiencies of herbaceous plants commonly found growing on sloping banks of River Gomati in North India. Besides grasses a few dicots are also found to be very effective soil binders such as *Phyla nodiflora* (C_v = 94%) and *Croton bonplandianum* (C_v = 68%).

Thus, if plantation of *Saccharum munja* and *Cynodon dactylon* are encouraged, at least on broad marginal lands between agricultural fields on slopes, much conservation of soil could be achieved (Ambasht, 1970). Further their underground parts are found to be far more effective in binding the soil and hence. *Cynodon* could be profitably used as fodder and *Saccharum* shoots should be used for thatching huts, roofs of mud wall houses and preparing cot knitting ropes to a fairly good extent without adversely affecting their conservation values.

On the hilly slopes of Vindhyas, Ambasht and Misra (1980) have made extensive field studies on the soil conservation values of grassland communities under different levels of grazing. They find that under protection the soil loss is less than two tons per hectare during one year's period, but under moderate grazing it rises to about 8 tons and under prevailing intensive grazing to about 21 tons. If the hill soil is scraped thoroughly and kept free of plants, with the same rainfall it rises to about sixty tons. This shows that plants have very significant role in the stabilization of the habitat. How much effectively each species binds the soil has also been quantified by Singh, Misra and Ambasht (1980) and values above 95% are recorded for dominant grasses.

In deserts, especially in western Uttar Pradesh and Rajasthan, the erosion is due to wind. Rainfall in these parts is low and the limited native vegetation has been reduced to a minimum due to excessive exploitation. The economy of the entire belt is adversely affected and in recent years measures for checking the march of desert (erosion and transport of sand to new areas) are being taken on an extensive scale. Strip cropping of trees requiring low water may be done between agricultural plots. This reduces the wind velocity considerably.

Plantation of trees in short blocks are called *Wind breaks*. Extensive plantation of trees are called *Shelter belts*. These are planted in rows at right angles to the direction of the wind. If there are more than one direction of wind in different seasons, then there are plantations of trees also in more than one direction. Usually, when the length of the wind break is doubled, the area of protection against wind erosion is increased to four times as would be clear from Fig. 14.1. But this is applicable for short block plantation only. Further, the shelter belts

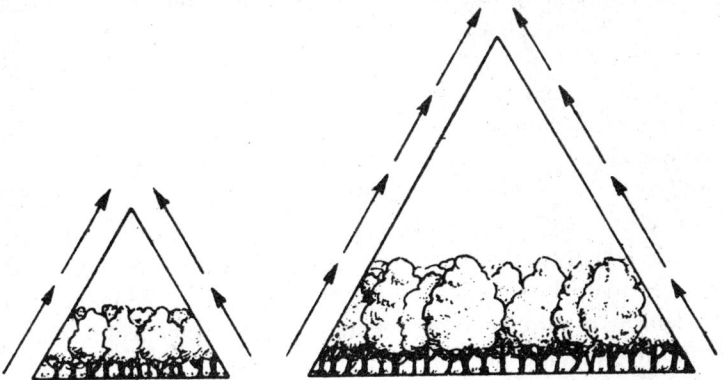

Fig. 14.1. Diagram showing that an increase in wind break length by twice increases the area of protection by four times. The arrows show the direction of wind and the triangles show the area of protection.

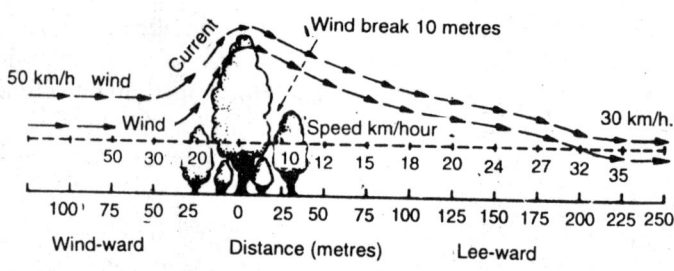

Fig. 14.2. A shelter belt of about ten metre high trees reduces the wind velocity of 50 km/hr speed to 30 and 20 km/hr on windward side and 10, 12, 15 km/hr on leeward side near the trees.

and wind-breaks reduce the wind velocity on wind-ward side upto about the distance equivalent to five times the height of trees and on leeward side to about thirty times the tree height. As it is evident from the Fig. 14.2, in a shelter belt of trees say of 10 metres height, wind coming at a speed of say 50 km/hr is reduced to 20 km/hr at 30 metres on the windward side and rapidly decreases to 12 km/hr at that distance on the leeward side. The wind very gradually gains its speed and the land is protected upto 300-400 metres distance. So on such lands there could be parallel windbreaks or shelter belts at a distance of half a kilometer and in between agriculture can be practiced without a serious problem of wind erosion. Both these types of plantations are being made in the region of Uttar Pradesh where desert is encroaching. The plantation is usually being done in two or three belts. Small sized plants planted on the windward side in the region are *Saccharum munja, Calligonum polygonoides, Laptadenia spartium, Cenchrus ciliaris, Balanites roxburghii, Kochia indica, Panicum antidotale,* etc. On the leeward side scrubs like *Acacia leucophloea, Acacia senegal, Ricinus communis, Tecoma undulata, Prosopis juliflora, Parkinsonia* sp. are being planted. These reduce the wind velocity considerably and the transport of lifted sand or soil particles is checked due to a physical barrier of trees. Ambasht (1982) has stressed on the significance of systems approach and interlocking mechanisms of different components in the management of resources. If we do not take note of side effect and go for any kind of manipulations in nature, we are likely to get ecological backlashes in which not only the projected gain is cancelled out in course of time but many new problems are created. For instance in Africa, construction of a dam on Zambezi river created an extensive area of marshy habitat in which fever causing *tse tse* flies grew in dense population causing disease and death to the tribal inhabitants. Excessive use of weedicides in Aswan dam to control weed infection has caused adverse side effects on the agricultural lands where the dam water was used for irrigation and also it has reduced marine fish production in the region of sea where such affected water was discharged. These unforeseen adverse effects that reduce the advantage on the one hand and cause adverse side effects on the other, are called *ecological backlash or ecological bomerang.*

We find from the foregoing descriptions that conservation of resources both renewable and non-renewable is of much importance for mankind. This can be done only with a thorough and extensive background knowledge of ecology especially of ecosystem (structure, function, dynamics and systems analysis).

Conservation of endangered genetic resources

Due to selective and excessive exploitation of a few selected species without proper management practices for their regeneration, and due to destruction of certain natural ecosystems, there is total extinction of a number of species. Once a species becomes extinct, that combination of gene pool is permanently lost and man can no more recover it for future use. Therefore, conservation of endangered plant resources is all the more important. Each disappearing species takes with it other dependent species of insects or higher animals etc. An endangered species is one in which the natural regeneration is not able to cope up or match with the prevailing exploitation level. There are several economic reasons also for the conservation of genetic resources as sometimes from among the wild varieties a few have proved to be extremely useful. Some are found to be of much medicinal importance while some others are good as fodder. Chang (1984) in his paper on 'conservation of rice-genetic resources : luxury or necessity' has shown how the newly discovered stunt-virus disease resistant wild species of *Oryza nivara* from India has been successfully used in cross breeding with appropriate rice cultigen *Oryza sativa* to produce disease resistant rice. This has tremendously increased rice production of the world. Without really fully realising their potential use through extinction we are rapidly loosing diverse genetic resources.

Further, conservation of wild-life, in national parks help in earning huge amount of foreign exchange by way of attracting tourists especially in African countries. On geological time scale the evolution of new taxa and extinction of the existing ones in somewhat balanced proportions have been taking place, but in recent years human activities by way of destruction of biota and natural habitats have accelerated the process of extinction to an alarmingly faster rates. But for a proper ecosystem balance, the preservation of species richness is necessary. For this the disruptive technologies have to be replaced by ecodevelopmental technologies. There have been some geological upheavals including glaciations through which mass extinctions took place, but currently the extinction rate due to man is far greater, being a few thousand species every year.

For the preservation of diversity of flora, there is urgent necessity of creating a series of nature reserves, biosphere reserves, national parks, sanctuaries, botanical and zoological gardens, etc. Germplasm banks of crop plants, medicinal plants, cash crops and important tree species also need to be established.

A large number of species are presently on the road to extinction. Their survival is *threatened or endangered* unless quick steps to protect them are taken. There are different stages of danger at which different threatened species lie.

Some were reported alive a few decades back but are not found alive now or they are kept alive under cultivated condition. (i) These are classed as extinct, (ii) The next category is of endangered taxa whose survival in the existing or prevailing stressed condition is not possible and they need to be protected against the stressed factor, otherwise they are likely to become extinct. The third category is of (iii) *Vulnerable* plants which if left to themselves under the prevailing level of exploitation are likely to soon pass into the endangered category. Then there are certain taxa whose number is already reduced to a low level and their area of distribution is very much restricted; they are classed in (iv) *rare* category. Both the vulnerable and the rare are likely to pass down to endangered category very soon. However *rare* ones are not necessarily under the threat of some environmental stress except that their number and area of distribution are restricted and in any catastrophic event they may be exterminated, while the *vulnerable* category are indeed under the stress of some known factor.

Commonest causes endangering the survival of species are (i) *overexploitation* for commercial or scientific or other purpose, (ii) *epidemics*, (iii) *habitat destruction* or modification and (iv) other *anthropogenic* and *natural stresses*.

The commonest recognised factors are (i) overgrazing, (ii) conversion of forests, savanna, grasslands and wetlands for agriculture, managed grass and tree farming, mining, dam construction, human settlement, industrialization etc., (iii) floods, (iv) droughts, (v) fire, (vi) air pollution, (vii) water pollution and acid rain, (viii) tourism, and (ix) botanical and zoological collections.

Among the overexploited species, some medicinal plants which were very abundant only a few decades ago are endangered now from their natural habitats. Examples are *Rauwolffia* sp., *Aconitum* sp. and *Podophyllum* sp. Non-medicinal plants are overexploited for other uses or commercial purposes. Botanical tours often lead to mass collection of rare plants from their small area of occurrence like *Psilotum* from Panchmarhi and *Nepenthes khasiyana* from khasi hills, in India are now endangered.

Some of the important plant species which are gravely 'endangered' are (i) *Rafflesia arnoldii* a root parasite with the largest flower in plant kingdom, (2) *Ramosmania heterophylla* (Rubiaceae) of which only one tree plant is known to be surviving in Rodrigues in an island in South-Indian Ocean. (3) *Diospyros hemiteles* is also represented by one male plant only in Mauritius. (4) *Saintpaulia ionantha* the African Violet of Tanzania, (5) *Paphiopedilum* spp. or Indian slipper orchids particularly *P. druryi, P. hirsutissimum, P. insignae, P. villosum,* (6) *Punica protopunica* (wild pomegranate), (7) *Senecio, handrosomus,* (8) *Ariocarpus agavoides* - a Mexican endangered cactus, (9) *Dicliptera dodsonii* a vine in Ecuador of which only one alive plant is known, (10) *Neoreitchia storkii* is a monotypic palm in Fiji, (11) *Microcycas calacoma* is a relict cycad in Cuba. Other less endangered plants are : *Psilotum nudum, Rhododendron nivalae, Rhododendron santapaui, Rhus hookeri, Saussurea bracteata, S. lappa, Camellia caduca, Colchicum luteum, Cyathea gigantea, Cyperipedium elegans, Osmunda regalis,* etc.

Environmental Pollution

Pollution of our environment has become a serious problem and the study of this aspect has assumed unusual significance in ecology. The air, water, light and temperature conditions, soils, mineral resources, industrial processes, dwelling conditions, crops, grasslands, forest and other useful plant communities, cattle, poultry, fishes and other animals etc. constitute the human environment. Pollution means the direct or indirect changes (usually but not always brought about by man) in one or more components of the ecosystem which are harmful to the system or at least undesirable for man. There are several kinds of pollution and the causes are also many. In any ecosystem the intensity of factors periodically change within a certain range and the organisms of the system are adjusted to such daily or seasonal changes. There are natural mechanisms of homeostasis through which the ecosystem self regulates and maintains the ecological balance. But if the forces of changes are too severe, the ecological balance gets disrupted leading to large scale ecological imbalances which may affect the organisms and the man. In the process of exploiting natural resources man did not understand the significance of ecological balances. So instead of adjusting his actions according to ecological principles, present day man through his technological activities has caused deterioration in his environment. The pace of environmental degradation is so rapid that the very survival of man may be at stake in the not too distant a future. Different aspects of pollution are described by Ambasht & Ambasht (2005) in their book *Environment and Pollution : An Ecological-Approach.*

Due to a rapid rate of increase in the human population the space on earth available to each man is getting smaller. The needs of modern man are also increasing both in quantity and complexity whereas the storehouse of natural resources are limited. In the process of the manufacture of finished goods some materials are invariably thrown out as wastes. The amount of wastes that are dumped in soil, water and air have reached such proportions that due to the limitation of spaces the waste dumping space of one section of population is the living space of another. This is to say that the living and dumping spaces have come to overlap. Thus pollution of soil, water and air is now found everywhere. Generally, the more industrially advanced a country is, the more polluted it is likely to be. This, however, is not always true for it is possible to make industrial developments without causing pollution, if some part of the profit is spent on antipollution measures and researches towards this. In fact many governments, have begun to insist on antipollution measures in their industries. Industrially less developed countries with a dense population are also getting polluted by the unhygienic way of living. The dwellings may have small, stuffy and poorly ventilated rooms; their surroundings have poor sanitation, full of organic wastes of garbage and filth and their drinking waters may not be clean specially in certain seasons. Therefore, it is of utmost importance that the poor people living in village be made to understand the value of maintaining a clean environment and the ill consequences of polluted surroundings.

We can understand the causes of pollutions, their extent and mechanisms to reduce or control them if we take up their different kinds separately as under :

AIR POLLUTION

The dry air contains 78.084% nitrogen, 20.9467% oxygen, 0.0314% carbon dioxide, 0.0018% neon, 0.0005% helium, 0.0002% methane, 0.0001% krypton, and in smaller traces, hydrogen, xenon, ozone, ammonia and carbon monoxide. Before the advent of life, however, the atmosphere consisted of 90% carbon dioxide, 1.9% nitrogen, and traces of oxygen and prevailing temperature might have been around 300°C. In this admixture of gases constituting the natural atmosphere there are a large number of other substances that keep on exchanging between the atmosphere and other natural and man-made sources.

1. Gaseous

In air the pollutants are like (i) CO, CO_2, (ii) hydrocarbons i.e. carbon and hydrogen containing compounds, oxygenated hydrocarbons emitted through combustion of fuel, (iii) sulphur compounds like SO_2 and SO_3 emitted by the burning of sulphur containing coal and other fuel and secondarily formed H_2S and sulphuric acid, (iv) nitrogen oxides like N_2O, NO, NO_2 and NO_3 and many malodourous amines, (v) ozone which is useful at higher level of atmosphere as a protective shield but is harmful pollutant at lower levels of atmosphere and (vi) hydrogen fluoride produced from industries like of aluminium.

Particulate matter which are small to very small in size that keep floating in the air. Particles around one micron size present in air in form of solid or liquid are generally called as *aerosol*. Particles smaller than *aerosol* assume the appearance of smokes and fumes while larger than the aerosol are dusts if the particles are solid and *mist* if they are liquid droplets. Other particulate pollutants are either living like bacteria, pollen grain, fungal and other spores, or non-living like flyash from burning coal and wood, chemical, compounds from metallurgical industries and lead combined with chlorine and bromine, etc.

We can take up the important pollutant gases one by one as follows: -

Carbon monoxide

Carbon monoxide gas is formed by incomplete combustion of fossil fuel like coal and petroleum, or other organic matter. Automobiles are the commonest source of carbon monoxide pollution in cities. Other common sources are oil refineries, metallurgical operations and other internal combustion engines. It is a major pollutant for man and other animals as it combines with blood haemoglobin more than 200 times faster than oxygen does. Therefore, in rooms, with burning coal even while the oxygen content may not be so deficient, carbon monoxide produced causes suffocation. It is estimated that this gas accounts for about half of the total air pollutants added to atmosphere. There are reports that in USA alone more than 65 million tons of CO are emitted annually, and in the city of Kolkata 450 tons are discharged every day. Smith (1984) has quoted Seiler (1974) the annual global input of this toxic gas is 6×10^{14} grams or 600000000 tons. Most of this emission is directly from anthropogenic or man operated sources from industrialised countries like Japan, Korea, USSR, UK, France, Germany, USA and Canada. Carbon monoxide is not toxic to plants as such and in fact green plants through its fixation (or conversion to CO_2) estimated to play an important role in significantly reducing its content. Vegetation and soil are regarded as a natural sink for carbon monoxide pollution.

Carbon dioxide

Carbon dioxide is a resource at its natural level of 0.03% as it is the raw material of photosynthetic reaction by which food is prepared by green plants and on this food ultimately all organisms including man depend. Carbon monoxide is a product of respiration and all living cells respire. The CO_2 so produced is naturally balanced by its photosynthetic utilization. However, two very important human activities : (i) the burning of fossil fuels, (ii) destruction of forests have resulted in an imbalance of CO_2 cycling. Its content in air is rising due to *higher release* and lower utilization and causing '*Green House Effect*' of warming up of air. Greenhouse gases in the atmosphere have the ability to allow solar radiations to pass down and reach ground surface, but on reradiation, the out going heat waves or infra-red radiations are held up or trapped by them. Greenhouse effect at its natural level is very essential for life to exist on this earth, but its increase, as is actually taking place in recent decades, is feared to cause adverse global climatic changes. C.D. Kellings has accurately measured CO_2 rise at Mauna Loa Observatory in Hawaii from 315 ppm or 0.0315% in 1958 to 345 ppm 0.0345% in 1985 with a narrow seasonal oscillation every year. From the isotopic carbon ratios, it is reported that the CO_2 content before the year 1850 was approximately 270 ppm or 0.027%. In nature, the rise in CO_2 is necessarily accompanied with decrease in O_2 content which has harmful effects on human health. Thus CO_2 increase has manifold adverse effects particularly on oxygen deficiency and green house effect on global weather and climate.

OTHER GREENHOUSE GASES

Besides carbon dioxide and atmospheric moisture, the other green house gases are Chlorofluorocarbons, methane, nitrous oxides, ozone and some other trace gases.

Chlorofluorocarbons or CFCs are chemicals synthesised by man for use in several kinds of industries including refrigeration and airconditioning and find their way into the atmosphere. CFC 11 or $CFCl_3$ and $CFCl_2$ or CF_2Cl_2 (trade name Feron 12 or F12) are commonest of the CFC. The US National Aeronautics and Space Administration (1986) have estimated that about 5% increase per year is taking place for greenhouse gases in the atmosphere. Both CFC11 and CFC12 are of much economic importance and their production and use have very rapidly risen in the past two decades. These occur in aerosol and nonaerosol forms. The aerosol form is suspected to damage the stratospheric ozone layer, and therefore environmentally alert and developed nations of the West have begun putting curb on the production of aerosol forms of CFC. CFCs are highly stable and non-destructive. They slowly reach the upper zone of atmosphere where ultraviolet radiation decompose them. In the present day computer based technology much air conditioning and refrigeration are needed and they require use of CFC12.

Their emission rate has to be curbed drastically even to maintain its concentration at the existing level in the atmosphere. It is estimated that if the CFC emission is totally stopped now, it will take about 100 years for its breakdown or in stopping its migration from troposphere to stratosphere, but there is little hope for reduction in its emission. The chlorine molecules combine with ozone to produce oxygen and ClO and ClO again combines with O_3 to produce $2O_2$ and Cl. Thus the chlorine of CFC acts as a catalyst to break ozone into oxygen.

Methane (CH_4) is another greenhouse gas which is naturally produced. In recent years its production and release into the atmosphere appears to have increased due to human influence. The main sources are biological processes such as enteric fermentation in cattle, sheep and other animals, anaerobic situations in wetlands and rice fields and burning of biomass and fossil fuel by man. Rasmussen and Khalil (1984) have worked on the methane content in past and present environment. Methane gas increases stratospheric water vapour on oxidation. The rise in water vapour is really more important source of greenhouse effect than the direct effect of methane gas. Hoffman and Wells (1987) estimate an input of 400 to 765 $\times$ 10^{12} g/yr of methane at the present times. It is difficult to project future picture for methane and its impact on greenhouse effect or warming up of the global environment. However, a rough estimate is that it will increase by about 1% per year from the existing level.

Nitrous oxide N_2O emitted from the burning of fossil fuel is another greenhouse gas. Its content is also reported to be rising all over the world. Correct rate of increase is about 0.2% to 0.3% per year. Some of the N_2O reaching into the air is from the breakdown (or denitrification processes) of manures and fertilizers. Hoffman and Wells (1987) consider that fertilizers and other natural processes account for 70-80% and fossil fuel combustion for 20-30% of the N_2O in the atmosphere. Once N_2O reaches in the air it does not change its form for many years. Only on reaching the upper layers it slowly reacts with atomic oxygen.

One fact about the greenhouse effect is that is a slow and imperceptible but over several years its impact would become alarming and irreversible. We do not have the technology to destroy or reduce the accumulating greenhouse gases. They break down very slowly by natural processes. All nations, irrespective of the fact that they emit these gases or not, will be equally affected. So the mere fact that some advanced countries are reducing emission of CFCs (to protect ozone layer) will not be sufficient. The environmental protection efforts have to be global without consideration for national boundaries. Unlike other pollution problems, the greenhouse warming effect cannot be reversed in a few years time by applying antipollution methods. The effect is long range lasting over to centuries.

Effects on forests

Warming up would lead to significant changes in the forest species. Change in natural growing condition may require modification in forest management plans with respect to timber trees particularly in tree breeding, and other forest technologies. CLIMAP (Climate : Long range investigation, Mapping and Prediction), COHMAP (Cooperative Holocene Mapping Project) and Atmospheric General Circulation Models (GCSs) and are being presently refined to yield reliable predictable picture for future vegetational changes due to warming. The prairies in USA may get converted into forest. For USA the general predictions are that Ponderosa Pine will increase in California and Oregon, but in Canada and Mexico it may decrease. Douglas-fir may not be much affected. Loblolly Pine distribution may shrink in Western USA but may increase in the East in New Jersey and Pennsylvania.

Effects on crops

Greenhouse effects on crop production with the rise of 1.5 to 4.5°C will be of varied types in different regions. Wheat and maize may suffer from moisture stress due to increased evapotranspiration. It is expected that CO_2 doubling will bring about a cascade of changes in (i) temperature regimes, (ii) water conditions and (iii) pest infestations. There may be many more unforeseen direct and indirect effects. So heat and pest tolerant varieties of crops will have to be evolved. There is some chance that soil moisture in crop fields may decrease because of rapid evaporational losses. So drought tolerant varieties will also have to be evolved. Reallocation of agricultural lands may also be required on account of global change in temperature and possibly rainfall patterns. Cycling of nutrient elements may be accelerated and this may upset the fertility status of soil. Warmer temperature may increase biological nitrogen fixation but due to reduced moisture the decomposition of organic matter may be retarded. A number of studies with GCM scenarios and with simulated CO_2 concentration levels in phytotrons and controlled chembers have been done during the eighties in USA and Europe, but there is a need of more comprehensive studies for reliable predictions.

Effects on water factor

Hydrological cycle is powered by sun through temperature changes, evaporation, wind movement and precipitation. Greenhouse gases are certainly going to alter the temperature regimes and therefore the hydrological cycles. It will also effect the infiltration and storage in soil pool. The future scenario on water resource is certainly going to have a drier summer and changed runoff patterns. In temperate belts winter runoff is likely to increase. Mather and Feddema (1986) have calculated changes in annual precipitation and evapotranspiration, water deficits and soil water surpluses using large scale water balance techniques. They conclude that some increase in precipitation and much more in evaporation leading to an annual water deficit in crop fields may take place.

Oxides of sulphur are the common pollutant being formed by sulphur containing coal and petroleum. Since the quantity of burning of such fossil fuels is fantastic. The quantity of sulphur dioxide emitted into the atmosphere is in the order of more than a hundred thousand tons each day. This gas is also quite harmful causing diseases of eye, throat and lungs. Smoke occurring in combination with fog is called *smog*. Under such a condition SO_2 reacts with O_2 and SO_3 is formed which in turn combines with water to form H_2SO_4. In such a case sulphur dioxide would be called as *primary pollutant* since this gas itself is a pollutant and H_2SO_3 and H_2SO_4 as *secondary pollutants*. Sulphuric acid corrodes many of the building and marble structures. The Taj Mahal of Agra is likely to be affected by the gases that would be discharged from the oil refinery in a nearby township of Mathura. Of course, this fact has been realised and antipollution measures are being taken.

Rao and LeBlanc (1966) and LeBlanc and Rao (1975) have done researches on air pollution due to sulphur dioxide and found that it causes bleaching of the leaf pigment. Firstly, this gas on entry into the plant leaf through stomatal pores combines with water to form sulphurous acid (H_2SO_3) in the intercellular spaces and this acid in turn damages the membrane activity and converts chlorophyll *a* into phaeophytin *a*. Thus, photosynthesis is affected. Leaves show necrosis on the margins and between the veins. However, the sensitivity to sulphur dioxide toxicity is widely different in different species. Lichens and some Bryophytes are found by Rao and LeBlanc to be extremely sensitive and could be regarded as indicators of SO_2 pollution. On the basis of the extent of damage to lichens the level of pollution could be guessed and infact they have given *Indices of Atmospheric purity* (IAP). On this basis Rao has observed that SO_2 damages the young leaves and flowers of mango trees around brick kilns and other places where coal is burnt in large quantities.

Pal (1974) has found that from an aluminium factory the gaseous discharge caused fluoride pollution of air which resulted in damage of the natural vegetation. Cattle feeding on these affected plants developed swollen knee bones, tooth decay and ill health. Lal and Ambasht (1981a and b) have found that flouride pollution results into the reduction of chlorophyll contents, biomass and size of leaves in an economically useful forest tree *Diospyros melanoxylon* in Mirzapur

District of Uttar Pradesh. The leaves of this plant is used as tobacco wrapper in the manufacture of an Indian kind of cigarette called *Biri*. This inexpensive smoke is most commonly used by millions of poor people. Thus, while fluoride gas reduces the leaf output, it may be reaching human lungs through Biri prepared from affected leaves.

In his studies on air pollution around coal unloading and loading stations, Rao (1971) has measured the quantity of coal dust that gets deposited on plants and dust traps. The dust settling from polluted atmosphere is a common experience in industrial cities like Kanpur. This causes damage to plants and diseases in man.

Large quantities of particles enter into the atmosphere through technological activities. Hydrocarbons, which are a variety of chemicals, are emitted in huge quantity by the burning of petrol in the automobiles. These are quite harmful and frequently combine with NO_2 and produce *photochemical smog*. N_2 causes toxic effects on human beings especially respiratory diseases. Some nitrogenous compounds produce dirty smell. Ozone is another gas that causes pollution (although it is often mistaken as air purifier) when occurring in concentration higher than 15 to 20 ppm. Lead which is mixed in petrol as an antiknock agent comes out from the automobiles and very badly affects human health. Another toxic pollutant is PAN (Peroxy acetyl nitrate), which is a bye product of automobile exhausts, which is reported to supress photosynthesis.

Sharma (1981) in his General Presidential address of the Indian Science Congress on the impact of development of science and technology on environment has given that the pollutant load entering in the air of Calcutta and Howrah city of India is as much as 1299 tons per day. Datta (1984) has quoted the National Environmental Engineering Research Institute (NEERI, Nagpur) findings that the gaseous pollutants entering Kolkata (900 tons) and Howrah (405 tons) twin cities is 1305 tons per day of which 560 tons are suspended particles, 450 tons of carbon monoxide, 123 tons of sulphur dioxide, 102 tons of hydrocarbons and 70 tons of oxides of nitrogen per day. The principal sources reported are industries (46% or 600 tons), transportation (28% or 360 tons) and thermal plants (15% of 195 tons). It is further reported that dust settling on the twin city is about 370 tons per day. Datta (1984) has compared the dust load of Kolkata atmosphere of 527 $\mu g/m^3$ with other metropolitan cities like Chicago with

280, Tokyo with 261, London with 221 and San Francisco with 104 $\mu g/m^3$, and Delhi with 700 $\mu g/m^3$. He has reported occurrence of '*acidic mist*' and inversions in different parts of Kolkata in certain months.

Air pollution control

There are various devices now developed to check particulate materials and harmful gases from being discharged into the atmosphere. If the gaseous waste is passed through a porous chamber, the particles are held up. The gases could then be passed through water and soluble pollutants are absorbed. *Spray Collector* or *scrubber* are also used in which the gas passes through a tower fitted with sprayer. The liquid flows down and the filtered gas let out from the top of the tower. *Bag filters, cyclone collectors*, and *Industrial Electrostatic Precipitators* are other devices to filter industrial fumes. Further as precautionary method, too much use of fossil fuels like coal and petroleum should be avoided as far as possible and instead hydroelectric and solar power should be harnessed more and more.

OZONE AND UV-B PROBLEMS

Ozone or O_3 gas is found in traces in air near the ground surface and concentrated in a layer at varying heights between 16 to 30 km at different latitudes. Ozone gas, although formed by three atoms of oxygen, is a toxic gas and adversely affects crop plants like *Glycine max, Triticum aestivum, Zea mays* and *Gossypium hirsutum*. Kohut et al. (1988) have reported leaf injury upto 21% in *Phaseolus vulgaris* on experimental exposure of 0.12 ppm O_3 for four hours every day (11 a.m. to 3 p.m.). Mulchi et al. (1988) have also recorded reduction in plant growth in 12 cultivars of soybean (*Glycine max*) due to ozone stress. Pell et al. (1988) found about 50% reduction in photosynthetic activity in potato and formation of poor grade smaller potato tubers. O_3 injures first the plasma membrane and then reacts with proteins and fatty acids. If the ozone stress is withdrawn, most plants, recover, but the exposure effect persists for long period. Ozone is a major pollutant and according to one estimate it is responsible for about three billion dollars loss in crop productivity in USA.

Ozone mantle or shield or umbrella as the layer is called around

16-40 km altitude in the atmosphere is extremely important as it absorbs the ultraviolet (UV) radiations. UV radiations are of short wavelengths between 200-400 nm, 200-280 nm is called UV-C; 280-320 nm UV-B and 320-400 nm as UV-A. Out of the three categories, UV-B is most dangerous to plants and animals.

Ozone layer is reported to be getting thinner due to its slow breakdown by a number of chemical radicals produced by anthropogenic activities which slowly move upwards and reach the ozone shield. The most serious culprits are chlorofluoro-carbons (CFCs), methane, nitrogen oxide and hydroxyl ions. Ozone layer has a certain natural concentration in stratosphere and is maintained at a balanced state due to its slow formation from oxygen and its equally slow breakdown into oxygen. This balanced state is getting depleted due to artificially added above named chemicals. Ozone levels in the stratosphere is measured in Dobson unit. The discovery of a steadily declining Dobson levels in Antarctics atmosphere every successive spring season (September and October) is alarming. This depletion has caused formation *'ozone hole'* of larger and larger sizes as reported for 1978 to 1993 as the ozone level fell from 225 Dobson units in 1978 to 109 units in 1987 (Watson, 1988). Measurements made by NOAA-11 satellite and TOMS instruments in Russion Meteor-3 satellite confirm further decline in ozone layer thickness (Gleason et al., 1993). It is predicted by Cicerone (1987) that a 10% reduction in ozone results in 20% increase in UV penetration. UV-A radiations are not so harmful while UV-B is really the problematic ultraviolet rays.

The mechanism of ozone depletion is mostly catalytic in nature CFCs have chloride and in form of chlorine it reaches ozone layer. CFC or Freon gas is used in airconditioning and refrigeration. It escapes in air in small quantities. Cl combines with O_3 to form ClO and O_2. ClO combines with O_3 to produce Cl and $2O_2$. Thus the same chlorine molecule breaks ozone molecules repeatedly and it reforms itself to act afresh. So each Cl molecule breaks hundreds and thousands of ozone molecules before itself getting converted into HCl on combining with water.

Similar catalytic effects are performed by OH, CH_4 and NO molecules. OH combines with O_3 to form HO_2 and O_2; the HO_2 reacts with O to form OH and O_2. Methane first combines with O to form

OH and CH_3. The OH then acts as above to break ozone molecules. NO also acts on O_3 to form $NO_2 + O_2$ and NO_2 combined with O to form NO and O_2.

Impact of UV-B and tropospheric ozone on plants

Results of researches on this aspect has been reviewed by Teramura (1983). It adversely affects photosynthesis, respiration, leaf expansion, discolouration of leaves in form of chlorosis, bronzing, glazing; germination of pollen grains, seedling growth and primary productivity. Ambasht, N.K. and Agrawal (1994), Ambasht (1988) and Ambasht and Ambasht (2005) have made excellent reviews of the UV-B impacts of agricultural crops. Krupa and Kickert (1989) have categorised plants into (1) UV-B *sensitive* in which primary productivity is clearly reduced, *tolerant* which bear it and maintain productivity at near normal level, and '*positive*' which show stimulatory or increasing effects. Several of the vegetables like cucumber, squash, okra, pumpkin, and watermelon are quite sensitive, some peanuts and many tuber crops are tolerant. But, by and large, the global agriculture will be adversely affected.

Depletion of good stratospheric ozone umbrella is responsible for generation of bad tropospheric (at ground level) ozone through the enhanced UV-B. Ozone at even 0.1 ppm level on 7 hours day^{-1} exposure would reduce crop production by 50%. The twin stresses of increased UV-B and ozone on plants, animals and human beings are creating serious adverse effects. Through Vienna convention, 184 countries of the world have taken extensive measures to protect the ozone layer. Tabazadeh et al. (2002) assign Polar Stratospheric Clouds (PSC) in September/October over Antarctica and near absence on Arctics as the cause of fast ozone hole formation in the Southern Pole areas. Adams et al. (1998) estimate that due to tropospheric ozone the total crop yield loss is of about 3 billion dollars. The damaging effect is further increased in presence of SO_2 and NO_x air pollution. Ozone enters leaf through stomata and generates free radicals whereas sulphydrils, cysteine and methionine serve as scavengers of free radicals. Ambasht and Agarwal (2003) found decline in biomass and seed yield of *Glycine max* on UV-B + O_3 treated condition by about 22%. In wheat plant, enhanced UV-B and O_3 (7.1 kJ m^{-2} UV-B and 0.07

μmol mol^{-1} ozone) reduced the biomass, yield, photosynthetic rate, chlorophyll, carotenoids and ascorbic acid content and catalase activity, but phenol content and peroxidase activity increased (Ambasht and Agarwal, 2003-04).

In rice plants on UV-B exposure the net production fell down by 13-17% (Ambasht and Agarwal, 1997). Stomatal conductance and chlorophyll a + b were reduced. Musil and Wand (1999) have made extensive field studies of several thousands km^2 of fynbos and karoo ecosystems in South Africa. On the basis of ecomodelling, change in UV-B enhancement for previous twenty and future fifty years, there is a reduction in the overall biodiversity of both mono- and dicotyledons at ecosystem level. Ambasht and Ambasht (2005) have comprehensively reviewed current literature on the UV-B and ozone impacts on plants.

WATER POLLUTION

Water resources of the earth are easily divisible into marine and fresh water. We can describe them separately.

Pollution of marine ecosystems

The vastness of expanse and depth of oceans with such a huge stock of water easily makes one to believe that any quantity of waste discharged here and there in sea gets thoroughly diluted and may not cause any serious pollution. But it is unfortunately not true. The main sources of sea water pollution is by oil. Sytnik (1985) in his edited book *Living in the Environment* has quoted a Russian estimate (1976) that the total influx of oil in the world's oceans is 4897 × 10^3 tons per year almost equally thrown out by ocean based sources (2407 × 10^3 tons) like tankers and transport ships and lands based sources (2490 × 10^3 tons) like automobiles, industries, oil refining that finally reach the sea. Accident of oil tanker (967,000 t) and transport ships (250,000 t) are major sources of oil spilling into sea water accounting for 25% of the total oil pollution input. Among the world oil polluted marine belts are the Baltic Sea and North Mediterranean Sea.

Besides oil and other petroleum products, there are numerous other kinds of toxic chemicals and garbage that are discharged into the sea and at regional levels where they become highly toxic. Further, through ocean currents, millions of tons of wastes from aluminium production plants and titanium dioxide producing chemical industries, domestic wastes, mining operations, and from power houses are annually reaching to the North Sea in Europe from Germany, UK, Netherlands etc. The Mediterranean waters are about six times more polluted than the world average for oceans. This has resulted into increased population of pathogenic bacteria; and at certain places, coastal people are suffering on this account. The problem of mercury pollutants has reached grave proportions in some European and Japanese sea water. At times fish samples have shown high mercury content to the extent as to cause serious disorders on consumption over long period and even have resulted into mass scale human deaths such as the *Minamata disease episode* in Japan. Mediterranean sea receives about one hundred tons of mercury besides 21,000 tons of zinc, 3,800 tons of tin, 2,400 tons of cadmium, etc. (Sytnik, 1985).

Large scale measures have been undertaken by various national and international pollution control agencies. The Environmental Protection Agency (EPA) of USA has evolved several control measures and set control standards. There are measures to recover the oil spills for recycling their use and controlling sea pollution. Application of technology for recovery of oil spills is the latest step to reduce pollution. There are cyclone recuperator and Nenufar recuperator which are capable of separating films of oil on sea surface.

There are exclusive bodies on oil pollution control measures. The new standards prescribed and enforced for tankers has reduced the oil pollution by accidents but the one due to washing of tankers and other ships has not been effectively controlled. There is indeed a most massive attempt involving 15,000 million dollars to prevent pollution of the Mediterranean sea from land based sources (accounting 85% pollution, Sytnik, 1985). In this programme the most-hazardous pollutants such as (i) organic substance containing halogen (chloride and fluoride), phosphorus, tin, mercury and cadmium compounds, (ii) nondegradable synthetic chemicals, (iii) radioactive isotopes and (iv) carcinogenous or similarly harmful materials are put in the "*black list*" and strictly prohibited from discharging into the sea. The next lists is called as "*grey list*" which covers somewhat less harmful materials like arsenic, zinc, antimony and their compounds, cyanides, organosilicon compounds, crude oil hydrocarbons, froth or foam forming detergents or surfactants. The release of these substances is not totally prohibited but very rigidly allowed in small quantities. The third list is called *"white list"* which includes solid wastes that enter the sea with the help of wind.

Pollution of freshwater ecosystem

Freshwater storages are *lentic* bodies like lakes, dams, ponds and pools, *lotic* like streams, rivers, irrigation canals, and channels on the surface. Huge quantity lies in form of ground water and in solid state in form of ice or snow. Of these, the surface storages of freshwater are most affected by pollutants. The physical, chemical and biological characteristics are easily affected by pollution and we can take them separately.

Colour of water easily gets affected. Organic wastes on decomposition produce humic and fulvic acids which impart yellowish brown colour. Industrial effluents give different shades of colour depending upon their chemical nature—some are dark black due to coal and oil rich slurry. *Odour* of water also indicates the kind and extent of pollution. Putrefying material and hydrogen sulphide give repulsive and offensive smell to some polluted pond waters. *Turbidity* is also imparted to dirty water due to suspended solids and excessive planktonic life forms. *Temperature* change, particularly the heating effects from power house water discharges cause thermal pollution. Higher temperature drives out the dissolved oxygen and fishes die due to oxygen deficiency. Migrating fishes passing through such a thermal block in a river also suffer heavily. Plant and animal life is drastically affected by the change of temperature. *Total solids* (T.S.) which include dissolved solids (DS) and suspended solids (SS) are also indicator of pollution. Some chemical pollutants deflocculate i.e. do not allow electrically charged clay and humus particles to aggregate and disperse them to the colloidal state. So they remain in suspended state and do not settle. Suspended material cut down light penetration, retard growth of macrophytes and phytoplankton and reduce oxygenation process of photosynthesis. *Alkalinity, acidity* and change of pH in natural waters are easily caused due to effluents entering to a water body and affect the aquatic biota. Dissolved oxygen (DO) is a very important aspect in aquatic pollution, DO is necessary for the respiratory needs of aquatic life, plants and animals alike. In natural waters some oxygen always remains dissolved and to quantity depends on temperature, turbulence, vegetation and photosynthesis. In sewage polluted water or other kind of organic wastes, decomposer microorganisms multiply in large numbers in presence of their food. They utilize all the available dissolved oxygen for their own respiration and cause oxygen deficiency for other aquatic plants and fishes. The oxygen demand of decomposers is called *Biological* or *Biochemical Oxygen Demand* or *BOD* and it is measured by first determinating the dissolved oxygen content of a polluted water sample and then after incubating at 25°C for 5 days in dark and again determining the DO to find out the quantity of DO used in known volume and known time period. BOD gives a good idea about the level of organic waste pollution. Similarly, the oxygen required in chemically oxidizing the organic matter in a polluted water (by chemical oxidants) is known as Chemical Oxygen Demand or COD. The carbon dioxide liberated by the plants, animals and chemical oxidants partly gets dissolved in free form and as bicarbonates.

Effluents and runoff water bring down a number of chemicals in the receiving water of rivers and lakes. Heavy metals, other metals, sulphates, nitrates, chlorides, phosphates, carbonates, ammonia, pesticides, phenols, detergents are the common chemical pollutants. There are a number of pathogenic microorganisms which cause water borne disease in man.

Among metals, the severe pollutants are mercury, lead, cadmium, arsenic, chromium, copper, zinc, sodium, manganese, iron, potassium, calcium and tin. Mercury, although a scarce metal, yet because of its widespread industrial use finds its way (through factory effluents) into river and coastal sea water. Methyl mercury is formed by the bacterial activity on mercury and this is a soluble in water and highly toxic to man. The Japanese *Minamata* disease, in which large number of coastal town people were killed, was due to eating fishes affected by methyl mercury. Mercury pollution is quite common around chlor-alkali industries, smelters and manufacturing units of batteries, thermometers, fluorescent lamps and caustic soda. *Lead* is highly poisonous and in soluble and colloidal state passes through filters and contaminates drinking water. Sources of lead pollution are manufacturing units of batteries, PVC plastics and paints. It is also mixed in petrol, and lead laden fumes settle on ground surface and get washed by rain to rivers and lakes. *Cadmium* in association with zinc are discharged from electroplating, chemical and metallurgical industries. Cadmium pollution causes hypertension, cancer of lungs and liver. The painful disease called *itai-itai* of bone in Japan is due to eating rice which is grown in fields irrigated by zinc smelter waste water. *Chromium* used in dyeing carpet industry is found to affect crops irrigated by such effluents (Mishra and Ambasht, 1988). *Copper sulphate* is extensively used as fungicide. Copper oxide is used in paints. Copper smelters are also a big source of copper pollution. The metal pollutants also show the common phenomenon of *biological magnification*. The concentration of the chemicals increase as it passes from water to the plant body. It further increases ten to hundred times or more as it passes from plants to herbivore fishes and again increases to carnivore fishes and then to birds and man. While the pollutant discharged may be at very low concentration or highly diluted level, quite safe for direct human consumption, but in course of years, the accumulated concentration in fish body or in course of accumulation in human body the effect may be severe and dangerous.

Phosphate is common and essential radical needed for biological energy storage and release processes through phosphorylation (ADP + P = ATP). But in higher amount reaching water bodies primarily from soaps and detergents, it causes excessive nutrient enrichment. The phenomenon is called *eutrophication*. This leads to explosive growth of plants (macrophytes and phytoplankton). This chokes the water body and hampers free flow of water and boats. On their death, huge load of dead organic matters causes oxygen deficiency and high BOD. This kills the fish life in ponds and water begins to stink. *Chloride* ions in low doses are not harmful but when it increases beyond 250 mg/l a salty taste is felt. In marine water, sodium chloride the common salt is about 35 g per litre or 3.5%. In drinking water supply, some chlorine is added to kill harmful bacteria. *Sulphate* and *Sulphide* are sometimes found in natural waters and sulphur springs. Sulphate causes hardness of water. Sulphate is received from SO_2 containing polluted air in form of acid rain. In *Nymphaea* sp leaf, on lower surface, there is a special tissue called hydropoten which selectively accumulate sulphates. *Nitrogen* in form of *nitrates, nitrites* and *ammonium* are found in water. In small quantity they are essential for a good growth of aquatic biota. But high concentration of nitrate in drinking water is decidedly harmful and cause a disease called *Methemoglobinemia* in which the skin turns blue. Nitrate pollution of water is going to become a big health hazard since we add nitrate fertilizers to our crops much more than what is consumed by crop plants. The excess run's off to rivers or infiltrates to ground water. *"Nitrate time bomb"* i.e. nitrate pollution may become most hazardous in years to come.

Pesticides are (i) organic phosphate compounds, (ii) chlorinated hydrocarbons, (iii) arsenic compounds. *Parathion* and *malathion* are organic phosphate compounds which affect nerve functions. The latter is safer since it rapidly breaks down into less harmful chemicals. *DDT, dieldrin* and *aldrin* are chlorinated hydrocarbons. DDT is the well known example of biological magnification. Through food chain it reaches to several non-target organisms. Indians are among the worst DDT affected people in the world. It gets accumulated in the fat tissues. DDT is not easily biodegradable and may remain intact even after 20-25 years of its application. DDT pollution has drawn a world-wide attention and its production and use has been prohibited by many countries. However advanced countries make use of pesticides much more than in India, but we use them most unscientifically. *Detergents* have also reached a high pollution load level in certain regions. The *alkyl benzene sulphonate* or ABS type of detergents do not degrade easily and remain on water surface in form of foam and froth. *Surfactants* are surface active agents that form a thin film over water surface and hamper normal air to water exchanges of gases and vapour molecules.

Pathogenic microorganisms are serious pollutants and cause epidemic diseases of cholera and hepatitis, etc. Common disease causing organisms in water are *Klebsiella* (a faecal coliform bacteria group), *Streptococcus* (Enterococcus), *Escherichia coli, Shigella dysentriae, Salmonella typhi*, etc. Besides bacteria, there are *enteric, polio* and *hepatitis viruses*, pathogenic fungi like *Exophiala* sp., *Trichosporium* sp., *Aspergillus fumigatus* causing pulmonary diseases. The sources of such pollution are human and domestic wastes.

Industrial wastes disposed in lakes and rivers are other important sources of pollution. Ambasht and Tripathi (1978) have studied the pollution level in wheat crops caused by effluents from fertilizer factory near Varanasi. They have shown that both the quantity and quality of agricultural crops are reduced and the soil properties are also adversely affected. Sodium, copper, chromium, cadmium and mercury are some of the more common effluent pollutants discharged from various kinds of factories. Metallic pollutants are very harmful for human health and therefore such affected waters need purification before supply by water works.

Ambasht (1989c) has briefly reviewed the water conservation aspects in respect to different types of pollution. Datta (1984) recorded an extremely bad situation about water pollution of Kolkata. About 120 million gallons of domestic and 50 million gallons of industrial water containing organic and inorganic substances flow out into Hooghly river. Among, organic wastes are raw cattle wastes and untreated domestic sewage and among industrial wastes are : sodium cyanide and barium salts from heat treatment furnaces, Cadmium, Chromium, Copper, Manganese, Nickel and Zinc salts from electroplating units, mobil oil from vehicle repair workshops, etc. The water quantity at water works intake points at Kamarhati, Serampore, Howrah and Garden Reach in Kolkata is reported to be highly contaminated with faecal-based organisms, unfit for normal chlorination and purification process.

Purification and waste-water treatment is an essential requirement for better human health, reducing pollution load and better management of water resources. Municipal water supply is filtered repeatedly to remove suspended materials, and then chemically treated, like chlorination to kill pathogenic organisms or application of $KMnO_4$ to wells. Sewage treatment plants have three main segments, the first is primary or mechanical treatment unit in which the waste water is passed through screens to remove big solid material, and then through grit chambers and trickling filters in which various dissolved organic compounds are broken down with the help of films of specialised microorganisms. Secondary or biological treatment unit in which there is activated sludge and trickling filters in which use various dissolved organic compounds are broken down with the help of films of specialised microorganisms. Finally the third stage is of advanced or tertiary treatment unit. The dissolved nutrients are eliminated by the use of autotrophs. Flocculation, adsorption and oxidation are common techniques to remove pollutants still persistent after the primary and secondary treatments.

SOLID WASTE POLLUTION

Most of our daily needs are obtained from the market in nicely packed containers made up of tin, polyethylene, plastics and glass. After the use of contents the packing material is usually thrown out as garbage. Many of the old used up things like automobile spares, machines, cycle parts etc. are also thrown out as junks. Some of the junk is readily degradable by the activity of microorganisms in nature and the materials are recycled but some are not easily degraded like metals, plastics, nylons and polyethylenes. The materials that have gone in the manufacture of these materials, therefore, remain locked up and out of the natural cycling of materials unless burnt. Material like glass cannot be burnt but recycled by melting and preparing new articles. Old paper is recycled to prepare fresh paper or other materials. Solid wastes are assuming alarming proportions in affluent countries like those of America and Europe, where labour charges of waste collection is high. In India however all old junks are purchased by professional hawkers from house to house. Old books, newspapers, tins, glass bottles, metallic junks, plastics are all purchased by them to be resold on marginal profits.

The control method of solid wastes could therefore, be : (1) by recycling, (2) by burning the wastes and utilization of heat, (3) composting of the organic wastes for preparation of manures and biogas, etc.

Xenobiotics

In the modern environment, plants are exposed to many foreign chemicals or mixture of chemicals which are biologically active and may cause damage or death of plants. These are collectively called as *xenobiotics*. Spray of insecticides, herbicides, fungicides, rhodenticides and many other biocides are done to selectively eliminate particular biotic component of a man-managed ecosystem. The desired species, like crop plants need to detoxify the chemicals to escape damage. There are only a few biological mechanisms of detoxification, whereas the number of chemicals now applied is increasing. Besides detoxification, on the application of pesticides, plants may produce certain defence proteins. These also provide resistance against pathogens. Xenobiotics that induce production of defence proteins are also put under *immunochemicals*.

From the atmospheric sources also xenobiotics like smog, ozone, acid rains, SO_2, HF are responsible for damaging the normal metabolism of plants. However, at very small concentration, some of them may become a resource for certain nutrients. In our modern drawing rooms, formaldehyde is emanated from furnitures, ply boards, carpets, curtains, foams, etc. and the concentration reach to harmful levels. Certain potted plants kept in drawing rooms or in the vicinity take up the excessive formaldehyde in the air and detoxify the place (phytoremediation). Sanderman (1994) has given a concept of "green liver" for such a function and Schulze et al. (2004) mention that *"Chlorophyton"* plants would detoxify a room of 100 m^3 containing twice the admissible level of formaldehyde in just six hours.

RADIOACTIVE POLLUTION

Ionizing radiations are other very important and harmful pollutants. Nuclear war materials and test explosions are principal sources of radioactive wastes in the atmosphere, soil and water. Already over two thousand detonations must have been done in underground, underocean and in atmosphere and the cumulative radioactivity level is rising particularly in oceans. Ionizing radiations cause mutations, abnormality and lethality in many organisms, including man. Cancer is commonly caused even under low level exposures. Radiation effects, persist for a very long period in the environment. Plutonium, the commonly used radioactive material in nuclear bombs has the half life of 24,360 years and its hazardous effect persists for a much longer period. Therefore, an utmost caution and completely foolproof technology is needed in handling such scientific activities to prevent radioactive pollutions.

Kormondy (1978) has regarded that 'fission byproducts both of nuclear detonation and water cooled atomic power reactors do indeed constitute more of a potential hazard than direct ionizing radiation, because they follow biogeochemical pathways. He has cited many case studies where radioactive isotopes reach human being through food chain. For example, radioactive iodine produced as a byproduct of nuclear detonation in 1954 test of atomic bomb on Bikini (Marshall Islands) which on falling on ground should not have entered plants due to its short half-life of 8 days, fell on plant leaves and reached to human beings through; milk of cattle feeding on such affected plants.

Children of the region had developed thyroid abnormaloities and cancerous growth. Kormondy has also cited reports of J.K. Miettinen of University of Helsinki that in the relatively simple ecosystems of Tundras, the human population of Laplanders have the highest radiation exposures found anywhere in the world because of the fact that a big portion of radioactive fall out of radioactive strontium, cesium, polonium, etc. are absorbed by the *lichen Cladonia* (reindeer moss) which are eaten by reindeers, which in turn are the food of men (Eskimos) of the region. A big proportion of strontium-90 gets trapped into the bone of reindeers and does not pass further to human system.

There are many plants which concentrate radioactive isotopes in their body from the environment. Use of such plants can be made to harvest out the pollutants in order to keep ecosystems clean. The affected harvested plants or animals could be appropriately disposed. The uptake of a harmful radioactive substance by plants can also be reduced by the application of some safe elements. For example, if calcium is present in sufficient quantity in the soil, radioactive strontium 90 is not taken up by plants but in absence of calcium, strontium is absorbed in its place. This could also be a very good line of management of ecosystems affected by radioactive isotopes.

NOISE POLLUTION

Unlike all other pollution causing components of environment, sound is not an element, compound or substance which can accumulate and harm future generation. It is a special kind of wave action usually transmitted by air in form of pressure waves and received by the hearing apparatus present in body of animals. Audible sound is one which our ears can perceive and a very wide range of frequency of sound waves is audible to human ear.

Our ear has an external broad lobed part called external ear or auricle or pinna which converges the sound waves to an inner tube called meatus or auditory canal. The tube ends at tympanic membrane or ear drum which is very thin and tough of about 1 cm diameter. Sound waves create vibration in the membrane and the vibratory motion is transmitted inside by three small ossicles to the cochlea. The mechanical impulse reaching the cochlea are converted into electrical impulses through numerous cilia and get transmitted through auditory nerve to brain. The ear is connected to nasal passage through eustachian tube to adjust the atmospheric pressure to a steady force on the ear drum. There is a semicircular canal which controls the equilibrium or balancing mechanism. Thus high pressure sound can cause damage in a number of ways on our hearing ability, *brain* and *balancing* mechanism. As a result of evolutionary process our ear has built in mechanisms to perceive from very feeble sound to hear the powerful thunder. With aging there is a gradual loss in hearing ability. Persons working in noisy places loose it faster. Thus the hearing ability of an average village dweller is much better than the corresponding age group person living in a noisy city or working in a factory.

Loss of hearing on account of noise pollution is monitored by screening to audiometric tests. For temporary threshold shifts (TTS) measures are taken at the end of day's work and then again at the start of next day work to find out the recovery level. If the recovery is complete, it would mean that the subjected level of noise pollution is not of permanent nature, while if there is some deficiency in the recovery of hearing ability, it would mean that in course of time the man is going to become deaf under the prevailing noise level at his work place.

Hearing aids help partially deaf persons to hear properly, if the auditory nerve and cochlea are not damaged.

Intensity of sound is measured in a scale called decibel or dB scale. It begins from zero which represents the faintest sound which is audible to a normal human ear. In the decibel scale each ten fold increase is represented by 10 db. The faintest audible sound is 0 dB. Ten times more intense sound is 10 dB, 100 times of sound intensity factor is 20 dB (i.e. 10 + 10 dB) and 1000 times more is 30 dB (i.e. 10 + 10 + 10). Likewise 10,000,000 times it is 70 dB, 1,000,000,000,000,000 = 150 dB and 1,000,000,000,000,000,000 = 180 dB. The faintest audible sound of sound intensity factor (SIF) 1 is 0 dB, of resulting of leaves, of 10 sound intensity factor is 10 dB, of very quiet place 100 SIF is 20 dB, of libraries with soft whispers 1000 SIF is 30 dB, of average living room 40 dB, light traffic noise 50 dB, during normal conversations 60 dB, cars, motorcycles, trucks, household machines between 70-80 dB, of jet planes upto 300 m height 100 to 110 dB, of truck horns near it 110-120 dB, jet plane at take off point 150 dB and rocket engine 180 dB to 195 dB (saturn rocket).

Noise is the unwanted sound, usually of high intensity and it causes irritation and discomfort. Sources of noise are very many, but it is always higher in urban and industrial areas than the rural. Industries expose their workers to high noise load for long periods of work every day. Road traffic particularly during peak hours is another noise pollution source to travellers as well as to the shopkeepers and residents of the affected area. Use of loudspeakers on almost all kinds of occasions like festivals, elections, worships (temples and mosques), during advertisements, are common almost all the year round. During election campaign too many loudspeakers make the worst kind of noise pollution around residential areas. Mining operations, use of bulldozers, dynamites to break rocks, drillers, are other important source of noise pollution. There are three main kinds, (1) with *intermittent* noise or *nonuniform*, (2) continuous or *uniform noise*. The third one (3) instantaneous or *impulsive* such as the explosions, gun shots, thunder, etc. The first two are quite annoying and fatiguing. Entertainment establishments and public address system often create prolonged noise and badly affect the hearers. Schools and common vegetable and fish markets are most noisy. Even the household gadgets such as a vacuum cleaner, TV and radio, Stereo music players, grinder, mixer, create too much noise and we feel so much of relief when their noise is stopped.

Fast moving jets create pressure waves that hit objects on ground and cause rattling of window panes. These sound waves are known as *Sonic booms*. Some booms are expressed in *mach* unit. Mach 1 is equal to the boom created by an object moving at the same speed as of sound. A supersonic jet moving at twice the speed of sound creates a boom of 2 machs. Mach 3 represents the boom created by an object moving at thrice the speed of sound, and so on.

NOISE POLLUTION IMPACTS

Noise affects human body in a number of ways. Blood vessels get constricted, breathing rate is affected, muscle tension changes, gastrointestinal motility and glandular reactions get affected. A general annoyance is felt at 75-85 dB. At higher *impulsive* noise pollution the pulse rate and blood pressure changes, stored glucose from the liver is released into the blood stream and there is an increased production of adrelin. The brain begins to show distorted electroencephalographic brain wave records and vision gets affected around 125 dB.

Loss of hearing is not the most important effect. Excessive anxiety, fatigue, fright and change in heart beat rates, dilation of pupil of eyes, constriction of blood vessels, damages of brain and liver, emotional disorders are more dangerous consequences of noise pollution.

Besides dB (sound intensity) and mach (booms), there are a few more units used in noise such as *phon* representing loudness (and not the sound pressure), *sone* also representing loudness equalling 40 phons (= 1 Sone). Noisyness is represented in *noy*.

The problems of pollution in the Indian context has been discussed by Ambasht (1972), by Sharma (1981) and Singal (2000). Possible methods are given for proper sewage treatments in urban settlements which may do away with such pollution and provide in addition to fertilizers and recovered water for irrigation. Through the construction of treatment tanks using biological means of decomposing organic wastes this problem can be overcome with advantage. In warmer climates as in India, these tanks function very efficiently with least expense as the prevailing temperature is suitable for microbial activity.

The following are some of the principal means to check pollution: (i) generating awareness about the causes of pollution so that preventive measures may be taken; (ii) monitoring the level of atmospheric pollution at weather stations in all principal cities and industrial areas; (iii) limiting the level of pollutant discharges; (iv) application of suitable antipollution measures by industries; (v) management of aquatic bodies; (vi) construction of treatment tanks for treatment of organic wastes; (vii) restriction of the use of non-biodegradable chemicals; (viii) research efforts to locate and replace pollution sources with safe methods etc. Turk, Turk and Wittes (1972) and Dasmann, Milton and Freeman (1973) have given excellent accounts on the ecology of environmental pollution and ecological principles in relation to economic development.

Singal (2000) has described noise pollution in respect of some Indian cities. He has compiled data from different sources. For market place the noise level reported in Delhi was between 86-102 dB, Kanpur 89-98 dB and Lucknow 67-99 dB. Even in certain functions, religious occasions and melas the peak noise level may exceed 100 dB. Loud speakers, fire works and high sound crackers are worst offenders. These are serious threats to human health, but very loud crackers are burst or fired on Deepawali, Durga puja and other festive nights. There appears a kind of competition for bursting the loudest cracker bombs. Most of the countries have set a maximum level of noise and it is 85 to 90 dB. In Indian cities noise pollution control norms and laws are in general not observed or enforced.

16 Ecotoxicology, Bioindicators, Biomonitoring and Bioremediation

Study of toxic materials in environment and their effects on biota and human health is *ecotoxicology*. Some plant and animal species or communities indicate the presence of certain stresses, pollutants, toxic chemicals, valuable mineral resources or factors like fire, flood, landscapes, etc. and they are called *bioindicators*. Using bioindicators to monitor the speed and quantum of changes in an ecosystem due to stresses (pollution, toxicity, nutrient enrichment, etc.) is called *biomonitoring*. Certain plant species are known to hyperaccumulate toxic chemicals from affected soil or water and thereby reduce the toxicity of the medium. Use of such plant or animal species to harvest out excess quantity of toxic material in the environment is called *bioremediation* (or phytoremediation). Hyper-accumulation is regarded when the concentration of the chemical becomes very high. *Phytomining* is a new term given for using bioaccumulators to obtain minerals on commercial scale. Iodine has been obtained through sea weed bioaccumulators like *Laminaria*. Presently for obtaining other minerals, sea weeds are being explored e.g. arsenic. This is doubly advantageous as it cleans the water or soil of the toxic effects at one place and supply these in useful forms for use at another. Bioharvestors can be used for double advantage, one for removing the toxic level of certain chemicals or heavy metals from affected ecosystem and second, using the harvested biomass for composting and using in appropriate agricultural fields to supplement macro and micronutrient elements and trace quantities of heavy metals in useful form and desired concentration. Ecotoxicology and bioremediation are highly applied in nature and concern scientists, technologists and medical scientists alike. Ecotoxicology has been defined variously by different scientists. We can define it as *"the science dealing with impact of environmental chemicals on organisms in populations and ecosystems, prediction of impacts, recognition of causes and remedial measures."*

Ecotoxicology involves all classes of organisms from microbes to man and all levels of biological organisation from organelle and cell to ecosystems. Markert and Oehlmann (1998) have reviewed literature on ecotoxicology and have regarded it as interdisciplinary science and a subdiscipline of biological sciences. They have emphasised that ecotoxicology should be looked in a wide perspective of environment and communities at ecosystem level rather than at individual level. They regard "toxic end points" as the concentration or level at which the test organisms succumb or die. Many toxic chemicals entering the aquatic environment on reaching human beings affect hormonal systems and reproductive cycle. These are called as *'endocrine disrupters'*. They are reported to cause testicular tumors and other urogenital disorders. EE2 or ethinyl estradiol, commonly used in oral contraceptives now-a-days by women, is excreted 90% unchanged. This ultimately reaches receiving water bodies where the reproductive cycle of fishes gets disrupted through sex change. Tributyltin (TBT) a biocide is the most toxic chemical produced and released in environment to prevent growth of undesirable animals. TBT also acts as a serious ecotoxicological agents disrupting sex-related hormones and sex organs. The prosobranchs among snails (molluscs) are most sensitive and therefore can be used as ecologically early warning bioindicators. Large number of herbicides, pesticides and fungicides as surplus from the place of application finally reach with runoff water to some wetland site and create ecotoxicological problems.

Oertel and Salanki (2003) have recently discussed biomonitoring and bioindicator aspects of aquatic ecosystems and John (2003) has described phycoremediation i.e. remediation with the use of algae in water bodies. Common toxic material occurring in nature and toxic metals like Hg, Cd, Pb, Cr, Al, Co, Ni, Cu, Zn, radioactive material and a great many types of industrially produced chemicals like chlorinated hydrocarbons, polycyclic hydrocarbons, aromatic hydrocarbons, cyanides and organic esters. Use of bivalves as *"mussel watch"* is advantageous as they have a high accumulation ability of a several toxic substances, the concentration factor being from 10,000 to 100,000. A luminous bacterium *Photobacterium phosphoreum* is used for non-specific toxicity in water samples. As the toxicity increases, the light emission by the bacteria decreases. This is called *Microtox test* and quickly gives directly the pollution level of the water sample. As a byproduct of Krebs cycle luciferase is produced which emits light energy. Same way, nitrifying bacteria are used to verify pollution load in river water. Ability to oxidise ammonium ions to nitrate in nitrifying bacteria is impaired by organic and inorganic pollutants. Decrease in nitrate level in water shows high pollution whereas an increase in nitrate shows the ecological recovery process is efficiently functioning.

Oertel (1994, 1996) has developed "*Dreissena basket*" to monitor river water using sets of zebra mussels in 8 mm meshes kept in perforated plastic tubes suspended in an outer container upto a depth of 30 cm. Oertel and Salanki (2003) regarded *Dreissena basket* method as reliable, cheap and easy to handle device of active and dynamic biomonitoring. *Daphnia* and some fishes are also commonly used in biomonitoring and the test is called '*Daphnia and fishes tests*'.

Biomonitoring techniques can be used for static and dynamic natures. In static methods the concentration of toxic substances is measured in test organisms to get a fairly accurate idea of environmental contamination level of toxic materials. The method is suitable so long as the toxicity is not beyond lethal level. Good biological indicator species are good accumulators of toxic metals and retain it for long time in their body. A good bioindicator should be readily available in wide range of places and in sufficient quantity. *Cladophora*, a common green algae, is a good accumulator or biomonitor for silver, cadmium and cobalt (Oertel and Salanki, 2003). In dynamic measurements, observations are made repeatedly at suitable time intervals and the difference values provide dynamics in pollution or toxicity status. For biologically early warning systems (BEWS) '*fish monitor*' and '*Daphania monitor*' have been used due to their sensitivity even at low toxicity levels. *Dreissena polymorpha* and *Unio pictorum* are good mussel monitors for BEWS also.

17

Flora and Vegetation of India

The terms *flora* and *vegetation* refer to the kinds of plants of a region. The two terms however differ in their meaning. The flora refers to the botanical composition of a place, i.e. the names of different species. It also indicates the way different species have come to occur at a place. Vegetation means the totality of forms in which the emphasis is not of different plants but their life forms, number, coverage, etc. Therefore, the floristic regions are generally similar to geographical regions whereas vegetational classification is on the basis of plant formations like different types of forests, grasslands, deserts, etc.

The Indian subcontinent was under forest vegetation for quite a long time, but the activities and demands of ever increasing human population have brought more of the area under the plough except on the hilly tracts, slopes, arid regions and other uncultivable areas. About 15,000 species of flowering plants and 600 species of pteridophytes are reported to occur in India of which a good number is thought to have reached India from surrounding countries in the geological and recent past and to have got naturalised here. The Orchidaceae, Leguminosae and Poaceae are the most dominant families in the Indian flora.

The Himalaya with its lofty heights has a large proportion of endemic species native of the region.

Indian subcontinent has wide variety of climates ranging from the cold and dry of the Western Himalayas, cold and wet of the Eastern

Himalayas, hot and very dry regions of Rajasthan and part of Punjab and Western Uttar Pradesh, hot and less dry region of Eastern Uttar Pradesh, part of Madhya Pradesh and regions of Gujarat, warm and wet regions of Assam, Bengal and Western Ghats in Maharashtra and Kerala and warm and semiwet regions of West Bengal, Bihar, Orissa, Andhra, etc. The flora has been described under separate botanical regions or provinces and different phytogeographers and ecologists have divided the country somewhat differently.

BOTANICAL REGIONS

India may be divided conveniently into the following nine botanical provinces : (1) Western Himalayas, (2) Eastern Himalayas, (3) West Indian Deserts, (4) Gangetic plains, (5) Assam, (6) Central India, (7) Malabar, (8) The Deccan, (9) Andamans and Lakshdweep Islands (Fig. 17.1).

1. The Western Himalayas

The Himalayas which form the highest range of mountains are one of

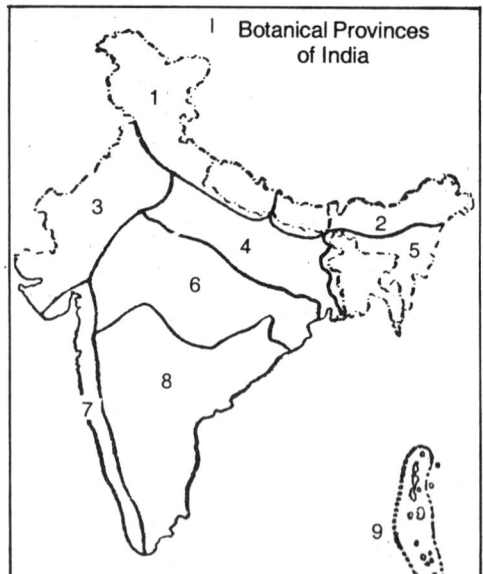

Botanical Provinces of India

1. Western himalayas
2. Eastern himalayas
3. West Indian desert
4. Gangetic plain
5. Assam
6. Central India
7. Malabar
8. Deccan
9. Andamans etc.

Fig. 17.1. Map of India showing nine botanical provinces of the country.

the most important botanical regions of the world with climate and vegetation ranging from truly tropical to temperate to arctic types from low to high altitudes in successive belts. The Western Himalayas ranging from the Central region of Kumaon to the North Western region of Kashmir are somewhat wet in the outer southern ranges and rather dry in the inner northern regions. The vegetation is divisible into : (a) submontane or lower regions up to about 1500 metres, (b) temperate or montane zone between 1500 to 3500 metres altitude and (c) the Alpine zone above 3500 metres up to the line of perpetual snow (Fig. 17.2). Singh and Singh (1987) have reviewed the forest vegetation of the Himalaya.

(a) Submontane or lower regions

This zone consists of the Himalayas from about 300 metres to 1500 metres altitude above sea level in the regions of Siwalika and adjacent areas. The forest is dominated by timber trees of *Shorea robusta* in regions receiving over 1000 mm, rainfall. In riverain regions *Dalbergia sissoo* (or Shisham) trees dominate while in more moist soils *Eugenia jambolana, Cedrela toona* and *Ficus glomerata* assume dominance. The grassy areas with less moisture have isolated trees of *Acacia catechu* and *Butea monosperma*. In the dry belt in the west *Shorea robusta* is replaced by xeric elements like *Zizyphus, Carissa, Acacia, Mallotus* etc. with patches of thorny succulent *Euphorbias* on hill slopes. On some elevations *Pinus roxburghii* (chir) begins to appear and around 1000 to 1500 metres altitude it assumes quite a characteristic position in the vegetation. The forest floor is often clear and ground vegetation is scanty.

(b) The temperate or montane zone

It ranges from about 1500 to 3500 metres altitude above sea level. *Pinus roxburghii* is gradually replaced by *Pinus excelsa*, the blue pine, at about 1600 metres. *Cedrus deodara*, the Deodar tree is quite abundant and forms almost pure forests in extensive ranges at about 1600 to 1800 metres. *Quercus incana* or oak tree also grows abundantly usually in separate patches at about these altitudes.

In the inner Himalayas in Kashmir, *Betula* (birch), *Salix* (cane) and *Populus* (Poplar) are abundantly common on certain soil types. At higher altitudes the horse chestnut (*Aesculus indica*), *Quercus*

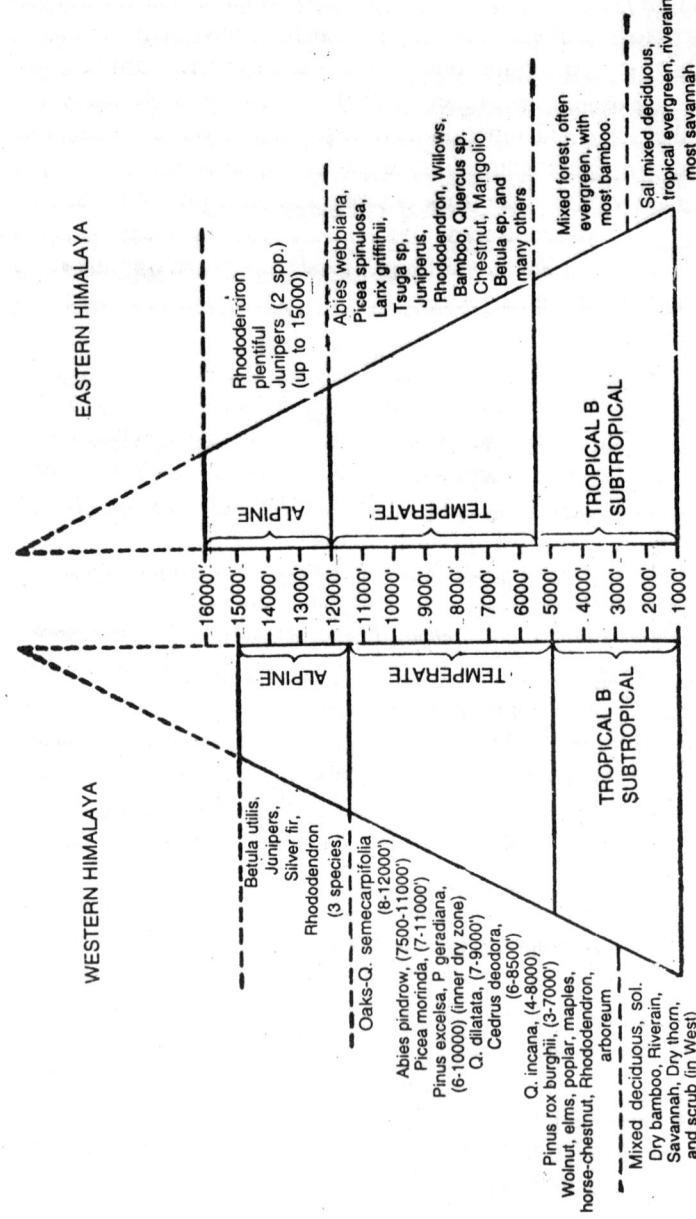

Fig. 17.2. Altitudinal Zonations and distribution of important species on the Western and Eastern Himalayas.

semecarpifolia, Q. dilatana along with conifers like *Abies pindrow, Picea morinda, Cupressus torulosa* and *Taxus baccata* grow commonly in the Kashmir Himalaya. *Rhododendron companulatum* grows on higher altitudes. In the inner valleys on dry mountains *Pinus gerardiana* (*Chilghosa*) is also present. The cultivable land in the dry region of Punjab is used for wheat and barley while in the west Kashmir valley it is used for paddy cultivation. Other important plants grown for economic purposes are saffron (*Crocus sativus*), apples, peaches, walnuts, almonds and a variety of other fruits. Wild pomegranate (*Punica* sp. or Anaar) are also found extensively.

(c) The alpine zone

It represents almost the limit of tree growth at about 3500 metres (*timber line or tree line*) at which the height of the plant is greatly reduced. Low shrubs and grassy meadows are more common. With increase in altitude the plant shape becomes more cushion-like and small. At about 5000 metres and above, the snow remains perpetually the year round and plant growth is almost nil. This altitude is called the *snow line.*

On lower levels of the alpine zone certain *Rhododendrons, Betula utilis* and small *Juniperus* are found. Above this zone, a large variety of herbaceous plants, many with beautiful flowers are found.

These have rather a short period of vegetative growth and flowering. Common among these are *Primula, Delphinium, Potentilla, Polygonum, Geranium, Saxifraga, Aster, Astragalus, Arenaria,* etc.

2. The Eastern Himalayas

The Eastern Himalayas consist of regions of Sikkim and to the east up to NEFA. This vegetational region is very similar to the Western Himalayas in vertical zonations in climate and vegetation. The chief differences are due to higher rainfall and warmer conditions in the Eastern Himalayas the tree line and snow line are higher by about 300 metres than the corresponding zone in Kashmir Himalaya. Species diversity and vegetation density are also higher in the East. This region can also be divided into : (a) submontane, (b) temperate, and (c) alpine zones according to the altitude (Fig. 17.2)

(a) The Submontane Eastern Himalayas

This zone, due to warm and humid conditions, is typically tropical with dense forests of *Shorea robusta*. It extends from the plains at the foot of the hill ranges to about 1800 metres altitude. The riverain region has forest growth of Shisham and Khair (*Dalbergia* sissoo and *Acacia catechu*). Mixed forests of deciduous trees like *Terminalia* sp., *Anthocephalus cadamba, Lagerstroemia, Toona ciliata, Bauhinia, Stereospermum*, etc. are predominantly present. Some tall trees of *Albizzia procera, Salmalia, Artocarpus chaplasha, Michelia champaca* and the bamboo *Dendrocalamus* are the other important plants.

(b) The Temperate Eastern Himalaya

It ranges roughly between 1800 to 3800 metres altitude. This is usually divided into lower temperate and upper temperate zones. The lower temperate has dicot elements like several species of oaks (*Quercus lemellosa and Quercus lineata*), *Michelia, Syzigium* sp., *Cedrela* etc. The upper temperate is cooler and is dominated by conifers like *Juniperus, Cryptomeria, Picia, Abies, Tsuga* etc. A bamboo *Arundinaria* is also common. Certain *Rhododendrons* are present on higher elevations.

(c) The Alpine Eastern Himalayas

It is above 3800 metres altitude where the vegetation is devoid of trees. Shrubby growth of *Juniperus* and *Rhododendron* are found in grassy areas.

On the whole the Eastern Himalaya has more tropical elements, a greater variety of oaks, rhododendrons and fewer conifers than the Western Himalayas. The Eastern Himalayas are regarded as the meeting point of several foreign elements in the flora like some Chinese and Japanese species. This region is also regarded as one of the richest vegetational regions of the world.

3. The West Indian Deserts

Part of Punjab, Rajasthan, Cutch, Delhi and part of Gujarat comprise this vegetational province. The climate is characterised by a hot and dry summer and a cold winter. Rainfall is low, usually less than 70 cm and in certain regions as low as 10-15 cm only. Historical evidences indicate that the area was under forests some two thousand years back but man gradually destroyed them for agricultural purposes and much of the land has become desert due to excessive dryness. The shrubby species of the zone are xeric such as *Acacia arabica* along rivers, *Prosopis spicigera, Prosopis juliflora, Salvadora oleoides, Salvadora persica, Tecomella, Capparis aphylla, Tamarix dioca,* isolated growth of *Zizyphus nummularia* etc. In the hilly ranges of Aravallis around Mt. Abu the vegetation is more dense with *Boswellia serrata* (salai), *Sterculia urens* and *Anogeissus pendula* as quite characteristic trees. On dried habitats *Acacia catechu* (Khair), bright red flowered *Butea monosperma, Euphorbia* sp. and *Acacia senegal* are common. The ground vegetation is mostly constituted by small *Calotropis, Panicum antidotale, Tribulus terrestris, Suaeda fruticosa* etc. Recently irrigation facilities have increased in this region and as a result agricultural crops are increasingly occupying greater areas. Grapes are also being cultivated. Areas of advancing deserts and lands facing strong wind erosion are being planted with *shelter belts* and *wind breaks* in several rows ranging from grasses, shrubs to trees. The common species used in plantations are the following :

Saccharum munja, Panicum antidotale, Cenchrus ciliaris, Capparis aphylla, Tamarix articulata, Parkinsonia aculeata, Prosopis spicigera, Prosopis juliflora, Acacia leucophloea, Acacia senegal etc.

4. The Gangetic Plain

This is one of the most fertile tract of land extending from Western Uttar Pradesh to Bihar and Bengal. The climate including temperature fluctuation and rainfall differences, in these regions together with other prevailing conditions of environment are responsible for the development of widely distinct types of vegetation. That part of Western Uttar Pradesh at the foot of the Himalayas where the river Ganga comes down to the plains near Hardwar is rather dry and comparatively cool. *Dalbergia* sp. and *Acacia* sp. grow along the river bank with occasional patches of *Tamarix* sp., *Shorea robusta* trees are also common. At the southern region of Uttar Pradesh there are desert conditions in the highly eroded ravines of the Jamuna river. *Capparis aphylla, Saccharum munja, Acacia arabica* are some of the characteristic species of this region. North of Vindhyas in Eastern Uttar Pradesh the hill slopes are under forests of usually open, scrub types.

The common trees are *Butea monosperma* (Dhak), *Terminalia arjuna* (Arjun), *Buchanania lanzan* (Chiraunji), *Diospyros melanoxylon* (Biri Ka Patta or Tendu), *Madhuca indica* (Mahua), *Cordia dichotoma* (Lisora), *Sterculia urens*, *Boswellia serrata* (Salai), *Flacourtia ramontchi*, *Acacia leucophloea*, *Acacia catechu* (Khair), *Emblica officinalis*, etc. The forests rich in shrubby growth of *Woodfordia fruiticosa*, *Zizyphus* sp. (Ber), *Wrightia tinctoria*, *Carissa spinarum*, *Ixora* sp. etc. Most of the Gangetic plain is however, under cultivation of wheat, barley, jowar, pulses etc., in the west and also rice in the east. The common village sides trees are *Mangifera indica* (mango), *Ficus bengalensis* (Bargad), *F. religiosa* (Peepal), *Azadirachta indica* (Neem) etc. and the wasteland weeds and grasses are *Xanthium strumarium*, *Cassia tora*, *Argemone mexicana*, *Amaranthus* sp., *Peristrophe bicalyculata*, *Dichanthium annulatum*, *Bothriochloa*, *Demostachya bipinnata* etc. In Bengal, *Lantana camera* on land and *Eichhornia crassipes* in water are the most outstanding noxious weeds. *Borassus* and *Nipa* are the common palms. The Gangetic delta region is extremely swampy and a special type of vegetation of halophytes called mangrove vegetation is found in this region of Sundarban. The plants have a variety of special features of adaptation against highly saline, swampy and anaerobic condition. The common species of sunderban are : *Rhizophora conjugata*, *R. mucronata*, *Kandelia rheedii*, *Ceriops roxburghiana*, *Bruguiera gymnorhiza*, *Avicennia alba*, *Avicennia marina*, *Sonneratia acida*, *Sonneratia apetala*, *Acanthus ilicifolius*, *Nipa fruticans*, *Excoecaria agalocha*, *Phoenix paludosa*, etc.

5. Assam

This botanical province receives the heaviest rainfall, and Cherrapunji in this region often with more than 1000 cm of rainfall is one of the rainiest places in the world. Excessive wetness and high temperature are responsible for the development of dense forests. The hilly tract has further rendered agriculture less possible and as such most of the area is under dense forests of a variety of broad leaved angiosperms besides some conifers like *Pinus khasiya* and *Pinus insularis*. The important forest trees of Assam are the evergreen tall *Dipterocarpus macrocarpus*, *Mesua ferrea*, *Michelia champaca*, *Shorea robusta*, *Endospermum chinense*, *Polyalthia jenkinsii*, *Dillenia indica*,

Artocarpus chaplasha, Alstonia scholaris, Sterculia alata, Morus laevigata, Sterospermum chelonoides, etc. Other characteristic plant species like bamboos—*Bambusa pallida, Dendrocalamus hamiltonii, Calamus* and grasses—*Imperata cylindrica, Saccharum arundinaceum, Themeda* sp., *Phragmites* sp., the insectivorous *Nepenthes* and on northern cooler region—*Alnus nepalensis, Rhododendron arboreum, Betula* sp., etc. are widely common.

On the whole the forests are of the broad leaved tall ever green type in central and southern Assam and of eastern Himalayan type in the Northern Assam.

6. Central India

It comprises of Madhya Pradesh, part of Orissa and Gujarat. The area is rather hilly and depending upon the quantity of rainfall, forests have developed into thorn, mixed deciduous and sal types. The forest vegetation is commonly dominated by *Diospyros melanoxylon, Butea monosperma, Terminalia tomentosa, Tectona grandis, Anogeissus latifolia,* or along streams by *Terminalia glabra, Ficus racemosa* with *Lagerstroemia.* The biotic disturbance specially due to grazing is quite severe and this has led to the development of open spiny thickets in place of dense forest. In such areas *Carissa spinarum, Mimosa rubicaulis, Zizyphus rotundifolia, Acacia leucophloea, Acacia catechu* are quite common.

7. Malabar

This region comprises of the western coast of India extending from Gujarat in the North to Kanya Kumari in the South. The region receives heavy rainfall from the moisture laden wind from the Arabian Sea. There are : (a) *tropical* moist evergreen forests, (b) mixed deciduous forests, (c) *subtropical* or *temperate* evergreen forest and (d) the *mangrove* forests.

The tropical wet evergreen forests are indeed very luxuriant in plant diversity and multistoreyed growth. The tallest trees of *Dipterocarpus indicus* and *Sterculia alata,* reach great heights. Other tall trees are *Cedrela toona, Tectona grandis, Dalbergia latifolia.* The bamboos are *Dendrocalamus strictus* and *Bambusa arundinacea.*

On the Nilgiri hills due to high altitudes subtropical and temperate conditions are met. Common trees are *Eurya japonica, Gordonia obtusa, Michelia nilagirica.*

8. The Deccan

The region is comparatively drier with rainfall of about 100 cm. The central region with mountain ranges has forest of *Boswellia serrata, Tectona grandis* and *Hardwickia binnata.* The common species of dry lands are *Santalum album* (chandan), *Cedrella toona, Soymida febrifuga, Capparis, Phyllanthus, Grewia, Euphorbia neriifolia, Borassus flabellifer, Phoenix sylvestris, Randia, Diospyros* etc.

9. Andamans

The Andaman group of islands have a wide range of spreading coastal vegetation like mangroves, beech forests and in the interior evergreen forests of tall trees. There are some pockets of dry regions also. The important species of the islands are *Rhizophora, Mimusops, Calophyllum, Dipterocarpus, Lagerstroemia* and *Terminalia.* Most of the area has been cleared and cultivation of paddy and sugarcane is done on most of the land.

VEGETATION OF INDIA

The Indian subcontinent has been under the influence of human culture for a very long time and as such much of climax formations have been altered or destroyed for the purpose of agriculture and human settlement. The remaining forests also are under heavy biotic pressures and management practices. The vegetation types of any region are the result of total effects of plants, animals, soil, climate and human influences and hence the Indian vegetation has also been classified by various workers taking some of the above mentioned ecological factors into consideration.

Of all the factors, the most important are rainfall, temperature biotic influence and plant life forms. The two most important types of plant formations are the forest and grassland. Forests are spread over 75 million hectares of Indian territory, i.e., on about 23% of the area. But much of this forest area is in a highly degraded state due to massive destruction of trees. Puri et al. (1983) have given a comprehensive review of Indian forest ecology.

FOREST VEGETATION

The Indian forests can be classified on the basis of temperature into four major types : (1) tropical, (2) montane subtropical, (3) temperate and (4) alpine, which represent from hot to progressively cooler conditions.

1. Tropical forests

Throughout the warmer plains of Indian tropics, forests of many kinds have developed. It is only on the hills and mountains, usually above 1000 metres altitude that the other categories of forests are formed. The tropical forests range from very dense, multistoreyed forests of diverse trees, shrubs and lianas in areas of high rainfall to dry, scrub jungles of thorny bushes in isolated patches on dry areas. Therefore, it is desirable to distinguish them into *moist tropical* and *dry tropical categories*.

A. *Moist tropical forests* : The moist tropical forests are principally of three types, again classified on the basis of relative degree of wetness.

  (i) *Tropical wet evergreen forests* develop in those very wet parts receiving over 250 cm of annual rainfall on the Western Coast, Assam, Bengal, and in the Andamans islands. These are regarded as climax formations. The diversity of species is high and trees uaually attain great heights (50 metres or more). Small trees, shrubs, epiphytes, lianas and dense ground vegetation are adjusted into so many storeys making the entire ecosystem impenetrable. The canopy remains evergreen all the year round. The tree trunks at base are usually buttressed. Grasses on the ground are almost absent whereas palms, canes and bamboos are often present. In the southern wet evergreen forests the dominant trees are *Dipterocarpus grandiflorus, D. pilosa, D. indicus, Hopea odorata, H. parviflora, Artocarpus chaplasha, A. hirsuta, Mesua ferrea* etc. On most of the trees epiphytic orchids of diverse types grow. In the second and third storeys also there are many species, chiefly *Mangifera, Emblica officinalis, Michelia* sp., *Syzigium* sp., *Ervatamia heyneana, Lagerstroemia speciosa, Strobilanthes, Ixora,* etc. The common climber species are *Ventilago, Jasminum, Calamus, Smilax, Pothos, Caesalpinia, Rubia and Gnetum.*

(ii) *Tropical moist semi-evergreen forests* are better developed in the northern than in the southern region of the country. Dominant trees usually shed and leaves for a brief period. In the north, these forests have developed in northern Assam and Bengal and parts of Orissa receiving heavy rainfall. There are some elements of evergreen nature like *Artocarpus, Michelia* and *Eugenia.* The principal deciduous species are of *Terminalia, Tetrameles* and often *Shorea.* Other species which occur in selected regions are *Odina wodier, Dillenia pentagyna, Sterospermum* sp., *Amoora rohituka,* etc.

(iii) *Tropical moist deciduous forests* have a number of tall trees which shed their leaves for a brief period and some other species are evergreen and semi evergreen. They are common in moist areas of Kerala, Karnatak and southern M.P. in the south and parts of northern MP, UP, Bihar, Bengal and Orissa in the north. These forests have tall trees (30 to 40 metres or even more) forming a closed canopy. In southern India, the moist deciduous forests are dominated by *Terminalia cernulata, Grewia* sp., *Garuga pinnata, Salmalia malabaricum, Terminalia paniculata, T. bellerica, Tectona grandis, Pterocarpus marsupium, Adina cordifolia, Lannea grandis,* etc. Teak and sal usually grow in separate stands, the former being a calcicole and the latter probably a calcifuge. However, in Baster area the two species are found to grow together. The sandal trees in Karnataka state grow in areas receiving 100-250 cm rainfall between 700 m to 1500 m altitude. They often grow in association with *Artocarpus, Melia, Albizzia, Dalbergia* etc.

In the northern half, *Shorea robusta* reaching around 30 to 40 metres is the dominant plant in forests of Gorakhpur and Tarai regions of UP, Khasi Hills (Assam) and northern Bengal. The other more common associates of sal are *Terminalia tomentosa, Dillenia* spp., *Eugenia* sp. and *Boswellia* sp. Both the sal and the teak forests are under intensive management practices of the forest department.

B. *Tropical dry deciduous forests* of India are composed mostly of such trees which remain leafless for several weeks in the dry season. The tropical dry deciduous forests can also be distinguished into the northern and southern regions.

The northern deciduous forests are extensively distributed in the Punjab, UP, Bihar and Orissa in regions which are neither wet nor too dry. The trees are of moderate height wigh a sparse canopy. Thorny scrubs, grasses and some bamboos are also present in many region. In the Punjab and western UP forests *Anogeissus latifolia, Acacia catechu, Terminalia tomentosa, Boswellia serrata* are dominant with subdominants and societies of *Dendrocalamus strictus, Emblica officinalis, Woodfordia floribunda,* etc. *Shorea robusta* forests are also scattered on somewhat wet regions. In many of the UP, Bihar and Orissa forests intense biotic pressure has degraded them into thorn scrub forests with extensive open areas dominated by grasses and thorny scrub on forest margins. In the interior *Shorea robusta, Terminalia arjuna, Boswellia serrata, Buchanania lanzan, Diospyros melanoxylon* assume dominance. The southern tropical deciduous forests are located in the dry areas of peninsular India in the States of Maharashtra, Tamil Nadu, Karnataka and Madhya Pradesh. The forests are of mixed type composed of deciduous trees with scattered patches of densely growing grasses intermixed with shrubs. *Terminalia, Anogeissus latifolia, Pterocarpus marsupium, Tectona grandis, Ougenia dalbergioides, Stephegyne parviflora, Boswellia* sp. form the top canopy followed by smaller plants of *Dendrocalamus, Bambusa, Lantana, Helecteris, Woodfordia,* etc. Common grasses are *Andropogon, Panicum* and *Heteropogon.*

2. Subtropical montane forest

These are cooler than the tropical and warmer than the temperate forests restricted to the hill of the Nilgiri, Mahabaleshwer and Pachmarhi and extensively on the Himalayas up to 1500 metres altitude. The southern forests have rather dense growth of trees of low stature with a number of ferns in their shade. The periphery and forest openings are occupied by shrubs, dicot weeds and grasses. Common trees are *Eugenia, Actinodaphne, Canthium, Memecylon, Mangifera* and *Ficus.* The climbers are *Piper trichostachyon, Gnetum scandens, Smilax macrophylla* and *Vitis elongata.* The northern subtropics has rather tall trees with an open canopy which makes possible the growth of a second storey of trees. In the eastern Himalayas, due to higher humidity, bamboos and many epiphytes, including orchids and ferns, are more abundant. Most of the trees are evergreen. Conifers and *Quercus* usually form separate stand. For the floristic description of this kind of forest the section on the Eastern and Western Himalayas under the Botanical Regions given earlier may be referred.

3. Temperate forests

These forests in India occur usually above 1600 metres altitude chiefly on the mountains of the Himalayas and Nilgiris. The Himalayan temperate forests have oaks and conifers in abundance. Conifers in the region are regraded to be in seral or successional stages. These are more common or northern slopes. Oaks form relatively stable evergreen pure stands on the southern slopes. The conifer trees here usually reach upto 20 to 25 m. The altitudinal zonations and floristics have already been described earlier (Fig. 14.1). In the temperate Himalaya successional trends starting from either hydrarch conditions through *Salix* or from xerarch starting with lichens and mosses through grasses and *Pinus roxburghii*, usually ends up in the climax oak forests.

The southern temperate vegetation is principally represented by the Sholas or extensive growth of grasses and evergreen forests on the Nilgiri and other hills on altitudes usually above 1300 metres. The forests are very dense because of heavy rainfall. Between 1000 to 1300 m alt tall trees of *Balanocarpus utilis, Hopea parviflora, Artocarpus hirsuta, Salmalia malabaricum, Hardwickia binnata*, and many others form dense closed canopy forests. Climbers are *Piper nilghirianum, Hoya* sp., *Jasminum* sp., *Dioscorea* sp., *Thunbergia* etc. In extensive areas there are stable grasslands which are regarded to have been formed secondarily under the influence of fire and biotic effects.

4. Alpine vegetation

The word alpine is derived from the word *alp* meaning high mountain. This type of vegetation is distributed extensively throughout the

Himalayas well above 3000 metres, tree heights become less and less with increasing altitude and around 400 m trees are replaced by a sparse growth of small plants of *Sedum, Primula, Saxifraga* and patches of lichens. The floristic details are given earlier in this chapter.

GRASSLAND VEGETATION

The Indian grasslands are not climax formations but have developed secondarily after the destruction of forest. In most cases they are maintained in their present seral stage due to biotic influence. They are spread in all the major bioclimatic regions of the country. Misra (1983) has regarded all tropical grasslands of India as a savanna. Savanna is a kind of grass dominated land beset with isolated growth of shrubs or trees at wide intervals. Indian savannas during the past four centuries have changed from mesic to xeric in nature and the common shrub elements now are *Acacia arabica, A. senegal, A. catechu, Calotropis gigantea, Mimosa rubicaulis, Phoenix sylvestris* and *Zizyphus nummularia* (Misra, 1983). Indian grasslands are tentatively divided into eight major types by Whyte (1957). These are as follows :

1. *Sehima-Dichanthium* type of grasslands are widespread in the black soils of Maharashtra, Madhya Pradesh, south-western parts of Uttar Pradesh and parts of Tamil Nadu and Karnataka. The dominant grasses are *Sehima sulcatum, S. nervosum, Dichanthium annulatum, Chrysopogon montanus, Themeda quadrivalvis.* Other common grasses are *Ischaemum rugosum, Eulalia trispicata, Isilema laxum* and *Heteropogon contortus, Themeda* and *Heteropogon* are more extensive on hilly tracts.

2. *Dichanthium-Cenchrus* type is by far the most extensively distributed type of grassland on the sandy loam soils of the plains of the Punjab, Haryana, Delhi, Rajasthan, Saurashtra, Eastern UP, Bihar, Bengal, Eastern MP, Coastal Maharashtra and Tamil Nadu. Both the dominant species viz. *Dichanthium annulatum* and *Cenchrus ciliaris* are most important fodder grasses. In dry areas of Rajasthan, Saurashtra and Western MP, after severe grazing these are replaced by sparse populations of annuals. Other characteristic perennial grasses are *Bothriochloa pertusa, Heteropogon contortus, Cynodon dactylon* and the annuals,

Eragrostis tennela, E. tremula, E. viscosa, E. ciliaris, Aristida adscensionis, Dactyloctenium aegyptium. On wet soils with good drainage, *Desmostachya bipinnata, Dichanthium annulatum* are more common.

3. *Phragmites-Saccharum* type in marshy localities of the Terai (moist) areas of northern UP, Bihar, Bengal and Assam and in the swamps of Sundarbans and Kaveri delta of Tamil Nadu, *Phragmitis karka, Saccharum spontaneum, Imperata cylindrica* and *Bothriochloa* are the main grasses.

4. *Bothriochloa* type on high rainfall paddy areas of Monavala tract of Maharashtra is a localised grassland of pure dense growth of *Bothriochloa odorata.* This is sweet scented perennial grass used in thatching huts but it is a poor fodder grass.

5. *Cymbopogon* type is restricted to the low hills of the Western Ghats, Vindhyas, Satpura, Aravali and Chhota Nagpur. Other associated grasses are *Themeda, Heteropogon* and *Aristida.*

6. *Arundinella* type on the other hand is on high hills of the Western Ghats, Nilgiris and throughout on lower Himalayas from east in Assam to west in Kashmir. On the Himalayas between 1500 m to 2000 m alt. *Arundinella nepalensis, A. setosa* with *Themeda anathera* form extensive stands with sporadic growth of *Chrysopogon* spp.

7. *Deyeuxia-Arundinella* type is on temperate regions of the upper Himalayas between 2000 m to 3000 m from Assam, Bengal through UP to Punjab and Himachal Pradesh. These are mixed types of grasslands characteristic of temperate climates. These are *Deyeuxia, Arundinella, Brachypodium, Bromus* and *Festuca.*

8. *Deschampsia-Deyeuxia* type of grasslands are restricted to the Himalayas above 2500 m alt in the alpine to subarctic type of region. The climate is very cold and soils are thin. *Deyeuxia, Deschampsia, Poa, Stipa, Glyceria* and *Festuca* are more common genera. Some of the grasses extend even beyond 5000 m such as *Deschampsia* and *Trisetum spicatum.*

Phytogeography

The plant communities are never identical at two different places. Even in a small place if you move from one part to another you will notice some difference in structure, composition and physiognomy of the communities. With distances and difference in topography, climate and biotic influences the vegetation changes markedly. If you happen to travel from Varanasi in the Gangetic plains to Madhya Pradesh in the south you will come across clearly different types of vegetation ranging from cultivated fields in plain level lands to a scrub type to open forest in the heavily grazed dry hills in Vindhyas to dense forest of sal in central Madhya Pradesh. On the other hand, if you go the north to the Himalayas you will come across different types of vegetational change with change in latitude. As you move up the Himalayas the change in vegetation is again clearly marked but this is in respect to altitudinal distribution. Similarly the animals also differ from place to place, e.g., Bengal tigers are found in India, giraffes are found only in Africa and kangaroos only in Australia. Plants and animals of any region are collectively called the *biota*. The study of world biota with regard to their origin, environmental interrelationships and distribution etc. is called *biogeography*. The science of biogeography has two major aspects, the historical dealing with the origin of the biota of a region and ecological dealing with their environmental interrelationships. Historical biogeography takes into account geological features like origin of life, movements of land masses and possible climatic condition in the geological past. Ecological biogeography dealing only with plant communities and vegetation is called *Phytogeo-*

graphy. The classical approach of phytogeographical studies has been towards enumeration of the taxa of a region and on the basis of broad floristic differences botanical regions have been recognized. With the aid of such information the causes and mechanism of evolution of different types of floras in different regions are also being studied. Thus, phytogeography has two major approaches of study : (A) *descriptive* or static phytogeography dealing with description of flora or vegetation of different botanical area and (B) *interpretive* or *dynamic phytogeography* dealing with interpretations of causes of plant distribution. The interpretive aspects of studies are based on certain basic phytogeographical and ecological principles.

(A) DESCRIPTIVE OR STATIC PHYTOGEOGRAPHY

The term static is rather unfortunate in the sense that vegetation and the environment are changing at every place especially with increasing biotic activities.

The earth is divided into the following broad vegetational belts : (1) The Arctic, (2) the North Temperate, (3) the Tropical and (4) the South Temperate.

These belts are also geographical and have different climatic conditions. Climate and vegetation go hand in hand. Within these four belts there are further subdivisions. The whole region of north temperate, subarctic and arctic zones is called Boreal or Holarctic.

1. ARCTIC ZONE

This zone is divided into the two (1) Arctic proper around the North Pole, and the (2) subarctic, a less defined part south of the pole.

(i) The *Arctic Proper* is covered with ice all the year round. There is very little biological activity and only highly specialized plants like some algae are found there. Somewhat further south of 80° latitude grow large varieties of flowering annuals for a few weeks of summer when the ice melts temporarily. This zone is also called the tundra zone. Mosses and lichens form a thick mat on the frozen soil, Tundra is a vast areas of barren land where life remains inactive over long periods in winter. As soon as the snow melts, animals come out of their hibernation or arrive from the south. Small grasses and some of the Rosaceous family members lying dormant quickly grow and produce flowers and fruits. The variety of plant species and their gregariousness increase as we move south to the subarctic zone. There is no clear cut demarcation between these zones because of different temperature conditions available in different parts due to extension of the sea on the European side. The vegetation of the zone as a whole is chiefly constituted by mosses *Polytrichum* and *Erytrichum*, lichens, some prostrate growing grasses, cranberries, *Rhododendron* sp., *Salix* spp. etc. Most of these perennate through underground root stocks that lie dormant for most of time. July is the month of maximum exposition of life. Men living in the tundra region are called *Eskimo*. They are dependent upon a common animal *reindeer* for most of their requirements including food. Reindeer in turn mostly eat lichens *Cetraira islandica* and *Cetraria cucullata* and 'tundra moss' *Cladonia* which is a lichen.

(ii) *The Sub-Arctic Zone* extends (from north to south) from southern arctic zone to the northern limits of temperate zone. The vegetation is similar both in North American and Euroasian regions of subarctic zone. The region is again very cold. Bogs are abundant, trees are of low height and shrubs and herbs are more characteristic in June and July. Tree species are mainly firs, *Pinus* spp. and small *Juniperus* among conifers and *Betula* spp., *Salix* spp., *Populus* sp., and some oaks and chestnuts among angiosperms are common. Many arctic species of *Rhododendron* are also found in this region. The ground is covered with *Lycopodium, Equisetum, Pyrola,* several orchids like *Goodyera,* marsh marigold, insectivorous *Drosera* etc. Mosses and lichens are also abundant. In the Alaskan region there are dense forests of tall evergreen trees of spruce. *Tsuga heterophylla, Chamaecyparis nootkatensis.* At its coastal region the giant sea weeds—*Macrocystis and Nerocystis* are abundantly found.

2. THE NORTH TEMPERATE ZONE

The north temperate zone extends roughly between 30°N lat. and 55°N lat. On account of some differences in vegetation and geography the north temperate zone is divided into two major sections : (i) the Old world or Eastern hemisphere consisting of Europe, part of

North Africa and Northern Asia and (ii) the New World or Western hemisphere comprising of northern parts of North America.

(i) The north temperate of the eastern hemisphere

The vegetation of the north temperate belt of the eastern hemisphere may be divided as follows :

(a) *Western* and *Central Europe* constitute a natural botanical region demarcated in the north by the subarctic and in the south by mountain barriers like the Alps. The British Islands are rather less cold due to the Gulf stream, a warm water current. The forests of western and central Europe are dominated by several gymnospermous tall trees like *Pinus sylvestris, Picea excelsa, Abies pectinata* and to some extent *Taxus baccata.* Among angiosperms the oaks like *Quercus pedunculata, Quercus robur* and *Quercus sessiliflora,* ash tree (*Fraxinus* sp.) maple (*Acer platanoides*), chestnut (*Castania* sp.) are more important. Among the ground vegetation, *Hieracium,* thistle, *Salvia,* several species *Companula, Viola* spp., *Dianthus* and some orchids are commonly found besides several wild roses, anemones and buttercups. At high altitudes the tree populations decrease and grassy expanses are common with anemones, primroses, buttercups and many other beautiful flowers that grow along with grass species forming thick cushions in the months of June and July.

The British Islands with a warm climate for its high latitude has somewhat different types of vegetation although the species content is much the same as described above for Western Europe. A few Mediterranean elements of the south are also found here like strawberries. Moors, bogs and peats are common the the UK.

(b) *The Mediterranean flora* extends between 30° and 40° N latitudes south of mountain ranges in Europe and in Asia around the Mediterranean Sea. The climate of this region is a rather warm temperate type highly suited for the growth of several economically important fruit trees. *Quercus ilex, Pinus pinea, P. pinaster, Populus* sp., and olives are the common trees. Various nut trees and oranges are also common. Many foreign elements like palms, cacti, *Acacias* and beautiful flowering species are now commonly found in this region. In the Asian region of Mediterranean as in Arab countries, there are high mountains and expanses of sandy deserts on account of the low rainfall. Vegetation is rather poor. *Artemisia tridentata, Atriplex* sp., *Alhagi* sp., *Polygonum* sp., *Phoenix dactylifera* (date palm or khajur) etc. are more conspicuous. In this region human culture had developed in very ancient time and wild varieties of many important crops like wheat, barley, grape and pomegranate are still to be found. Walnuts (or Akhrot, *Juglans regia*) are cultivated for economic purposes.

(c) *Northern Africa* : This region is essentially similar to the Mediterranean of Europe and consists of northern parts of Morocco, Algeria, Libya and Egypt. The Moroccon region has the high mountains of the Atlas. On the whole the rainfall is scanty and the vegetation is sparse. In cooler regions on mountains conifers like *Pinus halepensis, Callitris quadrivalvis, Cedrus atlantica* are common besides the broad leaved oaks (several species of *Quercus*). Several herbaceous and shrubby species occur in deserts. *Stipa tenacissima*—a grass in this region is used in the manufacture of paper. Succulent xerophytic *Euphorbia* spp. and *Mesembryanthemum*, and hard woody *Acacia* sp. are common. The Sahara desert is a strikingly barren expanse without plants for miles, around springs or oases there grow *Phoenix dactylifera, Carissa, Astragalus* etc. Around the Nile river and its delta rich crops of rice, wheat, legumes, vegetables and good quality cotton and many edible fruits are grown. *Salix* and *Acacia* also grow there.

(d) *The Himalayas, Eastern Asia* and *Japan* are the other parts of the eastern hemisphere temperate region. The vegetation of the Himalayas, the highest range of mountains in the world, is described in details separately under the title "Flora and Vegetation of India".

Tibet, China and Japan have a very diverse type of vegetation. China, being a country of dense human population and a very ancient human culture, has lost much of its original vegetation due to extensive cultivation of land. The conifer trees of China and Japan are *Cryptomeria, Sciadopity, Cuninghamia, Cephalotaxus, Torreya,* etc. Maiden hair tree *Ginkgo biloba*—the only survivor a vast group of dominant trees of Mesozoic (some fifteen crores of years ago) is still found growing naturally in China. *Cycas* is also common. Among angiospermic trees several *Rhododendron* spp., *Citrus*, palms and bamboos are quite characteristic in some regions. Camphor (*Cinamomum camphora*), *Mangolia, Begonia, Lilium auratum*, beautiful *Pittosporum tobiva* and many varieties of lilies are other important species. In fact horticulture has received highest attention in Japan.

(ii) The north temperate of the western hemisphere

It consists of parts of United States and Canada lying mostly between north latitudes 30° to 55°.

The Eastern coastal region of the United States and Canada in the temperate belt have some characteristic species not met with in the interior or the continent, like *Shizaea pusilla*—a tropical fern. The forests are composed mostly of conifers and deciduous tree like *Acer saccharum, Betula* sp., red spruce, *Pinus strobus, Abies balsamea* and *Thuja occidentalis*. The coastal land is rocky. On lower altitudes *Epigèa repens, Myrica carolinensis*, some wild cherries, plums, roses, and a number of orchids like *Cypripedium acaule* are abundant. In coastal waters species of *Fucus* and *Ascophyllum* are quite characteristic. In the New England region trees of *Ulmus americana* and *Castania dentata* are abundant in the inland wet areas. *Typha* and *Zazania aquatica* (wild rice) grow abundantly along lake margins. Lake vegetation, consisting chiefly of *Potamogeton, Vallisneria*, and *Elodea* is very much the same throughout the world. In the southern parts of the United States, temperate region some rich forests have developed. Some larger trees are *Liriodendron tulipifera, Liquidamber styrauciflua, Magnolia grandiflora, Magnolia acuminata*.

Other important North American tree species of the temperate belt in the East are *Taxodium distichum, Pinus rigida, P. palustris, P. caribaea, Quercus rubra, Q. macrocarpa, Carya microcarpa, C. alba, Fraxinus sambucifolia* (Cotton wood tree), *Tilia americana*, a mangrove tree *Rhizophora mangle*, royal palm *Oreodoxa regia* etc. The ground vegetation has at places *Dionea muscipula* (venus-fly trap, an insectivorous plant), *Arundinaria macrosperma*, royal fern *Osmunda regalis, Viola* sp., buffalo grass, *Buchloe dactyloides*, some species of *Bouteloa, Stipa* and cycadophyte *Zamia floribunda*.

To the west on the Rocky mountains and slopes of the Pacific side the entire area is, indeed, a vast expanse of rugged mountains, high peaks, covered with forests of different types. Some lowlying areas are even below sea level. There is a vast expanse of desert in southern Arizona and south eastern California. The coastal region has an equable climate as against the extremes of cold and hot seasons found in the interior.

The major forest trees in different parts are *Pinus ponderosa* associated with *Pseudotsuga* at about 2000 metres, *Picea* sp., *Pinus flexilis* at 4000 metres, and *Abies laciocarpa* at even higher altitudes. On the humid Pacific coast are tall *Larix occidentalis, Abies grandis* and *Taxus brevifolia*. In Northern California there exists a long stretch of forests of the world's tallest trees—the *Sequoia sempervirens*. These trees are even over 100 metres tall and with over 6 metres diameter of trunk, *Sequoia* or redwood trees form almost pure stands as other species scarcely withstand in competition for light against such gigantic trees. *Pasania densiflora, Arbutus* sp. are the chief associated by *Abies grandis*. Further towards the coastal region are *Rhododendron californicum* and *R. occidentale*. In the absence of Sequoias in open spaces *Quercus* and *Aesculus californica* also occur.

The ground vegetation also differs from place to place. In salt marshes *Salicornia herbacea* and *Rumex maritima* are common. In peaty soil *Monotrapa uniflora, Pyrola, Goodyera,* etc. grow profusely. At high altitudes trees are replaced by meadow lands with several species of *Saxifraga, Epilobium latifolium, Castilleia, Mimulus* sp., *Primula farinosa* etc.

In the Colarado desert of Arizona and south eastern California there are large varieties of xerophytic plants. *Larrea, Parkinsonea* and *Fouqneria* form characteristic bushes. The Californian fan palm *Washingtonia filifera* commonly grows in these regions. Cacti of wide varieties are most characteristic in deserts, *Agave, Ephedra* and *Prosopis* are also common. *Cereus giganteus, Ferocactus* and *Echinocactus* are some of the cacti predominantly found in Arizona.

3. THE TROPICAL ZONE

The tropical zone is also broadly divided into the (i) *Palaeotropics* or the Old World or Eastern Hemisphere tropics and the (ii) Neotropics or the New World or Western Hemisphere tropics.

The Palaeotropics have two distinct botanical areas : (a) *Tropical Africa* and (b) *Tropical Asia* comprising of India, Pakistan, Burma, Thailand, Indonesia, etc.

Tropical Africa is a large landmass of uneven topography with the greater part of the area at a relatively high altitude or over 1000 metres above sea level making the climate somewhat subtropical. The Sahara desert receives very little or no rainfall while some other regions receive high rainfall. Thus a variety of vegetation patterns from very dense and diverse to scanty and sparse types are met with in this belt. In equatorial regions on coastal land the mangrove plants like *Rhizophora mangle, Avicinnia nitida*, etc. grow. On less swampy but highly saline soils *Caesalpinia crista* and *Cassytha filiformis* occur. Gradually from the coast to the inner region of the continent there appear *Pandanus* sp. *Phoenix spinosa* and some leguminous shrubs and the oil palm *Elaeis guieneensis*. In the interior very dense forest of all trees supporting lianas like *Landolphia kirkii* (yielding rubber), *Quisqualis indica, Clerodendrum splendens*, etc. grow. Important species are *Ficus, Bombax* sp., *Khaya senegalensis, Diospyros ebenum* (mahogany) and several leguminous trees. On the ground *Canna indica. Cyperus papyrus, Zingiber, Phragmites* and *Saccharum* occur abundantly. In eastern Africa *Sterculia tomentosa* is quite common. The most remarkable of all plant species found in the South Western Coast of Africa is *Welwitchia mirabilis* a member of Gnetales in the Gymnosperms. It does not occur anywhere else in the world and is restricted to a small region of Africa. In East Africa many plants common to India also grow. These are the Indian fan palm *Borassus flabelliformis, Tamarindus indica, Ficus* sp., *Asparagus, Clematis, Phaseolus, Cassia fistula, Erythrina*, etc. Several species of *Acacia, Albizzia, Zizyphus, Bauhinia*, etc. also occur in open forests. In rain forests *Syzygium* trees are common.

The Asiatic tropics : Important botanical regions of the Asiatic tropics are Arabia, part of Pakistan, India, Bangladesh, Burma, Sri Lanka, Thailand, Indonesia, Philippines, etc. The vegetation of India is described separately. In Arabia the rainfall is extremely low and temperature is high except along the high mountains where it rains abundantly. Otherwise, it is all desert condition and plants adapted to extreme xeric conditions grow. Several *Acacia* species and *Prosopis*, are quite common. *Coffea arabica*, the coffee plant is supposed to be a native of Arabia. Date palms are also found near water bodies. Sri Lanka is very rich in diversity and density of plant life. The climate is equatorial, i.e., warm and humid which favours high plant productivity all the year round. Most of the ground is under intensive cultivation of crops like rice and sugarcane, and fruits of *Eugenia*, banana, papaya and mango. Hill slopes and many other regions are under the cultivation of tea. To some extent coconut and rubber are also grown commercially. Very little area is left under natural vegetation. Ferns of a wide variety like *Ophioglossum pendulum, Lygodium, Helminthostachys, Gleichenia* sp. and *Botrychium* are commonly found. In the famous botanic garden at Paradeniya, a wide variety of tropical plants like giant bamboos and branching palms grow. A number of orchids, beautiful flowers of roses, violets and fuchsias are also abundant.

Burma, Thailand and other areas around them are mostly under the cultivation of rice. The common trees are mostly of Jack fruit (*Artocarpus integrifolia*), orange, banana and mango. *Areca catechu* (supari) yielding betel nut is another beautiful plant of this region.

In Malaysia and the group of islands of Indonesia as Java, Sumatra and Bali have very high rainfall and rich soil that bears one of the most luxuriant vegetation to be found anywhere in the world. Large varieties of palms like *Nipa fruticans, Onchosperma horrida, Crystostachys* sp., *Arenga saccharifera, Caryota urens*, etc. are common. Ferns are also abundantly widespread. *Durio zibethinus* a fruit tree belongs to this region. Its fruit is regarded as one of most tasteful but at the same time it smells very offensively bad, but local people eat it with gusto. Forest trees are *Albizzia, Diospyros, Eugenia*, etc. infested with lianas. *Nepenthes* or pitcher plant is commonly found in Malaysia. *Dendrocalamus giganteus* grows to tall heights on hill slopes. In Borneo almost everywhere sago palm, bananas and coconuts are grown. Many ferns and beautiful orchids grow epiphytically. *Wormia pulchella* a beautiful yellow flowered shrub is found growing almost everywhere. Java is regarded as the richest place from the point of view of vegetation. The soil is exceedingly rich because of its volca-

Fig. 18.1. A mature crop of rubber plantation at Bogor Indonesia. Rubber is tapped from the cuts made in the main stem (Photograph : R.S. Ambasht).

nic origin. Rice is extensively cultivated besides rubber (Fig. 18.1), coffee, condiments and spices, sugarcane, tobacco, *Cinchona* etc. and exceedingly large variety of trees grow in Java. Important ones are *Albizzia, Pterocarpus, Tamarindus, Cassia, Bombax, Durio* and *Artocarpus, Dendrocalamus* is also very common.

The Neotropics is constituted of Mexico and a major part of South America. The temperature, moisture and topography are very similar everywhere in the western hemisphere and hence the vegetation is fairly homogeneous. Around the equator in South America there exists one of the densest and largest expanses of forest in the Amazon basin.

The Mexican region is quite hilly and the plateau is high. In regions of low rainfall there grow a number of xerophytes like tall cacti (*Pachycereus*), *Agave* and *Yucca*. Much of the land is under cultivation of crops of wheat and maize and a variety of fruits and vegetables. Comparatively cooler region at higher altitudes are full of trees of *Pinus, Spruce, Quercus, Populus* etc. Mountain peaks are under grass vegetation containing several grasses, sedges, members of Compositae, Rosaceae and Cruciferae. The wet lower regions are more tropical in look with abundant growth of mosses, bamboos, palms and epiphytic orchids. The vegetation of South America is very dense and most extensive due to high rainfall, rich alluvial soil around the river Amazon and its tributaries, and the equatorial climate. A large expanse of forest is of the *flood forest* type which remains under inundated condition for the greater part of the year. In the ecotone region between flood forests and uplands the soil has a better combination of soil moisture and air and it bears a thick forest. In drier regions savanna type of grassland is found.

The most widespread trees are *Bertholletia excelsa, Maximiliana regia, Eukylisia* sp. etc. In mangrove conditions *Rhizophora* is most extensive. *Lacythis* sp. and *Bombax* sp. are other trees found fairly extensively. A large number of epiphytes of Bromeliaceae, Araceae and Orchidaceae with hanging roots are conspicuous. A variety of ferns, bananas, *Zingiber, Canna* and arrow-roots are extensive on the ground. A species of travellers tree *Ravenala guyanensis* is also found here. On less wet areas forests of a large variety of leguminous trees like *Cassia, Bauhinia* and *Inga* etc. are abundantly found besides *Ficus* spp., *Artocarpus* sp. etc. Cultivation of para rubber (*Hevea braziliensis*) is being done on an extensive scale.

4. THE SOUTH TEMPERATE ZONE

Some of the extreme southern region of Africa is in a temperate belt with vegetation showing a transition of tropical and subtropical elements into the temperate type. The fernlike gymnosperms—*Encephalortos* and *Stangeria* are native of this region (Natal). On the hills of Kilimanjaro the temperature is low and conifers like *Podocarpus* and *Callitris* are dominant. On lower wet regions *Salix* and *Phragmites*, and in dry regions *Andropogon* and *Panicum* grasses and *Acacia giraffe* are common.

Australia and New Zealand being isolated from the rest of the land mass through oceans have a large variety of species endemic or specific to this region. There is very little scopes of altitudinal distribution of climatic zones in Australia due to the absence of any high mountain there and the general climate is rather dry.

In the northern part of Australia, floristic elements are similar to those of South East Asia such as palms *Caryota, Borassus* and betel nut *Areca*. Towards the south, trees of *Araucaria* are very common. *Eucalyptus* attaining great height is a typical Australian tree occupying large areas. A large variety of *Acacias* are also abundant. Another very characteristic tree of this continent is *Casuarina*. Among ground vegetation *Drosera*, some orchids, a *cycad-Macrozamia*, some *Lycopodium, Psilotum, Tmesipteris* etc. are of great botanical interest. New Zealand on the other hand is more hilly and forests are abundantly composed of conifers like *Agathis, Podocarpus, Dacrydium* and others. Many ferns like *Dicksonia squarosa, Hemitelia smithi, Trichomanes reniformae, Todea superba* and *Cyathea medularis* are abundant. The only palm in New Zealand is *Rhopalostylis* sp. which often supports epiphytic growth of *Astelia solanderi* a liliaceous plant. A large number of species of *Metrosideros* of the family Myrtacaece is another characteristic of the island. In the matter or bryophytic flora New Zealand is probably richest with exceptionally gigantic sizes of *Dawsonia superba* (a moss) and *Monoclea foresteri* (a liverwort).

B. INTERPRETIVE OR DYNAMIC PHYTOGEOGRAPHY

Based on the knowledge of many biological and earth science a large number of phytogeographers have laid certain principles as the basis or foundation of dynamic phytogeography. Good (1931, 1947), Mason (1936) and Cain (1944) are the important phytogeographers who have elaborated and discussed this subject in detail and have in essence laid thirteen basic principles, divisible into four categories, pertaining to (i) environment, (ii) plant responses, (iii) migration of floras and (iv) climaxes. The thirteen principles as given by Lawrence (1951) on the basis of works of Good, Mason and Cain are as follows :

BASIC PRINCIPLES

(I) Environment

1. Of the several aspects of the environment, the *climate* is regarded as *the primary factor* that controls the nature of the flora of any region. Other environmental factors like soils, biotic conditions,

etc. are also responsible for the development of flora but they are treated as secondary causes, subordinate to the climate.

2. The second principle stresses the fact that climate, which is the master factor in the evolution of floras *has varied in the past* and as such the interpretation of present day flora of any region should be viewed with a background knowledge of climate of the region in the past.

3. Another principle pertaining to climate is that in the geological past the *relations of land and sea have varied* and they were much different from what they are at present. This fact is responsible for commonness of flora of lands now widely separated by the sea but having belonged to the same land mass in the geological past. Further the relation of sea to land has its own effects on the climate of any region.

4. The soil conditions : Their origin, chemical and physical nature, depth and texture with soil moisture etc. also control the nature of vegetation but the *edaphic factor is regarded as secondary to climate*.

5. Biotic factors like association of other species, presence or absence of certain parasites, soil microbes, role of man and domesticated or wild animals etc. have also important effects on the evolution of flora.

6. Another principles of dynamic phytogeography is that all the three—climate, biota and soil conditions operate on the development of floras and stability of communities in an *integrated* or *holocoenotic fashion*.

(II) Plant responses

7. The ranges of plants for their spread and evolution in any region are limited by the degree of their ecological tolerances.

8. The degree of ecological tolerance, that is tolerance in a range of environmental conditions are governed by the genetic make up of the species.

9. In the ecological life cycle, different ontogenic phases have different levels of tolerances. A plant species may easily survive

through in some stage like seeds or diaspores but may not tolerate the same in seedling stage.

(III) Migration of floras

10. Migration of floras have taken place on a large scale in the past due to a variety of factors like movement of land masses, movement of ice (glaciation), and through human activities.
11. Successful migration involves transport of propagules from one place to another and its successful establishment at the new place.

(IV) Perpetuation and evolution of floras

12. The success of perpetuation of a flora largely depends upon migration and species evolution in newer environments.
13. Perpetuation further depends upon natural selection of some out of a lot of migrated and evolved flora in a region. All these principles have been reviewed by Wulff (1943).

AGE AND AREA THEORY

Willis (1922) has propounded dynamic phytogeographical theory called 'Age and area' theory. He is of the opinion that the area of distribution of a species is directly related to the age of the species. That is, a species which evolved a long time ago has a large area of distribution and a young or recently evolved species is distributed in a small area. He has explained this theory with examples of several species of *Coleus* growing on plains and hills in Sri Lanka. The philosophy behind this theory is that in advancing age of the species its area of occupation also goes on increasing. This may be true in certain specific cases but when we look to geological evidences and many other aspects of biology the theory of *age and area* fails miserably. For example the dwindling species like *Ginkgo biloba* has at present a very narrow area of its growth and according to Willis' theory the species should be regarded as young. But paleobotanical evidences clearly demonstrate that the species is actually old and existed in much wider areas several millions of years ago. Similarly due to their economic importance many crop species, although relatively young, are occupying most of the land under cultivation in the world. Therefore, merely the area of spread of a species which depends upon a number of environmental, genetical and evolutionary conditions cannot be regarded as a sole criterion to assess the age of species.

ENDEMISM

Species differ in the area of their distribution. Some are spread over a wide area. These are called *cosmopolitan*. Most of the species which are not cosmopolitan have still quite wide regional distribution. Some species are, however, restricted to a small region. These are *endemics*. *Endemism* is the phenomenon of restriction of species or taxa in a small region. The distribution of species is dependent on ecological and geographical factors and the age of the species. In young species for instance the distribution is narrow in the beginning and it is likely to grow in its area in course of time. These are called *progressive* or *expanding* endemics. On the other hand certain old species on account of gradual dwindling become restricted to a small region. These are called *retrogressive* or *contracting* endemics.

Endemics may be due to poor adaptability of a species to a wide range of ecological conditions. It may also be due to geographical barriers such as sea, high mountains etc. Evaluation of age of the endemic species helps greatly in determining its nature. If the species is of great geological antiquity with extensive distribution in the past but in the present time restricted to a narrow region, it may be due to geological revolutions and geographical and climatic changes. This is actually a relic of past extensive flora and, therefore, this is called *relic endemism*.

In certain regions around which the changes in environment are abrupt as in climatically isolated regions like islands, mountain tops etc. the number of endemics are more frequent. Wulff (1943) has shown that the Alps mountain in Europe have 200 endemic species. The islands of Madagascar have 66%, New Zealand has 72% and Hawaii 82% endemics.

The endemics of ancient origin are called *relic* or *conservative* or *ancient* or *paleo-endemics* and the newly developed endemics are called *secondary, progressive or neoendemics*. Even among endemics some are restricted to very localised spot and these are called *local endemics*. Sometimes here and there a few mutants appear which do not compete successfully and, therefore, disappear quickly. These are often referred to as *pseudo endemics*.

In India the high mountain ranges of the Himalayas form a range of distinctive climate especially at high altitudes. Chatterjee (1939) has estimated that as many as 3169 dicot species, or about 28% of the Himalayan dicots, are endemic to the region. Some of the well known endemic tree species of India are *Ficus religiosa, Ficus benghalensis, Feronia elephantum, Aegle marmelos* which are incidentally of some religious importance. Some important species like *Piper nigrum, Elettaria cardamomum* and oil crop *Sesamum indicus* are also endemic to India.

CENTRE OF ORIGIN

From the foregoing description and theories of phytogeography it is evident that plant migrations have taken place on a large scale in course of evolution of floras. Many plants have travelled almost round the world and are cultivated everywhere while some others have not moved much from the place of their origin. For every taxa there is some place where it came into existence for the first time, this region is known as its *Centre of origin*. Evolution of most of the angiospermic taxa has taken place in the geological periods when man was not evolved and hence inorder to find out the centre of origin, some direct and indirect evidences and criteria are used. Important criteria on the basis of which it is assumed that a genus or species has evolved are based on a number of features. An area where the maximum number of species of a genus is found, or the place where most primitive species are in abundance, or where the species are least affected by diseases and pests, or where the size and productivity are at their best are regarded as possible centres of origin of the taxa. From the centre of origin some clear cut relationship with the migration routes of the taxa is also of importance. Vavilov (1926) the famous Russian botanist, has made extensive studies on the centre of origin of cultivated plants. It is thought that India is the centre of origin of many crop plants like *Oryza sativa* (rice), *Saccharum officinarum* (sugarcane) and *Cicer arietinum* (gram). Other important species of Indian origin are *Cocos nucifera* (coconut), *Cajanus cajan* (arhar). *Emblica officinalis* (aonla), *Areca catechu* (supari) etc. In the Malaysian region *Musa paradisiaca* may have originated. Wheat, barley, onion, pomegranate, grapes and pea appear to have their centre of origin in the Middle East and Mediterranean regions. *Litchi chinensis* (lichi), *Camellia sinensis* (tea), *Solanum melangena* (brinjal) and *Papaver sominiferum* have originated in China. *Zea mays* (maize), *Gossypium hirsutum* (cotton), *Capsicum annuum* (chilli) and tomato are thought to belong to Central America and South Mexico while *Arachis hypogea* (groundnut) and *Hevea braziliensis* originated in Brazil. Cinchona, tomato and tobacco are also thought to have originated in South America.

PLANT MIGRATION AND BARRIERS

In nature, there are a variety of forces or agencies which tend to transport the seeds, fruits and other diaspores from one place to another and also there are variety of hurdles that prevent such transfer. The movement of floristic elements from one place to another is known as *migration* and the hurdles as *barriers*. Successful migration depends upon a number of factors such as an *effective agency of transport* like (i) wind or water current, (ii) through moving animals mostly grazers from one region to another carrying seeds either on their coat or through their dung, (iii) migratory birds carrying seeds in mud sticking to their legs, (iv) man transporting useful grains and inadvertently several weeds in grain stocks or purposely for plantation purposes, several variety of trees, shrubs and crop plants, etc. The transported diaspores do not establish everywhere because of a variety of unfavourable conditions like soil pH, presence of pests, absence of symbionts or appropriate pollinators, adequate photoperiod necessary for flowering and seed production, etc. which all act as barriers to successful migration. The opposite is also sometimes true such as the population of cactus in its native place had a regulated population but when it migrated to India, its population grew much more extensively in absence of insects that fed on cactus. Their population was ultimately biologically controlled by importing the insect as well in India.

For migrations on a large scale, absence of barriers between the centre of origin and places of distribution is essential. There are many parts of different continents which are separated by effective barriers like deep oceans or high mountains, but have many common plants (not brought by man) in present and recent geological past periods. Several theories have been propounded to explain such discontinuous distributions. Wegener (1915, 1924) has regarded that in early Paleozoic era there was one landmass only which gradually separated into two halves almost 200 million years ago into the northern *Laurasia* and Southern *Gondwana land* with Tethys sea in between. The process of land mass drifting continued and a big part of Gondwanaland from African block drifted to north and joined the Asian block. At the point of joining of these two land masses, gradually due to pushing force Himalayan mountains were formed and the southern part formed the Indian peninsula. The narrow strip of Tethys sea of this part got filled up by soil eroding from the Himalayas and formed the Indo-Gangetic plains. The hypothesis is supported by the commonness of fossil flora of Gondwanalands of African and Indian regions. Even now, it is supposed that there is slow drifting of landmass and rising of the Himalayas are taking place. Croizot (1952) has regarded that angiospermic flora have mostly migrated from the southern to the northern hemispheres through several points of land connections which he calls as *gates of angiospermae*. He has in all recognized thirteen main routes through which angiospermic flora have moved northwards.

Barriers are mainly due to land features, oceans and extensions of other water bodies, climatic conditions and biotic forces.

Land features

There are many kinds of land features such as occurrence of chain of high mountains and absence of frequent passes or low valleys, extensive deserts or dry sands, absence of adequate soil thickness or presence of pebbles and stony surface, occurrence of marshes or saline and alkaline lands (userlands), etc. that do not allow migrating diaspores to pass through or prevent their germination. There are many such barriers found in different parts of the world such as the Himalayas and the Alps mountains and Sahara desert.

Oceans

This is one of the most effective barrier across which most island originating flora have failed to cross until in recent years when man has bridged this gap through frequent travels by large ships transporting selected natural wealth and unknowingly carrying many inconspicuous biological materials contaminating resources. Even attached to the ship many aquatic plants have reached distant lands. Because of oceanic barriers we find the preponderance of endemics on isolated islands. Large water bodies other than oceans also act as barrier on a small scale. Migratory fish eating and other water loving birds bridge such lakes and act as agents of migration.

Climatic barriers

In course of evolution, the life of any taxa gets strongly conditioned by the climate of its centre of origin. Some of them are very rigidly fixed in their photoperiodic and thermoperiodic needs, and fail to produce flower and fruits at new places even if their vegetative growth is vigorous in absence of required day length or alternating range of diurnal temperatures. Thus they die out in course of time. Due to actions and reactions between the flora and climatic, edaphic and biotic forces, there has been evolution of ecological races or ecotypes adjusted to new climatic regimes.

Biotic forces

A variety of biotic forces also act as barriers for the migrating plant diaspores. Many animals selectively eat them. Many natural communities are so dense as to prevent germination and perpetuation of migrating populations under their canopy. There are often absence of necessary biotic symbionts like mycorrhizae, actinorhiaze, pollinating insects on the routes of migration. There could be many other competitive organisms as well that do not allow growth of new arrivals. Infact, there is a constant interaction between forces of migration, evolution of floras and barriers on a global scale.

Experiment No. 1. Determine the minimum size of a quadrat for the study of grassland vegetation by the species area curve method

Things required : Metre scale, thread, nails and graph paper

Quadrat is a sampling area of any shape or size but usually it is square. The size of sampling area should be such that with minimum efforts of sampling, maximum information is achieved. As we increase the size of the quadrat the number of species falling within it increases initially but later with the size increase there is no more corresponding increase in the number of species. We have to determine that size where there is a diminishing return with respect to effort input and information gained.

Lay a quadrat of 10 × 10 cm in a grassland and note the name and number of species falling within this size. Now increase the size to 20 × 20 cm and look for new or additional species that come(s). Likewise increase the quadrat size to 30, 40, 50, 60, 70, 80, 90, 100, 120, 140, 160 × 160 cm size and note the additional number of species. Record the data in a table giving quadrat size on one side and the number of species on the other. Draw a curve with the quadrat size on X axis and the number of species of Y axis. Take care that the graph is showing the area and not the length of quadrat arm. The curve initially rises but at some point it flattens, indicating that the number of species is no more correspondingly increasing. This point may be regarded as the minimum size of the quadrat required for the study of the grassland. Refer to chapter 10 and Fig. 10.3.

Experiment No. 2. Find out the frequency of different plant species found in grazed and ungrazed grassland or in a playground and outside it. Prepare their frequency class diagram and compare them with Raunkiaer's Normal Frequency diagram

Things required : Quadrat frame of 1 × 1 m size

Frequency is a measure of dispersion of individuals of a species, well dispersed species are likely to be found in most of the quadrats. Their percentage occurrence or the frequency will be high.

Selecting two contrasting grassland habitats, lay a quadrat and record the name of the species present. Lay the second quadrat at random elsewhere and record the name of the species. Similarly at random lay 10 quadrats in each of the two sites and record the presence or absence of the constituent species. Tabulate the data. A species occurring in 7 out of ten quadrats will have 70% frequency and another species occurring in four of the ten studied quadrats will have 40% frequency. Against the name of each species write the frequency value.

Raunkiaer gave the concept of frequency class by dividing the species into five frequency classes; A for those species whose frequency values are less than 20%. Frequency class B 21-40%, C is 41-60%, D is 61-80% and E is 81-100%. Now note the number of species falling in A, B, C, D and E classes and calculate their percentages. Raunkiaer prepared a World Normal Frequency Diagram giving the standard share as A-53%, B-14%, C-9%, D-8% and E-16%. Compare your result with the standard values and discuss the results. Please refer to Chapter 10.

Experiment No. 3. Find out the density and abundance values of herbaceous species in a grassland

Things required : Quadrat frame of 1 × 1 metre or of 50 × 50 cm

Both the density and abundance are expressions of number of individuals of a species per unit area. Density refers to the number per unit area which is studied while the abundance is the number per unit area of quadrats in which it occurred and not quadrats studied. Refer the formula given in Chapter 10.

Lay at random at least 10 quadrats as in the previous experiment but instead of only recording the presence or absence of occurring species, record the number of individuals of each species. Total the number of individuals of a species found in all the studied quadrats and divide it by the number of quadrats studied and it will give the density i.e. individuals per quadrat. If the quadrat size is 1 square metre then the density is so many per square metre. If the quadrat size was 50 × 50 cm or 0.25 sq.m. then multiply the average number of individuals by 4 or 16 to get per sq.m. If instead of dividing the total number of individuals of a species by the number of quadrats studied, it is divided by the number of quadrats in which it had occurred, then the value obtained is for abundance. For example if a species has 128 individuals in the ten studied quadrats, its density would be 128/10 or 12.8 per quadrat, while if the species was present in 8 quadrats only and absent in two the abundance value would be 128/8 or 16 per quadrat. Of course the values need to be expressed in per square metre.

Experiment No. 4. Find out the basal cover and canopy cover of the herbaceous plants in a given habitat

Things required : Quadrat frame, calipers or graph paper, scale

Plant occupy space on ground surface by the emerging stems and it is referred as the basal area and also occupy space is the aerial space by their foliage and it is referred as the canopy cover. Cover values are of significant importance in phytosociology as it gives an insight into the dominance status of different species, level of adjustment and competition between the component species. In grassland, foliar cover is also called herbage cover.

Take 50 × 50 cm quadrat frame and lay it at random in the study size. The purpose is to faithfully chart the vegetation on a graph paper so as to give the area covered by the foliage and the bare area. From the graph paper the area covered can be found out directly or by the use of a planimeter. This is the canopy cover. Now clip all the emerging stems. Measure the diameter of each stem by a calipers or by placing the stem pieces closely adpressed to each other on a graph paper and find the combined diameter of several stems and then convert it into area. For a faithful recording, the quadrat may at first be divided into small segment of 10 × 10 cm subsquares by using threads. Both the canopy and the basal cover are to be expressed as percentage of the ground area covered. ·

Experiment No. 5. Find out the relative frequency, relative density, relative basal cover and compute the ecological Importance Value Index of the component plant species in a grassland

Things required : Quadrat frame, graph paper, calipers, scale, thread

We have learnt to measure frequency, density and basal cover which are expressions in respect to the ground area. Relative values differ from the above as these have nothing to do with per unit area, but with respect to all other associated species. Relative frequency, relative density, and relative cover are the percentages for one species out of sum of all frequency values, density values and basal cover values of all species respectively.

Select a study area dominated by grasses or forbs and lay atleast 10 or preferably 20 quadrats. Measure the three study parameters as done in the earlier experiments. Tabulate the results. Make a few additional columns in the table. Add all frequency values of all species and note. Likewise add density values as also all basal cover values. Find out the percentage relative frequency of each species by dividing the number of occurrences of the species by the total number of occurrences of all species and multiplying this by 100. Similarly, the relative density is obtained by dividing the total number of individuals of the species in all quadrats and multiplying the resultant by 100. Relative cover is likewise obtained dividing the total basal area of the species in all quadrats by the total basal area of all species in all quadrats and then multiplying it by 100. Enter these values in the additional columns against each species in the table. Refer to Fig. 10.7 to prepare polygraphs to show the relative values in the scale of 100. Add all the three relative values to obtain the ecological Importance Value Index of each species out of 300. The fourth radius in the polygraph has to be divided in 300 parts and IVI is shown on this radius. Now arrange the species according to their IVI values.

Experiment No. 6. Find out the Simpson's Index of Dominance in a grassland community

Things required : A quadrat frame

This is a reflection of relative importance of component species in respect of the number of individuals. In any plant community there are a large number of species but there may be a few with much larger number of individuals. So their index of dominance will be high. The Index is out of a maximum of 1. Simpson has given the formula e = Σ $(n_i/N)^2$ where c is the Index of dominance, n_i is the number of individuals of a species and N is the number of individuals of all species and Σ refers to summation. Other parameters than the number of individuals can also be considered, e.g. the biomass or productivity.

Lay 10 quadrats in a crop field and 10 in a play field. Record the number of individuals of each of the component species found in them. Apply the above formula to find out the Index of dominance for the different species. In the crop field the index may be very high for the sown crop, say about 0.8 or more. For example if in the field were 10 species and total 100 individuals in a quadrat with 91 of the sown crop and one each of the remaining nine weeds then the Index of Dominance of the crop would be $(91/100)^2 + (1/100)^2 + (1/100)^2$ + ... upto 9 times like this. It comes to 0.829. If in the other field all the species have equal number of individuals, say 10 individuals of each of the ten species present them the index of dominance of every species will be :

$(10/100)^2 + (10/100)^2 + (10/100)^2 + (10/100)^2$ ... ten times and it comes to 0.1.

Experiment No. 7. Find out the similarity and dissimilarity index between two stands of grasslands (protected and grazed)

Things required : A quadrat frame

Sorensen has given the formula S = 2C/(A + B) where S is the similarity index, A is the number of species per unit area in site one and B is the number species in the second site. C is the number of species common to the two sites.

Lay 10 quadrats at random in field one and 10 quadrats in the second field, all at random and note the name of the species present in each of the quadrat. Apply the above formula and find out the similarity index. If there are many common species the similarity index will be high i.e. 0.6 or more.

Dissimilarity index is 1-S. So if the similarity index is 0.7 then the DI will be $1 - 0.7 = 0.3$.

Experiment No. 8. Find out Margalef's and Menhinick's Diversity Index in a vegetation stand

Things required : A quadrat frame

Diversity refers to variety and it can be at different biotic levels but the number of species is regarded as an important parameter. The simplest index is of Margalef who has given the formula : $d = s/1000$, i.e. the number of species per 1000 individuals in a vegetation stand. For this lay quadrats at random and record the name of the species and number of individuals. As soon as the combined number of all species reaches 1000, count the number of species encountered upto the stage.

Menhinick has given the formula : $d = S\sqrt{N}$ where d is the index of dominance, S is the number of species and N is the number of individuals of all species.

For Shannon's Index of Diversity refer to Chapter 10.

Experiment No. 9. Estimate the standing crop herbage biomass and the total biomass in a grazed and ungrazed field

Things required : 25 × 25 cm quadrat frame, trowel or khurpi, tray to carry and washing monoliths, oven, balance

Select the study site and lay at random ten quadrats. For herbage biomass the aerial portion has to be clipped close to the ground surface. Samples from each of the quadrat may be separately oven dried at 80ºC for 48 hours. The herbage biomass is expressed as grams or kilograms per square metre or tons per hectare.

For measuring the total biomass the entire plant sample including the roots is to be taken out from each quadrat for oven drying. In view of the difficulty of digging and transporting, smaller quadrats of

15 × 15 cm may be taken. Lay the quadrats at random. Dig out upto 25 cm depth, take out the monolith, wash the soil in a tray with water carefully avoiding breakage of fine roots. The plant material of each quadrat may be oven dried together to obtain standing crop biomass per metre square. If desired the roots and shoots may be separately weighed. Dominant and rest of the species may also be separated for detailed analysis.

Experiment No. 10. Measure the net rate of primary production by short term harvest method

Things required : Quadrat frame, oven at 80°C and weighing balance

The rate at which the biomass is built in unit area over a known period of time is called productivity. Therefore, this experiment has to be carried out at short period intervals of 15 days or one month for one year or more. For classwork, you may sample crop plants sown at 15 days intervals beforehand, so that at the time of practicals, samples of 15 days, 30, 45, 50, 75 and 90 days old plants are available at one time. Take out for biomass samples in monoliths, wash carefully, oven dry and weigh. In such a sown crop there is every chance that there will be increase in the standing crop biomass in successively older samples. The difference in biomass from 30 days and 15 days $(T_2 - T_1)$ gives the rate of productivity during the period. Like that the productivity at short intervals are obtained. Summation of successive biomass difference values gives the productivity over the studied period. In field conditions the differences in biomass is not always positive. For example, in a grassland the biomass in April may be more than in dry month of May and so also in June, but with rainfall the biomass may again increase in July and August. In such cases for calculating annual productivity the positive biomass difference values are considered for summation and the negative values are ignored and not deducted.

For example if the biomass g/m² values for January to next January i.e. 13 months are 45 grams, 51, 68, 72, 50, 30, 60, 83, 88, 80, 70, 68, 47 gram respectively, then the 12 difference values would be 6, 17, 4, –22, –20, 30, 23, 5, –8, –10, –2, –21. The annual productivity will be sum of positive values (6 + 17 + 4 + 30 + 23 + 5) i.e. 85 g/m²/yr.

Experiment No. 11. Study the biomass profile of the plants in a herbaceous ecosystem

Things required : Trowel, quadrat frame, scissors, oven and balance

Take out soil monolith with herbaceous plants, wash carefully and bring the material to laboratory. The shoots and roots may be clipped at 10 cm height and depth profile. Oven dry each of the segment separately. Record the data of biomass per quadrat and then convert to per square metre. Express the results in height profile of first 10 cm, 10-20 cm, 20-30 cm and so on and for roots surface to 10 cm depth, 10-20 cm depth and 20 to 30 cm. This can also be illustrated in form of a diagram on a graph paper.

Experiment No. 12. Measure the Leaf Area and Kemp's Constant for a dicot and monocot plant and the Leaf Area Index of a Cropland

Things required : Plainmeter or graph paper and balance. Fresh leaves of plants to be studied

Leaves are not rectangular or perfect circular. Therefore some corrections to Length × Breadth has to be made depending upon the shape and this is called K or Kemp's constant. The value of K varies for different leaf shapes. The leaf area index is the total area of leaf over per unit ground area. The area of these leaves can be measured by a simple instrument - *plainmeter* or a modern but expensive instrument - *leaf area metre*. It can also be measured by placing the leaves on graph paper, drawing its outline and then counting the squares and the area is found out. Alternatively the leaf sample is superimposed on a paper and along the margin the paper is cut. This paper of shape and size equal to leaf is weighed. Another piece of the same paper is cut out say of 10 cm × 10 cm and weighed. Suppose the leaf shape and paper weight is 3 g and of 100 sq.cm. paper is 2 g then the leaf would be 2 g is equal to 100 sq.cm. 3 g paper weight will equal to $\frac{100 \times 3}{2} = 150$ sq.cm.

After actually measuring the leaf area; measure the length and breadth (at the broadest point) of the leaf.

Actual leaf area = L × B × K, or K = $\dfrac{L \times B}{\text{Actual leaf area}}$

In narrow long leaves of wheat or rice the K value is about 0.9, in maize it may be 0.8 and in broad leaves like mango or guava it may be 0.6.

For measuring the LAI select a crop field, lay a quadrat of 25 × 25 cm, collect all leaves occurring within this, find out the area of all leaves. LAI is equal to total leaf area/total ground area. Repeat it in about 10 quadrats. LAI may be 2, 3, 4, 5 or even more. Primary productivity increases with increase in LAI upto 4 or 5 and then levels off due to too much of self shading effects.

Experiment No. 13. Prepare the shoot profile of a tree stand along a line transect

Things required : Graph paper, pegs, string, metre scale

In a forest, the trees, shrubs and herbs adjust themselves in the available space on ground and in the aerial space. The result is both due to competition and adjustments. Lay a line transect by taking two points between which the profile diagram is to be drawn. On a graph paper also demarcate these points at the base of the paper. The real position of each emerging stem can easily be marked on the graph paper by measuring with a metre scale. Upto about two metres height also we can use the scale, but at higher levels the height has to be determined by the use of klinometer. There are other methods also be measure tree heights.

Experiment No. 14. Prepare a shoot and root profile diagram on a bisect in a grassland

Things required : As in experiment No. 13, trowel

Select a site, and lay a transect of about 2 metres. Cut carefully a soil profile of about 50 cm depth about 20 cm away from the transect. Now with the help of trowel or khurpi and knife work your way to expose the roots without disturbing the shoot or cutting the root. Once the roots are exposed the bisect is ready for its faithful recording on a graph paper. Bisect diagram gives a good idea of adjustment in the space, both aerial and underground.

Experiment No. 15. Prepare the biological spectrum of your college campus or Botanical Garden using Raunkiaer's life form classification

Raunkiaer had given a classification of plants based on the position of buds that perennate during the adverse season (described in Chapter 10). He classified plants into Phanerophytes, Chamaephytes, Hemicryptophytes, Cryptophytes and Therophytes.

Demarcate the study area to be covered. Walk along a straight line and record the name and life form of the plants encountered. Repeat the process by walking on another line a few metres away from the first. You may also move at random or in crisscross way. The idea is to record the name of all species present in the study campus and to note its life form. All shrubs and trees have perennating buds located well above the ground as the branch tips remain alive all the year round. They are Phanerophytes (Ph.). Many of the cold country plants during severe winter perennate by buds located upto 25 cm height and the parts above which die. These are Chamaephytes (Ch) and those in which the perennation is done by parts (buds) located at the soil surface or just below it are called Hemicryptophytes (HC). Those plants which survive by underground parts such as bulbs are called Cryptophytes (Cr) or Geophytes. All annuals which die in seasons adverse to them and regenerate by seeds are called Therophytes (Th).

Raunkiaer gave a normal biological spectrum of the world flora based on a very extensive study. The values are Ph 46%, Ch 9%, HC 26%, Cr 6%, and Th 13%.

Prepare the biological spectrum diagram and compare with the above values and discuss the results.

Experiment No. 16. Make detailed study of external (morphological) features of adaptation in surface floating plants. Collect samples, draw their diagrams and record features which are favourable for free floating plant life

Experiment No. 17. Make detailed study of the features of adaptation in rooted hydrophytes with floating leaves

Experiment No. 18. Make detailed study of the submerged hydrophytes

Experiment No. 19. Bring out features of adaptation in emergent wetland species

Experiment No. 20. Make detailed anatomical study of different category of hydrophytes for the ratio of lacunate tissues (aerechymatous tissue) with other tissues

For all these adaptation experiments, student will collect different material and the results will differ from plant to plant and to some extent on the type and position of wetland and the season of collection (for details see Chapter 4).

Experiment No. 21. Collect different types of succulent xerophytes and study the features of ecological interest. Draw their diagrams

Experiment No. 22. Make anatomical studies of different kinds of xerophytes such as *Nerium, Casuarina, Acacia, Euphorbia,* etc.

Details can be found in Chapter 4.

Experiment No. 23. Compare the soil properties of grassland, cropland and drying pond bed for texture, pH, carbonate, nitrate, and base deficiency by spot test

Things required : China clay plate (or cavity tile), test tubes, pH paper or Universal indicator, dil. HCl, 0.002% diphenylamine in conc. H_2SO_4, alcoholic saturated solution of ammonium thiocyanate, H_2O_2.

Collect soil samples from the above mentioned sites :

  (i) For texture rub the soil sample between the thumb and forefinger. If the feeling is of coarse type, it is sandy, if extremely fine and smooth it is clayey, if only somewhat rough it is silty and if a mixture of the above, it is loamy.

 (ii) For pH take a small quantity of soil sample (about 2 g) in a test tube and add about five times of its volume distilled water. Shake it well and leave to settle it for a few minutes. Test with a pH paper or by adding a few drops of Universal Indicator. Compare the colour developed with the colour chart on the pH booklet or indicator bottle.

(iii) For carbonate, take a pinch of soil sample in the test tube and add dil. HCl. Effervescence indicates the presence of carbonate. An arbitrary scale of 1 to 5 from trace to rapid effervescence is used to record carbonate. 1 represents trace, 2 slightly high, 3 is moderate, 4 is high and 5 is very high.

(iv) For nitrate take small quantity of soil samples on cavity tiles separately and add to each a few drops of 0.002% diphenylamine in H_2SO_4. In presence of nitrate blue colour develops. Here also a qualitative value of 1 for feeble, 2 for moderate, 3 for good, 4 for rich and 5 for very high nitrate content and the blue colour intensity increases from very light to deep blue.

(v) For base deficiency test for iron is done as its presence indicates the deficiency of bases like Ca, K, Mg etc. Take a little quantity of soil in test tube, add saturated alcoholic solution of ammonium thiocyanate, and shake well. Red colour develops due to ferric iron. Add a new drops of H_2O_2 and the red colour may intensify as ferrous iron also gets converted into ferric. Light red to intense red grade is divided into 5 levels where 1 represents least deficiency i.e. rich soil, and 5 as highly base deficient. Drying pond bed soil may show maximum red colour or highest base deficiency.

Experiment No. 24. Estimate the moisture content of soil samples collected from surface, 15 cm and 30 cm depth

Collect soil samples from surface and freshly cut soil profile at 15 and 30 cm depths without allowing moisture loss. Collect in polythene bags, bring to laboratory, weigh about 50 grams of it and place in oven at 105°C for 48 hours. Weigh the soil after drying. Calculate the moisture content on the oven dry weight of the soil e.g. if the 50 g wet sample on oven drying became 45 g the moisture content % will be calculated as 45 g of dry soil contains 5 g of water, hence 100 g of dry soil will contain 5/45 × 100 or 11.1%.

MEASUREMENTS AND CONVERSIONS

Some measurement prefixes and their values :

Geo.....	10^{20}	(geo)	Deci.....	10^{-1}	(d)
Tera.....	10^9	(T)	Centi.....	10^{-2}	(c)
Mega.....	10^6	(M)	Milli.....	10^{-3}	(m)
Kilo.....	10^3	(K)	Micro.....	10^{-6}	(μ)
Hecto.....	10^2	(h)	Nano.....	10^{-9}	(n)
Deca.....	10	(da)			

1. Length

1 cm = 0.01 m = 0.3937 inches

1 m = 0.001 km = 39.37 inches

1 km = 1000 m = 0.62137 miles

1 inch = 2.54 cm = 0.0833 ft

1 ft = 30.48 cm = 12 inches

1 yard = 3 ft = 36 inches = 0.9144 m

1 mile = 1760 yd = 1.60935 km

2. Weight

1 mg = 0.01 g = 0.0154 grain = 0.0000353 ounces

1 g = 15.43 grain = 0.035 ounces

1 kg = 1000 g = 35.274 ounces = 2.05 pounds

1 pound (lb) = 453.59 g = 16 ounces

3. Volume

1 ml = 0.001 litre = 0.002113 liq. pints.

1 lit = 1000 ml = 61.025 cubic inches

1 gal (US) = 0.8269 gals. (British) = 3.7853 lit.

1 liq. pint = 0.473167 lit. = 16 ounces.

1 gal. (British) = 1.2 US gal.

4. Area

1 sq cm = 0.155 sq inch

1 sq m = 10,000 sq cm = 1550 sq inches

1 ha = 10,000 sq m = 107600 sq ft = 2.47 acres
1 sq km = 100 ha = 247 acres = 0.3861 sq miles
1 sq mile = 640 acres = 259 ha = 2.59 sq km
1 sq yr = 9 sq ft = 0.846 sq m

5. Pressure

1 atm. = 1.01325×10^6 dynes sq cm = 760 mm of Hg
1 cm of mercury (at 0°C) = 0.013158 atm.

Other miscellaneous equivalents :

1 Å (angstrom) = 10^{-10} m = 3937×10^{-12} inches
1 barrel = 42 gallons
Temp. degree centigrade = (°C × 9/5) + 32 = °F
Temp. degree fahrenheit = (°F − 32) × 5/9 = °C
1 chain = 792 inches = 20.12 m = 22 yd
1 Dalton = 1.650×10^{-24} g
1 erg = 10^{-7} Joules = 1.341×10^{-10} horsepower = 10^{-10} kilowatt
1 footcandle = 10.764 lumen/sq m or lux
1 furlong = 220 yd = 660 ft = 40 rods
1 joule = 10^7 ergs = 0.7376 ft pounds = 2.39×10^{-4} kg cal
1 km/hr = 27.78 cm/sec = 16.67 m/min = 54.68 ft/min = 0.6214
    miles per hour
1 light yr = 5.9×10^{12} miles = 9.464×10^{12} km
1 lux = 0.0929 ft candles
1 mile/hr = 88 ft/min = 1.609 km/hr
1 kg/sq m = 10 t/ha
1 sq m = 10^{-4} ha = 10.76 sq ft
1 sq ft = 0.0929 sq m = 0.111 sq yd
1 quadrant = 90 degrees angle = 5400 minutes (angle)

Glossary

Abiotic : The non-living components in an ecosystem.

Abundance : The number of individuals per unit area calculated on the basis of number of quadrats of occurrence.

Abyssal : Deep sea region ranging between 2000-6000 m depth.

Acclimatization : Adaptation to a changed or new environment.

Accretion : Gradual addition in reference to land, inorganic bodies or minerals in soil.

Acre : Land area 43.560 sq ft, Acre feet = 1 ft thick column of water over 1 acre area.

Adaptation : Structural, functional and or behavioural changes for a better adjustment to its surroundings.

Adsorption : Settling or adhesion of a substance on the surface of another substance, usually at very small particles or ionic levels.

Aeolian : Soil transported by wind action (also eolian).

Aerosol : Suspended fine particle, usually liquids in the atmosphere.

Aquaculture : In specially created or managed ponds growing fishes, prawns and shellfishes.

Alien species : A species occurring at a place outside its native or natural place, *exotic* species.

Allelopathy : Inhibitory activity of one plant species commonly through exudation of chemicals from their roots on the seed germination or growth of other associated species. Also known as allelochemic interaction and *antibiosis*.

398

Alluvium : Soil carried and deposited by water current.

Alpine : Very cold climate or referring to high mountains such as Alps or Himalaya.

Ambient : The outside air.

Amphiphytes : Plants that grow in two contrasting conditions such as on land and in water.

Aquifer : Underground natural water storages, sometimes flowing out through artesan wells due to sufficient pressure on the storage water.

Associates : Species occurring together.

Atmosphere : The gaseous envelope around the earth, the air. It has many gases but nitrogen, oxygen, carbon dioxide and water vapour are most important. It exerts a pressure of 14.7 pounds per sq inch or in terms of column of mercury equivalent to 760 mm.

Autecology : Study of a species or its population with an emphasis on ecological life-history.

Autotrophic : Related to producing its own food such as the green plants.

Barrier : Hurdles in the migration of flora on geographical scale such as mountains and oceans.

Bathyal : Ocean zone between 200-2000 m depth.

Beaufort scale : Wind velocity scale named after the inventor Sir F. Beaufort of the British Navy.

Bench mark : A reference point of permanent structure to measure elevation.

Benthic : Of bottom water body; benthos for organisms living on the bottom of lakes, river or sea.

Biodiversity : Sum total of variability among living organisms at genetic, species and ecosystem levels.

Biological control : Use of some organisms to control the growth of some other pests or of weeds, etc.

Biological magnification : Successive increase of concentrations of some chemicals through different food level organisms such as of DDT from water to algae to herbivore fish to carnivore fish to fish eating birds and to man.

Biological spectrum : The range and ratio of different life forms in an area such as of phanerophyte, therophyte, etc. or the range of different levels of biotic organization from sub-cellular to cellular, to organ, individual, population and community.

Biome : A large unit of climax communities usually representing well defined climatic zone.

Biomonitoring : Using bioindicator organisms to monitor the speed and quantum of change in ecosystem (due to pollution toxicity or nutrient enrichment etc.).

Biomonitors : Certain species whose presence indicates particular kind of toxic material, pollutant, stresses or resources in the surrounding environment.

Biosphere : All parts of earth where organisms live.

Biosphere reserve : Major vegetational zone or characteristic habitats set aside and protected against human disturbance to act as a reference area of natural habitat.

Canopy : The foliage cover at the top layer of the plant.

Carnivore : Organisms feeding on other living organisms or flesh.

Carrying capacity : Maximum number of a particular species which could be maintained indefinitely on an area or renewable resource stock.

Climax : The community that perpetuates itself on a habitat indefinitely, i.e., in balance with the environment.

Climograph : Rainfall-temperature relationships diagram.

Coliform bacteria : Those infesting the intestine of man or other animals. Fecal coliform bacteria occur in fecal material : *coliform index* is for expressing the relative purity of water based on these bacterial counts.

Colluvium : Soil deposits at the base of hills, transported by gravity.

Commensalism : Association of two kinds of organisms in which one or both are benefitted.

Community : Group of organisms and their populations living together with interacting influences.

Compensation point : At certain depth in a water column where the reduced light intensity is just sufficient to allow only as much photosynthetic production as is completely used in the respiratory process.

Competition : Act of using and defending a resource in a way as to reduce its availability to associated individuals or species.

Conservation : Managing any segment of biosphere for preservation and greatest sustainable use and yield for present and future generations of mankind.

Contour : Lines on map connecting points of equal value of depth, ups and down on a land surface or the bed of an aquatic body.

Cryogenic conservation : Preservation of microbes, embryos, seeds, diaspores, semen etc. at very low temperature below freezing point. These remain alive but fully dormant.

DBH : Diameter (of a tree) at breast height, i.e. at 1.3 m or 4.5 ft.

dB : Decibel, a unit of sound intensity on a logarithmic scale.

Delta : Alluvial sand deposits at the mouth of a river, usually triangular in shape and hence named after the Greek letter delta (Δ).

Density : Number of individuals per unit area.

Desalination : Removing salt from sea water.

Desertification : The process of desert formation; permanent loss of plant life in a region, loss of capacity to sustain good plant growth.

Diversity : Number of different species growing together.

Dredging : Removal of river or lake bottom mud to deepen the same.

DDT : One of the commonest chlorinated hydrocarbon, Dichloro diphyenyl trichloroethane which is used as an insecticide, particularly for mosquito control. It reaches non-target organisms such as man through food chain and accumulates in the fatty tissues.

Ecad or ecophene : It is a form variation within a species associated with some environmental condition but the character is not genetically fixed.

Ecological bomerang or back lash : The unforeseen cancelling of projected gain or unforeseen adverse or catastrophic reaction from a natural system in response to human interference, management or developmental activities.

Eco-development : Development taking prior consideration of the holistic impacts on the concerned system.

Ecological pyramid : Diagrammatic representation of numbers, biomass, energy or productivity of different trophic level organism with autotrophs at the base superimposed by those of herbivores, carnivores of the first order, second order and so on.

Ecological services : The human society and individuals receiving benefits from ecological processes and ecosystem functions, such as oxygen enrichment, CO_2 sink, absorption of pollution, litter decomposition, etc.

Ecosystem : The structural and functional entity of biotic communities and their environment. It may be of any size.

Ecotone : The transition zone between two different adjacent communities showing characteristic edge effect.

Ecotoxicology : Study of toxic material in environment and their impacts on organisms and human health.

Ecotype : Genetically distinct ecological races within the framework of species and variation in character is associated with variation in some environmental condition.

Effluent : Liquid discharges or flowing out materials that usually pollute the environment.

Electrostatic precipitator : Device to collect dust from factory chimney smoke by imparting electrical charge to dust particles to reduce air pollution.

Emission standard : The maximum permissible or safe limit of discharge of pollutants.

Endangered species : Species which are on the verge of extinction.

Endemism : Restricted area of distribution of a species to an isolated region.

Energy conserving efficiency (ECE) : Percentage of half of the incident solar radiation that is fixed through photosynthesis by plants in unit area and time.

Environment : The totality of all kinds of influencing forces that surround an organism, both non-living and living.

Eolian : Same as aeolian, wind transported soil.

Erosion : The physical removal of top soil by wind or water, wearing under the forces of wind, rain beating, water current, tramplings, scrapings, etc.

Ethnobotany : Studies about traditional use of wild plants mostly by tribals and village folk for rituals, medicines and food. Broadly, all human uses of plants.

Eutrophic : Lakes rich in nutrients and productivity, i.e. choked with rich growth of vegetation.

Fall out : The descending radioactive fission products from atmosphere to earth's surface.

Fault : Deep cracks in earths surface, fractures in rock layers, etc. due to geological events.

Flagship species : Popular and charismatic species that stimulate conservation awareness.

Flora : The botanical composition of a place or a book describing taxonomic character of plants of a region.

Fecundity : The rate of offspring production at individual level.

Feedback : The reverse direction flow (of information or any other things) from the products back to the reactants.

Fission : Splitting of atom into lighter fragments and release of energy and neutrons.

Flow metre : The instrument used to measure the quantity of water or effluents flowing out of a system over unit time.

Flu gas : The dust and other small fragments of burnt up material dispersed by gas mixture usually emitted through chimneys.

Food chain : The sequence of different trophic level organisms starting from primary producers to top carnivores.

Food web : The network of food chains in an ecosystem.

Fossil fuel : The fuel produced out of fossil plants and animals such as coal, petroleum and petroleum gas.

Fresh water : Drinking water, less than 0.2% salt content.

Fusion : Combining of nuclear or other things; combination of atoms such as of hydrogen forming helium and associated with release of energy (as in sun).

Gene bank : Where germ plasms are stored in *in vitro* culture of cryogenically.

Green house effect : The heating effect through reradiation. In glass house the sunlight enters, but on reradiation the heat is trapped inside the glass house thus increasing the inside temperature. Similar effect is caused by atmosphere, dust; water vapour, ozone, carbon dioxide, etc. which allow the sunlight to reach ground surface but trap the reradiating heat waves.

Gross production : The total increase in organic matter, energy fixation due to photosynthesis over a unit area (or volume) and time. If the loss due to respiration is deducted it is *net production*.

Guild : Group of species found together and sharing same food resources.

Half life : The time taken by a radioactive isotopes (or by pesticides) to come to its half strength or decay to its half level from the initial.

Halophytes : Plants adapted to grow in saline soils or sea coast marshes.

Heliophytes : Plants adapted to grow in open sun.

Holistic : Taking all the aspects and components of ecosystem into consideration.

Homeostasis : The capacity of ecosystem to resist changes due to disturbances or to return to balanced state.

Humic acid : The different kinds of acids formed due to partial decay of litter in the soil. These are decay resistant dark coloured acids.

Humus : Dark coloured fine powder of organic matter formed after the decomposition of litter.

Hydrology : The study of water storages, distribution and movement on earth's surface.

Hygrograph : Instrument to measure relative humidity.

Incineration : Complete combustion of materials in a controlled condition.

Indicator species : Species whose presence indicates certain status in ecosystem like some minerals and metals or fire, etc.

Infiltration rate : The maximum rate of water entering the soil system through soil pores when it is abundantly available at its surface.

Infrared radiation : Wavelengths just larger than the red wave between 0.7 to 1.0 μm. It has a heating effect.

Ionosphere : The upper most layer of atmosphere above the stratosphere, where ionization takes place. In day time its lower limit is

56 km and in night 96 km above the earth's surface. It reflects radio signals.

Isotherm : Lines on a map connecting places with equal temperature.

Isotope : Forms of an element with same number of protons but different number of neutrons. They differ in radioactive behaviour and atomic weights but have the same atomic number and chemical properties.

Keystone species : Species with disproportionate, far-reaching effects on associated species, whose loss causes major changes in eco-system.

LC 50 : Indicates the concentration of chemicals that would kill 50% of the target organisms.

Leaching : The process of downward movement of soluble nutrients and other salts through soil profile with moving water.

Lentic : Standing water such as in pond or lake.

Limnology : Study of fresh waters for its various physico-chemical and biological characteristics.

Littoral : Shallow zone of a water body, also on sea coast, the zone receiving low to high tides.

Lotic : Running water such as streams and rivers.

Lux : A measure of light intensity obtained at 1 m distance from a standard candle, 10.76 Lux is equal to 1 foot candle.

Meandering : Loop-like changing course of rivers from time to time.

Merological : Study in parts as against holological, i.e. holistic study.

Microcosm : A miniature ecosystem often controlled to study the cause-effect relationship of environmental factors on test organisms or populations. Often used in aquatic ecology.

Migration : Movement of plant diaspores or of animals such as birds to distant places along some set routes.

Mimicry : Adaptation of structure and colour to merge into its habitat in order to avoid detection by predators.

Mist : Fine droplets larger than fog (i.e. from 40 μm to 500 μm) of liquids in the air.

Thermocline : The sharp zone separating the warm epilimnion and cold hypolimnion in water body.

Trophic : Related to food and feeding habit.

Troposphere : The zone of atmosphere containing clouds and moisture.

Vulnerable species : Species which are on the verge of becoming endangered.

Watershed : The area drained by a stream or river.

Water table : The upper layer of ground water.

ZPG : *Zero Population Growth*, i.e. each couple producing only two children so as to stabilize the population at the existing level.

References

Abrol, Y.P., Wattal, A. Gnanam, Govindjee, D.R. Ort, A.H. Teramura, 1991. Global climatic changes on photosynthesis and plant productivity. Oxford IBH, New Delhi, 824.

Adams, R.M., Glyer, J.D., McCari, B.A. (1998). In Assessment of crop loss from air pollutants (Eds. W.W. Heck, D.J. Tingey and D.C. Taylor, Elsevier, London.

Agarwal, B. and Tiwari, S.C., 1988. Effect of prescribed fire on plant biomass, net primary production and turnover in a grassland at Srinagar, Garhwal Himalaya, India. Proc. Nat. Acad. Sci., India (B), 58 : 291-302.

Akkermans, A.D. and C. van Dijk, 1981. Non-leguminous root nodule symbioses with actinomycetes and *Rhizobium*. In W.J. Broughton (Ed.) *Nitrogen Fixation & Ecology,* 57-103, Oxford Univ. Press, London.

Ambasht, N.K., 1993. Effect of enhanced ultraviolet-B irradiation on crop-weed competition, Ph.D. Thesis, Banaras Hindu University.

Ambasht, N.K., 1998. Ozone depletion and UV-B radiation enhancement impacts. In : Ambasht, R.S. (Ed.) *Modern Trends in Ecology and Environment.* Backhuys Pub., Leiden, 307-317.

Ambasht, N.K. and M. Agrawal, 1994. Enhanced ultraviolet-B radiation and its impacts on agricultural crops—A review. Energy Environment Monitor, 10(2) : 141-146.

Ambasht, N.K. and Agrawal, M., 1994b. Effect of enhanced UV-B radiation on the physiology of field grown *Sorghum vulgare* plants.

Proc. VII IUAPPA Regional Conf. for Pacific Rim on Air Pollution and Waste Issues III : 111-119.

Ambasht, N.K. and Agrawal, M., 1995. Physiological responses of field grown *Zea mays* L plants to enhanced UV-B radiation. *Biotronics*, 24 : 15-23.

Ambasht, N.K. and Agrawal, M., 1997. Influence of supplemental UV-B radiation on photosynthetic characteristics of rice plants. *Photosynthetica*, 34 : 401-408.

Ambasht, N.K. and Agrawal, M., 1998. Physiological and biochemical responses of *Sorghum vulgare* plants to supplemental ultraviolet-B radiation. *Canadian Journal of Botany*, 73 : 1290-1294.

Ambasht, N.K. and Agarwal, M., 2003-04. Effect of enhanced UV-B and tropospheric ozone on physiological and biochemical characteristics of field grown wheat. *Biologia. Plantarum*, 47(4) : 625-628.

Ambasht, N.K. and Ambasht, R.S., 2003. Applied ecology of biodiversity. In : Ambasht R.S. and Ambasht, N.K. (eds.), *Modern Trends in Applied Terrestrial Ecology*. Kluwer Academic/Plenum Publishers, New York.

Ambasht, R.S. and Ambasht, N.K., 2003. Conservation of soil and nutrients through plant cover on wetland margins. In : Ambasht R.S. and Ambasht N.K. (eds.). *Modern Trends in Applied Aquatic Ecology*. Kluwer Academic/Plenum Publishers, New York, 269-280.

Ambasht, N.K. and Ambasht, R.S.. 2005. Plant responses to changing ozone and UV-B Scenario, A Review. Proc. Nat. Acad. Sci., India, 75 : 159-168.

Ambasht, R.S. and Ambasht, N.K., 2006. Conservation of wetlands— Global and Indian perspectives in advancing frontiers of ecological research in India (Ed. A.K. Kandya and A. Gupta), Bishen Singh, Mahendra Pal Singh, Dehra Dun, 247-257.

Ambasht, R.S., 1958. Studies on the underground parts of *Alhagi camelorum* Fisch, *Proc. Nat. Acad. Sci. (India)*, 28(B) : 106-107.

Ambasht, R.S., 1962. Root habits in response to soil erosion silting and inundation. *Ph.D. thesis, Banaras Hindu University (India)*.

Ambasht, R.S., 1963. Ecological studies on *Alhagi camelorum* Fisch, *Tropical Ecology*, 4 : 72-82.

Ambasht, R.S. 1963. (a) Ecological problems in the tropics—erosion in relation to vegetation cover in two areas of Varanasi. *Proc. Nat. Acad. Sci. (India)*, 33 : 158-162.

Ambasht, R.S., 1964. Ecology of river bank, *Proc. symp. Recent Advances in Tropical Ecology*. ISTE Varanasi, 466-470.

Ambasht, R.S., 1970. Ecosystem study of a tropical pond in relation to primary production of different vegetation zones. 1971, *Hidrobiologia*, 12 : 57-61 (Rumania).

Ambasht, R.S., 1970. Conservation of soil through plant cover of certain alluvial slopes in India. *Proc. IUCN XI Tech. Meeting*, 44-48 (Switzerland).

Ambasht, R.S., 1970. Conservation Ecology. *In Proc. School on Plant Ecology* (Ed. Misra and Das), Oxford and I.B.H., 271-276.

Ambasht, R.S., 1972. Pollution Problems in India Context. *Current Events*, 28(11), 35-37 (*Science News and features*).

Ambasht, R.S., 1974. Ecological implication in the control measures of aquatic weeds. Mimeographed. *Proc. S.E. Asian Workshop on Aquatic Weeds. BIOTROP, Indonesia.*

Ambasht, R.S., 1978. Observations on the ecology of various weeds on Ganga river banks at *Varanasi*, India. *Proc. S.E. Asian Workshop on Aquatic Weeds. BIOTROP, Indonesia.*

Ambasht, R.S., 1978. Observations on the ecology of various weeds on Ganga river banks at *Varanasi*, India. *Proc. VI Asian Pacific Weed Science Conference, Jakarta* (Indonesia), Vol. II, 109-115.

Ambasht, R.S. 1978. Ecology education and research in India in relation to weed problems *Proc. VI*, Vol. II, 679-684. *Asian Pacific weed science Conf. Indonesia.*

Ambasht, R.S., 1980. Grassland Resources : A Synthesis. In ecology and Development (Ed. J.I. Furtado), *ISTE Univ. of Malaya*, 131.

Ambasht, R.S., 1981. Plant Resources of Diverse Ecosystems around Varanasi. *Souvenir. Ind. Sci. Congr.*, 13-16.

Ambasht, R.S., 1981. Responses of Aquatic Plants to Pollution : Proc. *WHO Workshop on 'Biological Indicators and Indices of Environmental Pollution*, Osmania Univ., Hyderabad, India, 61-66.

Ambasht, R.S., 1982. India In *'Biology and Ecology of Weeds'* (Ed. W. Holzner and M. Numata), W. Junk & Co., Netherlands, 267-275.

Ambasht, R.S., 1982. Ecological Perspectives of Conservation of Nature and Natural Resources. *Prajna Science special, Banaras Hindu University*, 323-336.

Ambasht, R.S. River Ecology, *Nat. Conf. River Pollution and Human Health,* Nat. Env. Cons. Asso., India, Vol. I, 70-72.

Ambasht, R.S., 1985. Primary productivity and soil and nutrient stability an Indian Hilly Savanna Lands. In J.C. Tothill and J.J. Mott (Ed.), *Ecology and Management of World's Savannas.* Australian Acad. Sci. Canberra, 217-219.

Ambasht, R.S., 1986. Forest—the key of all natural resources. *Smarika* (Ed. A.K. Sinha), Vikas Bharati, Ranchi, 1-3.

Ambasht, R.S., 1986. Natural Resource Conservation in stressed environments. In Ambasht, R.S. (Ed.). *Recent Advances in Environmental Biology.* ERL, Botany Deptt., Banaras Hindu University, 14-21.

Ambasht, R.S., 1987. On the fragility of sloping land ecosystems. *Recent Advances in Plant Sciences*, (Ed. M.R. Sharma and B.K. Gupta), 344-354.

Ambasht, R.S., 1989a. Community structure, biomass, production and energy aspects of aquatic macrophytes and soil, water and nutrient conservation of Ganga River. Chapter 19, Ganga—A Scientific Study (Ed. C.R. Krishna Murty, K.S. Bilgrami, T.M. Das, R.P. Mathur), 141-149.

Ambasht, R.S., 1989b. State of Art of Flood plains of River Ganga and its tributaries and wetlands of North India. Proc. *International Conf. on Wetlands and River Corridor Management,* at Charleston, USA.

Ambasht, R.S., 1993. Conservation of some disturbed Indian tropical forest ecosystem : *Restoration of Tropical forest Ecosystem* : (ed. H. Lieth and M. Lohmann) : Kluwer Acad. publ., The Netherlands, 203-208.

Ambasht, R.S., 1992. Ecology of river corridors in India with special reference to soil conservation and pollution. *Natural History Research*, 2(1). 73-76 (Japan).

Ambasht, R.S., 1998. World water and wetland resources. In : Ambasht, R.S. (Ed.), *Modern Trends in Ecology and Environment*, Backhuys Pub., Leiden, 115-130.

Ambasht, R.S. (Ed.), 1998. *Modern Trends in Ecology and Environment, Backhuys Pub., Leiden*, 362.

Ambasht, R.S. 2003(a). Environmental awareness of water and wetland uses including riparian ecosystems. In Environmental care and sustainability (Ed. A. Kumar), SPMD College, Allahabad, 1-10.

Ambasht, R.S., 2003(b). Ecology and economy of water resources conservation and management In Eds. D.N. Singh, J. Singh and K.N.P. Raju, *Water Crisis and Management*, Tara Publ., Varanasi, 102-112.

Ambasht, R.S., 2005. Global Perspective of water resource and management options. In sustainable water management (Ed. Arvind Kumar), S.P. Mukherjee, Govt. College, Allahabad, 19-26.

Ambasht, R.S. and Ambasht, N.K., 1992. Pressing global issues of water resource management. *Indian Forester*, 118(5) : 348-351.

Ambasht, R.S. and Ambasht, N.K., 1998a. Ecology of India wetlands. In : Majumdar, S.K., Miller, E.W. and Brenner, F.J. (eds.), *Ecology of wetlands and associated habitats*. Pennsylvania Aca. Science, USA, pp. 104-116.

Ambasht, R.S. and Ambasht, N.K., 1998b. Biodiversity with special reference to soil management. In : Bagyaraj, D. et al. (eds.), *Modern Approaches and Innovations in Soil Management*. Rastogi Publication, Meerut, 241-250.

Ambasht, R.S. and Ambasht, N.K., 2003. *Modern Trends in Applied Aquatic Ecology*. Kluwer Academic/Plenum Publishers, New York.

Ambasht, R.S. and Ambasht, N.K., 2003. *Modern Trends in Applied Aquatic Ecology*. Kluwer Academic/Plenum Publishers, New York.

Ambasht, R.S. and P.K. Ambasht, 2005. *Environment and Pollution—An Ecological Approach*. 4/e. CBS Publishers & Distributors, New Delhi

Ambasht, R.S. and S.N. Chakhaiyar, 1979. Composition and productivity of weeds in an oil crop ecosystem. *Proc. VII Asian Pacific Weed Sci. Soc. Conf. Univ.* Sydney, Australia, 425-426.

Ambasht, R.S., R. Kumar and N.K. Srivastava, 1994. Strategies for managing Rihand river riparian ecosystem deteriorating under rapid industrialization. In : *Global Wetland—Old World and New* (Ed. W.J. Mitsch), Elsevier, Amsterdam, Elsevier, 725-728.

Ambasht, R.S. and B. Lal, 1979. Ecological researches on weeds of North India. Proc. *VII Asian Pacific Weed Sci. Soc. Conf. Univ.* Sydney, Australia, 339-341.

Ambasht, R.S. and B. Lal, 1978. Ecological Studies of *Chrozophora rottleri.* A Juss. *Vol. II, VI APWSS* Proc Jakarta, 153-163.

Ambasht, R.S. and A.N. Maurya, 1963. Ecology of the Banaras Hindu University grasslands, Abs. Proc. Nat., Aad. Sci. (India).

Ambasht, R.C. and A.N. Maurya, 1970. (a) Reproductive capacity of *Dichanthium annulatum* in relation to biotic factors. *Tropical Ecol.*, 10(2), 186-193.

Ambasht, R.S. and A.N. Maurya, 1970. (b) Effect of certain biotic factors on the vegetative reproductive capacity of *Dichanthium annulatum* Stapf. *Science and Culture*, 36 (6) : 354-356.

Ambasht, R.S., A.N. Maurya and U.N. Singh, 1972. Primary production and turnover in certain protected grasslands of Varanasi, India, *Tropical Ecology with an Emphasis on Organic Production* (Ed. F.B. Golley and P. Golley, Georgia Univ. Athens (USA), 43-50.

Ambasht, R.S. and K.N. Misra, 1980. Conservation studies of a Hilly grassland. In '*Tropical Ecology and Development*, Ed. J.I. Furtado, ISTE, Kuala Lumpur, Malaysia, 133-139.

Ambasht, R.S. and K.N. Misra, 1987. Structure and function of a *Bothriochloa grassland* ecosystem, *Indian Journal Range Management*, 7(1) : 11-18.

Ambasht, R.S. and T.N. Pandey, 1975. Net Primary Productivity of *Aristida cyanantha* dominated grassland communities in Varanasi division, India. *Abs. XII International Botanical Congress,* Leningrad, USSR.

Ambasht, R.S. and T.N. Pandey, 1981. Seasonal changes in the phytosociological and productive structures of two stands of *Aristida cyanantha. Geo. Eco. Trop.* Belgium, 5(1) : 45-56.

Ambasht, R.S. and K. Ram, 1976. Stratified primary productive structure of macrophytic weeds in a large lake : *Aquatic weeds* (Ed. C.K. Varshney and Rzoska), W. Junk & Co., Netherlands, 147-154.

Ambasht, R.S. and M. Shanker, 1992. Rehabilitation of River Corridors. In : (ed. J.S. Singh), *Restoration of degraded lands* : Concepts and Strategies. Rastogi Publ., Meerut, 192-209.

Ambasht, R.S. Shardendu and M. Sikandar, 1983. Climate and aquatic productivity in India—State of Art, *Internat. Biometeorological Conf. Souvenir* and State of Art, Vol., 86-104.

Ambasht, R.S. and Shardendu (1988). The state of art on ecology for riparan and diarah lands. *In wetlands* (Ed. R.K. Garg and L.N. Vyas), *Wetland Conservation,* EEC Publ., 175-187.

Ambasht, R.S. and Shardendu (1989). Conservation and management of Flood Plains of River Ganga. Proc. III Internat. Wetland Conference at Rennes, France (1988).

Ambasht, R.S. and Shardendu, 1989. Morphometry and Physicobiotic character of Varanasi Ponds. *Proc. Nat. Acad. Sci.,* India.

Ambasht, R.S. and Sharma, E., 1983. *Alnus nepalensis* the multiple use tree of the eastern Himalayas. *Science Reporter*, 20(10) : 612-614.

Ambasht, R.S. and E. Sharma (1988). Fifty years of Ecology Research at the Banaras Hindu University : *Perspectives in Ecology* (Ed. J.S. Singh and B. Gopal), Jagmander Book Agency, New Delhi, 495-510.

Ambasht, R.S., M.P. Singh and E. Sharma, 1983. Changes with plant age in fractionated phytomass, productivity and chlorophyll density of crops on a riparian agroecosystem. Proc. *IX, APWSS Conf.* Manila, Philippines, 491-501.

Ambasht, R.S. and A.K. Singh, 1980. Productive status of grassland in deciduous forests of Vindyan Hills. In *Tropical Ecology and Development.* Ed. J.I. Furtado. ISTE, Kuala Lumpur, Malaysia, 155-159.

Ambasht, R.S., A.K. Singh and K.N. Misra, 1982. Energy conserving efficiency and productivity of a gradient of communities in

Chakia forest ecosystem. Proc. *Internat. Forestry Seminar*, Univ. Pertanian, Malaysia (Ed. P.B.L. Srivastava et al.), 209-218.

Ambasht, R.S., M.P. Singh and E. Sharma, 1984. Soil water and nutrient conservation by certain riparian herbs. *Jour. Environmental Management*. Acad. Press, London, 18 : 99-104.

Ambasht, R.S., M.P. Singh and E. Sharma, 1984. An experimental study of soil and water conservation through herbaceous plants. *Nat. Acad. Sci. Letters*. May, 1983, 143-146.

Ambasht, R.S. and M.P. Singh, 1988. Conservation and management of riparian wasteland. In Problems of *Wasteland and Forest Ecology in India* (Ed. P. Singh), Centre for study of Environmental Science, Allahabad, 44.

Ambasht, R.S. and U.N. Singh, 1975. Monthly variations in the biomass and energy structure of *Heteropogon* and *Vetiveria* grass stands on Vindhyan Hills, Varanasi. Abs. *XII International Botanical Congress*. Leningrad, USSR.

Ambasht, R.S. and A.K. Srivastava (1988). Energy dynamics of macrophytes at up and down of river Ganga at Varanasi. *Proceedings of the National Seminar on Perspectives in Aquatic Biology,* Nainital, 171-176.

Ambasht, R.S. and N.K. Srivastava, 1991. Primary Productivity of Indian wetland and their future scenario. In : Y.P. Abrol, P.N. Wattal, A. Gnanam, Govindjee, D.R. Ort & A.H. Teramura (ed.), *Impact of global climatic changes on photosynthesis and plant productivity*, 605-624. Oxford IBH, New Delhi.

Ambasht, R.S. and N.K. Srivastava, 1994. Restoration strategies for the degrading Rihand river and reservoir ecosystems in India. In: *Global Wetland—Old World & New* (Ed. W.J. Mitsch), Elsevier, Amsterdam, Published, 483-492.

Ambasht, R.S., N.K. Srivastava, R. Kumar and A.K. Srivastava, 1992. Ecological characterization of Rihand River. Proc. All India Seminar on Protection of Freshwater Bodies from Pollution : publ. Institution of Engineers (Varanasi Chapter), 69-75.

Ambasht, R.S. and A.K. Srivastava, 1992. Ecology of Nitrogen fixing trees in the tropics. Proc. Internat. Symp. Rehabilitation of Tropical Rain Forest Ecosystems : Research and Development

Priorities (E.D.N.M. Majid, IAA Malek, M. Hanzah, K. Jusoff Univ. Pertanian, Malaysia, 57-65.

Ambasht, R.S., A.K. Srivastava and N.K. Ambasht, 1994. Conserving the biodiversity of India : An ecological approach. *Indian Forester*, 120 : 791-798.

Ambasht, R.S. and Alok, K. Srivastava, 1994b. Nitrogen dynamics of actinorhizal *Casuarina* forest stands and its comparison with *Alnus* and *Leuaena* forests. *Current Science*. 66(2) : 60-63.

Ambasht, R.S. and Alok K. Srivastava, 1994c. Tropical litter decomposition, an holistic approach. In : *Soil organisms and Litter Decomposition in Tropics* (ed. M.V. Reddy).

Ambasht, R.S. Srivastava, A.K. and Ambasht, N.K., 2000. Degradation and restoration processes for Tropical Ecosystem. In *Environmental hazards - plants and people* (Ed. Iqbal M., Srivastava, P.S. and Siddiqui, T.O., CBS Publishers & Distributors, New Delhi, pp. 289-296.

Ambasht, R.S. and B.D. Tripathi, 1978-79. Assessment of effluents of a chemical and fertilizer factory for irrigation of agricultural lands, *Jour. Sci. Res.,* Banaras Hindu University, 29 : 83-87.

Ambasht, R.S., B.D. Tripathi and M. Sikandar, 1984. Ecological Investigation on the plant biomass and energy used for the burning of human dead bodies at Varanasi Ghat. In Ambasht & Tripathi (Ed.) : *River Pollution and Human Health*, 62-69.

Ambasht, R.S., K.R. Verma, Shardendu, M. Shankar and A.K. Srivastava, 1987. Macrophyte ecology of North-Indian Freshwater—a review (abs.), Ed. A.R. Zafar, Limnology in India.

Attiwill, P.M., 1994. Ecological disturbance and conservative management of eucalypt forests in Australia. Forest Ecology and Management, 63 : 301-346.

Bacquerel, P., 1932. C.R. Acad. Sci., Paris, 194 : 2158 cited by Mayer and Mayber, 1963. The germination of seeds. *Pergamon Press,* London.

Bacquerel, P., 1934. La longevire des granies macrooci- otiques *C.R. Aad. Sci.*, Paris, 199 : 1662-1664.

Bar, W., Pfeiper, P., Dettner, K., 2000. Biochemiche Interaktionen Zwiischen Kalanchose Pflanzen. Biol. Unserer Zeit, 30 : 228-234.

Barbour, M.G., J.H. Burk and W.D. Pitts, 1980. II ed., 1988. *Terrestrial Plant Ecology*. Benjamin Cummings Publ. Co. Inc., London, 584.

Begon, M. and M. Mortimer, 1981. Population Ecology : A unified study of Animals and Plants. Blackwells, Oxford.

Bharucha, F.R. and D.B. Ferreira, 1941. The biological spectra of the Matheran and Mahabaleshwar Flora. *Jour. Indian Bot. Soc.*, 20 : 195-211.

Billings, W.D., 1964. Plants and the ecosystem. Macmillan & Co., London. 1970, *Plants, Man and the Ecosystem*, 2nd ed.

Billore, S.K. and L.P. Mall, 1977. Dry matter structure and its dynamics in *Sehima* Community II Dry matter dynamics *Tropical Ecology*, 29-35.

Biswas, D.K. and C.L. Trisal, 1993. Initiatives for conservation of wetlands in India. In : *Biodiversity Conservation* (Ed. Bob Frame and Joe Victor), British High Commission, New Delhi, 153.

Biswas, R., 1964. Ecology of medicinal plants—*Rauvolfia tetraphylla*, and *Rauvolfia serpentina. Ph. D. thesis,* Banaras Hindu University.

Bolin, B., Cook, R.B. (editors), 1983. The *Major Biogeochemical cycles* and their Interactions, Wiley Chichester, England.

Bond, W.J., 1994. Keystone species. In : Schulze, E.D. and Mooney, H.A. (ed.), *Biodiversity and Ecosystem Functions*. Springer Verlag, NY, 237-253.

Bormann, F.H. and G.E. Likens, 1967. Nutrient Cycling. *Science,* 155: 424-429.

Braun-Blanquet, J., 1932. *Plant Sociology : The Study of Plant Communities*. McGraw Hill Book Co., Inc, NY.

Braun-Blanquet, J., 1951. *Pflanzensoziologie*. Springer Verlag, Vienna.

Brown, N.A.C., 1993. Promotion of germination of fynbos seeds by plants derived smoke, *New Phytol*, 123 : 575-583.

Burkholder, P.R., 1952. Cooperation and conflict among primitive organisms. *Amer. Sci.* 40 : 601-631.

Cain, S.A., 1944. *Foundation of Plant Geography*. Harper and Brothers, NY.

Champion, H.G., 1936. A preliminary survey of forest type of India and Burma, *Ind. For. Rec.* (New Series) I(1).

Chapagan, A.K. and Hoekstra, A.Y., 2003. Virtual trade : A quantification of virtual water flows between nations to international trade of livestock and livestock products. In : Hoekstra, A.Y., 2003. Virtual water trade, IHE Report Series 12, Delft.

Chatterjee, D., 1939. Studies on the endemic flora of India and Burma, *Jour. As Soc. Bengal Sci.*, 5 : 19-67.

Chaturvedi, O.P. and J.S. Singh, 1987 (a and b). The structure and function of Pine forest in Central Himalaya. I Dry matter dynamics. *Annals of Botany*, 60. 237-252, II. Nutrient dynamics *Annals of Botany*, 60 : 253-267.

Chauhan, K.P.S., 1993. Research options and priorities for conservation of biological diversity in the Western Ghats. In : *Biodiversity Conservation* (Ed. Bob Frame and Joe Victor). British High Commission, New Delhi, 153.

Christensen, N. and C.H. Mullar, 1995. Effect of fire on factors controlling plant growth in *Adenostoma* Chaparral *Ecological Monograph*, 45 : 29-55.

Cicerone, R.J., 1987. Changes in Stratospheric ozone. *Science*, 237 : 35-42.

Clausen, J., D.D. Heck and W.M. Hiesey, 1940. Experimental studies on the nature of species. Carnegie Inst., Wash Publ., 520.

Clements, F.E., 1916. *Plant Succession, an analysis of the development of vegetation.* Carnegie Institute of Washington.

Connell, J.H. and Slatyer, R.O., 1977. Mechanisms of successions in natural communities and their role in community stability and organization. *The American Naturatist, III* : 1119-1144.

Conway, V.M., 1937. Studies on the autecology of *Cladium mariscus.* III, The aeration of subterranean parts of plants. *New Phytol.*, 36: 64-96.

Cottom, G. and J.T. Curtis, 1956. The use of distance measures in phytosociological sampling, *Ecology*, 37 : 451-460.

Dahlman, R.C. and Kucera, C.L., 1965. Root productivity and turnover in native prairie. *Ecology*, 46-89.

Dansereau, P., 1957. *Biogeography—An Ecological Perspective*, The Roland Press Co., NY.

Dasmann, R.F., J.P. Milton and P.H. Freeman, 1973. *Ecological Principles for Economic Development*, I.U.C.N., Wiley & Sons Ltd., 257.

Datta, S.C. and S.D. Chakrabarti, 1982. Allelopathic potential of *Clerodendrum viscosum* Vent, relation to germination and seedling growth of weeds, *Flora*, 172 : 89-95.

Daubenmire, R.F., 1966. Vegetation identification of typical communities, *Science*, 151 : 291-298.

De Santo, R.S., 1978, *Concepts of Applied Ecology*. Springer-Verlag, 310.

Dudley, J.L. and Lajtha, K., 1993. The effects of prescribed burning on nutrient availability and primary production in sandplain grasslands. Amer. Midl. Naturalist., 130 : 286-298.

Dwivedi, Ram Snehi, 1969. Evaluation of methods of measuring productivity in *Triticum aestivum and Dichanthium annulatum. Indian J. Agric. Sci.*, 39(11).

Dyer, M.I. and U.G. Bokhari, 1976. Plant-animal interactions: Studies of effects of grasshopper grazing on blue grama grass. *Ecology*, 57 : 762-772.

Eckardt, F.E. (Ed.), 1968. Functioning of terrestrial ecosystem at primary level *Proc. Copenhagen Symposium, UNESCO.*

Edroma, E.L., 1986. Effect of fire on primary productivity of grasslands in Queen Elizabeth National Park, Uganda. In : Joss, P.L., Lynch, P.W. Williams, O.B. (Eds.), *Rangelands : Response under siege*. Austr. Acad. Sci., Canberra.

Farnworth, E.G. and Frank, B. Golley, 1974. Fragile Ecosystems, Springer-Verlag, Berlin, NY, 258.

Forbes, S.A., 1887. The lake as a microcosm. Bull. Sci. A. Peoria. Reprinted 1925 in Ill. Hist. Surv. Bull. : 15 : 537-550.

Forcier, L.K., 1975. Reproductive strategies and the co-occurrence of climax tree species. *Science*, 189 : 808-810.

Fox, J.E.D., 1998. Role of fire in forest and grazing lands. In : R.S. Ambasht (Ed.), *Modern Trends in Ecology and Environment*. Backhuys Publ., Leiden, 251-276.

Gadgil, M. and W.H. Bossert, 1970. Life historical consequences of natural selection. *Am. Nat.*, 104 : 1-24.

Gadgil, M. and V.M. Meher-Homji, 1983. Ecological diversity. In G.C. David and J.S. Serrao, conservation in developing countries : Problems and prospects. Oxford Univ. Press, 175-198.

Gadgil, M. and O.J. Solbring, 1972. The concept of r and K selection: Evidence from wild flowers and some theoretical considerations, *Am. Nat.*, 106 : 14-31.

Gilmour, J.S.L. and J. Heslop-Harrison, 1954. The deme terminology and the units of microevolutionary change. *Genetica* 27 : 146-161.

Gleason, H.A., 1926. The individualistic concept of plant associations. *Bull. Torrey Bot. Club* : 7-26.

Gleason, J.F., P.K. Bhartiya, J.R. Herman, R. Mc Peters, P. Newman, R.S. Stolarski, L. Flynn, G. Labow, D. Larko, C.W. Planet, 1993. Record low global ozone in 1992. *Science*, 260: 523-526.

Goldman, C.R., 1966. Primary productivity of aquatic environment. Palanza Symposium, UNESCO Univ., Calif. Press.

Golley, F.B., 1960. Energy dynamics of food chain of an old field community. *Ecol. Monogr.* 30 : 187-206.

Golley, F.B., 1969. Calorific value of wet tropical forest vegetation. *Ecology*, 50 (Late Spring).

Golley, F.B. and Lieth, H., 1972. The bases of tropical production. *Tropical Ecology* (Ed. Golley and Golley), New Delhi symposium : 1-26.

Golley, P.M. and F.B. Golley, 1972. *Tropical Ecology with an Emphasis on Organic Productivity*, 418, Univ. Georgia, Athens, USA.

Good, R.A., 1937. A theory of plant geography, *New Phytol*, 30 : 149-171.

Good, R.A., 1947. The *Geography of flowering* Plants, 403, London.

Goodall, D.W., 1963. The continuum and the individualistic association, *Vegetatio* II : 297-316.

Gregor, J.W. Ecotypic differentiation. *New Phytologist*, 45 : 254-270.

Grime (1979). Plant strategies and vegetation processes. Wiley, NY.

Gupta, R.K., 1992. Restoration of degraded watersheds. In : J.S. Singh (Ed.). *Restoration of Degraded land : Concepts and Strategies*. Rastogi publ. Meerut, 321.

Gupta, R.K. and R.S. Ambasht, 1979. Use and Management. In *Grassland Ecosystems of the World* (Ed. R.T. Coupland), 241-244, Cambridge Univ. Press, UK.

Gupta, S.R., 1968. Studies on the growth and ecology of *Hygrophila auniculata Ph.D. thesis.* Banaras Hindu University.

Haeckel, E., 1969. Generelle Morphologie der organismen. George Reimer, Berlin (2 Vols.).

Harper, J.L., 1977. *Population Biology of Plants.* Acad. Press, NY and London, 892.

Harvey, H.W., 1950. On the production of living matter in the sea off Plymouth, *J. Mar. Biol. Assoc.*, UK (NS), 29 : 97-137.

Haskell, E.F., 1949. A clarification of social science. Main Current in Modern Thought, 7 : 45-51.

Hatch, M.D. and C.R. Slack, 1966. Photosynthesis by sugarcane leaves. A new carboxylation reaction in the pathway of sugar formation. *J. Biochem.*, 101 : 103-111.

Hejny, S., 1960. Okologische characteristik der Wasser-und Sumpflanzen in den slowakischen tiefebenen. US Akademic Vide Bratislava.

Heywood, W.H. and Watson, R.T., 1995. *Global Biodiversity Assessment.* Cambridge Univ. Press, 1140.

Hoekstra, A.Y. and Hung, P.Q., 2003. Virtual water trade. IHE Report No. 12, Delft.

Howe, H.F., 1995. Succession and fire season in experimental prairie plantings. *Ecology*, 76 : 1917-1925.

Hutchinson, G.E., 1957. *A Treatise in Limnology, Geography, Physics and Chemistry*, Vol. I, John Wiley & Sons, NY, 1015.

Hutchinson, G.E., 1965. *The Ecological Theatre and Evolutionary Play.* Yale Univ. Press.

Hutchinson, G.E., 1978. *An Introduction of Population Ecology*, Yale Univ. Press, 260.

Iverson, J., 1936. Biologische Pflansentpen als Hilfsmittel in der Vegetationsforshung. Ph.D. Thesis, University of Copenhagen.

John, J., 2003. Phycoremediation : Algae as tools for remediation of mine-void wetlands. In : Ambasht, R.S. and Ambasht, N.K. (eds.), *Modern Trends in Applied Aquatic Ecology.* Kluwer Academic/Plenum Publishers, New York.

Kamen, M.D., 1954. Discoveries in nitrogen fixation. *Scient. American*, 188 : 38-42.

Kaul, V., 1959. Physiologico-ecological studies of *Xanthium strumarium* L and *Croton sparciflorus* Morong. *Ph.D. thesis*, Banaras Hindu University.

Kaul, V., 1965. Physiological ecology of *Xanthium stramarium* Linn I seasonal morphological variants and distribution. *Trop. Ecol.* 6, 72-87.

Krupa, S.V. and R.N. Kickert, 1989. The green-house effect : Impacts of Ultraviolet-B (UV-B) radiation, Carbon dioxide (CO_2), ozone (O_3) on Vegetation *Environ. Pollut.*, 61 : 263-393.

Kohut, R.J., Laurence, J.A. and Colavito, L.J., 1988. Influence of ozone exposure dynamics on the growth and yield of kidney bean. In : Dempster, J.P. and Manning, W.J. (eds.), Spl. Vol. *Environmental Pollution*, 53 : 79-88.

Kormondy, E.J., 1996. *Concepts of Ecology*. IV Ed. Prentice Hall, NY, 559.

Kumar, R., R.S. Ambasht and N.K. Srivastava, 1992. Nitrogen conservation efficiency of five common riparian weeds in runoff experiment on slopes. *Jour of Env. Management*, 34 : 45-57.

Kumar, R., R.S. Ambasht and N.K. Srivastava, 1992. Conservation Efficiency of five common riparian weeds in movement of soil, water and phosphorus. *Jour. Applied Ecology*, 29, 734-744.

Laetsch, W.M., 1968. Chloroplast specialization in dicotyledons possessing the C_4 dicarboxylic acid pathway i. photosynthetic C fixation. *Amer. J. Bot.*, 55 : 875-883.

Lal, B and R.S. Ambasht, 1978. Growth of *Chrozophora rottleri*. A Juss in relation to different watering levels. *Ind. Jour. Ecol.* 5(2): 172-180.

Lal, B and R.S. Ambasht, 1979. Autecology of *Scoparia dulcis* Lin. *Proc. Ind. Nat. Sci. Acad.* B 45(4) : 368-374.

Lal, B. and R.S. Ambasht, 1979. Biomass productivity and growth analysis of *Scoparia dulcis*. Linn under varied light intensities. *Proc. Nat. Acad. Sci.* India.

Lal, B. and R.S. Ambasht, 1980. Effect of cement dust pollution on the leaves of *Psidium guava*, I—Pigment content and leaf biomass, *Ind. Jour. Environm. Health* : 22(3) : 231-237.

Lal, B. and R.S. Ambasht, 1981. Impairment of Chlorophyll content in leaves of *Diospyros melanoxylon* by fluoride pollution, *Water, Air* and *Soil pollution*. California, USA, 16 : 361-365.

Lal, B. and R.S. Ambasht, 1981. Mineral concentration of leaves of *Diospyros melanoxylon* tree near a thermal power plant. Proc. V. Internat. Symp. on Environmental Biogeochemistry (Stockholm), Abs., 60.

Lal, B. and Ambasht, R.S., 1981. Fluoride accumulation in a deciduous forest tree species in the neighbourhood of an aluminium factory *Ind. Jour. Forestry*, 4(4) : 262-264.

Lal, B. and Ambasht, R.S., 1982. Impact of cement dust on the mineral concentration of *Psidium guayava. Environmental Pollution*, UK, 4 : 241-247.

Lal, B. and R.S. Ambasht, 1982. Ecological studies on seed germination of *Leonotis nepetifolia* (L) in relation to environmental factors with emphasis on fluoride polluted soils, *Geo. Eco. Trop.*, 6(3), 229-237.

Langlet, O.A. Cline or not a cline—a question of scots pine. *Sylvae Genetics*, 8 : 13-22.

Larcher, W., 2003. Physiological Plant Ecology, 4th ed. Springer Heidelberg, 450.

Lawrence, George H.M., 1954. *Taxonomy of vascular plants*. Macmillan, NY.

Leblanc, F. and D.N. Rao, 1975. Effect of air pollution of Lichens and Bryophytes, *Response of plants to Air pollution*, Acad. Press, NY, 519-528.

Lerner, H.R., A.M. Mayer and M. Evenari, 1959. *Physiol. Plant*, 12 : 245 (cited by Mayer and Mayber, 1963).

Levy, E.B. and E.A. Madden, 1933. The point method of pasture analysis, *N.Z. Agri.*, 46 : 267-279.

Lewis, J.K., 1970. Production in grassland Ecosystem US/IBP. Grassland Ecosystem : A Supplement, Edited by R.L. Dix and R.C. Beidleman. Colorado State University, USA.

Lieth, H., 1964. Versuch einer Kartographischen Darstellung der Productivital der Pflanzendecke auf Erde In : *Geographisches Taschenbuch*, 1964/65, Franz Verlag, Weisbaden, 12-80.

Lieth, H., 1968. The measurement of calorific values of biological material and determination of ecological efficiency, *Copenhagen Symposium*, UNESCO (See Eckardt, 1968).

Lieth, H., 1968. (a) Continuity and discontinuity in ecological gradients and plant communities, *Bot. Rev.* (3) : 291-302.

Lieth, H., 1975. The measurement of calorific values : *Primary Productivity of Biosphere* (Ed. Lieth H. and R.H. Whittaker), Springer-Verlag, 119-130.

Lieth, H., 1975. Productivity of major vegetation units of the World, *Ibid*, 203-216.

Lieth, H. and E. Box, 1972. Evapotranspiration and primary productivity. C.W. Thornthwaite Memorial Model Publ., *Climatology*, 25(2) : 37-46.

Lindeman, R.L., 1942. The trophic dynamic aspect of ecology. *Ecology*, 23 : 399-418.

Lotka, A.J., 1925. Elements of Physical Biology. Williams and Wilkins, Baltimore, 460 (reprinted : Dover Pub., NY, 1956).

Lovelock, J.E., 1981. The Electron Detector Capture-A personal odyssey. *Chemtech* (Sept. 1981) : 531-537 (7-10).

Lovelock, J.E., 2002. The Evolution of the Earth. In : *A Better Future for the Planet Earth*, Vol. II, Asahi Glass Foundation, Tokyo, 326.

Luther, H., 1949. Vorsalag zueinerokologichen grundeinteilung der Hydrophyten. *Acta. Bot. Fenn.*, 44 : 1-15.

Mc Arthur, R.H., 1965. Patterns of Species Diversity. *Biological Review*, 40 : 510-533.

McCook, L.J., 1994, Understanding ecological community succession. *Vegetatio*, 110 : 115-147.

Mc Intosh, R.P., 1967. The continuum concept of vegetation. *Bot. Rev.*, 33 : 130-187.

McKinney, M.L. and Schoch, R.M., 1998. *Environmental Sciences : Systems and Solutions,* John & Bert.

Mc Naughton, S.J., 1966. Ecotype function in the *Typha* community type : *Ecol. Mongr.*, 16 : 297-365.

Mc Naughton, S.J., 1967. Relationships among functional properties of California grassland. *Nature*, 216 : 168-169.

Macfadyen, A., 1964. Energy flow in ecosystems and its exploitation by grazing. *Grazing in terestrial and marine environment*, Ed. D.J. Crisp. Blackwell Scientific Pub. Oxford.

Machta, L., Hughes, 1970. Atmospheric Oxygen, 1967-1970, *Science*, 168 : 1582-1584.

Majumdar, S.K., Miller, E.W. and Brenner, F.J. (eds.), 1988. *Ecology of Wetlands and Associated Systems*. Pennsylvania Acad., Sciences, Penn.

Margalef, R., 1968. *Perspectives in Ecological Theory*. Univ. of Chicago Press, 112.

Markert, B. and Oehlmann, J., 1998. Ecotoxicology. In : Ambasht, R.S. (ed.), *Modern Trends in Ecology and Environment*, Backhuys Pub., Leiden, 35-53.

Marwah, P. and R.S. Ambasht, 1972. Community architecture and productivity of wheat crop community. *Trop. Ecol.* 13(2) : 176-182.

Mason, H.L., 1936. The principles of geographic distribution as applied of floral analysis. *Modrono* 3 : 188-190 (cited by Lawrence, 1951).

Maurya, A.N. and R.S. Ambasht, 1973. Significance of seed dimorphism in *Alysicarpus monilifer*, D.C. *Jour., Ecol.*, 61 : 213-217.

Mayer and Poljakoff Mayber, 1963. The *germination of seeds*. Pergamon Press, London.

Meentmeyer, V., 1978. Macroclimate and Lignin control of litter decomposition rates. *Ecology*, 59 : 465-472.

Menhinick, E.F., 1964. A Comparison of some species diversity indices applied to samples of field insects : *Ecology*, 45 : 858-862.

Milner, R.E., R.E. Hughes, C.H. Giminham, G.R. Miller and R.O. Slatyer, 1969—Methods of *measurement of the primary, production of grassland*. I.B.P. Handbook No. 6, Blackwell, UK.

Mishra, K.N. and Ambasht, R.S., 1988. Physico-chemical nature of effluents from woollen cottage industries. Proc. Nat. Acad. Sci., 58 : 161-165.

Misra, K.C., D.N. Rao, R.S. Ambasht, K.L. Mukherjee, R.S. Dwivedi and P.V. Ananthkishman, 1970. *Ecology—Study of ecosystem*. A.H. Wheeler and Co., Allahabad, 311.

Misra, M.K. and B.N. Misra, 1981. Seasonal changes in leaf-area index and chlorophyll in an Indian grassland. *Jour. Ecol.,* 69 : 797-805.

Misra, R., 1938. Edaphic factors in the distribution of aquatic plants in the English Lakes, *J. Ecol.,* 36 : 441-451.

Misra, R., 1944. Variation of leaf form in *Potamogeton perfoliatus Jour. Ind. Bot. Soc.,* 23 : 44-52.

Misra, R., 1944. (a) The vegetation of Rajghat ravines *Jour. Ind. Bot. Soc.* 23 : 113-121.

Misra, R., 1945. The soil complex as studied in plant ecology, *Jour. Banaras Hindu University,* 9 : 12-16.

Misra, R., 1946. A study in the ecology of low-lying lands, *Ind. Ecologist,* 1 : 27-46.

Misra, R., 1952. The forest complex of Patharia Hills, *Jour. Ind. Bot. Soc.,* 31 : 155-10.

Misra, R., 1957. Plant Ecology, Progress of Science in India, *Nat. Inst. Sci.* India, 7 : 141-148.

Misra, R., 1969, *Ecology Workbook,* Oxford & IBH, Kolkata.

Misra, R., 1969. (a) Studies on Primary productivity of terrestrial communities at Varanasi, *Trop. Ecol.,* 10(1) : 1-15.

Misra, R., 1969. (b) Primary production of Chakia forest and IBP/PT study of organic production and nutrient cycling in monsoon forests, grassland and croplands, Proc. IUCN, XI, Technical meetings.

Misra, R., 1983. Indian Savannas. In *Tropical Savannas* : Ed. F. Bourlier. Elsevier Sci. Pub. Co., Amsterdam, 151-166.

Misra, R., S.S. Ramam and R.S. Ambasht, 1968. Floristics and ecology of Varanasi, *Souvenir,* Indian Science Congr., Varanasi session, 25-32.

Misra, R. and K.P. Singh, 1971. Ecology in India—A Bird's Eye view Adv. Printed J. Ind. Bot. Soc. Golden Jubilee Volume.

Misra R. and G.S. Puri, 1956. *Indian Manual of Plant Ecology,* English Book Dept., Dehradun.

Misra, R. and Siva Rao, 1948. A Study of autecology of *Linenbergia polyantha Royle. Jour. Ind. Bot. Soc.,* 27 : 168-193.

Mitsch, J., 1994 (Ed.) *Global Wetlands : Old World and New,* Elsevier.

Mitsch, W.J., Mitsch, R.H. and Turner, R.E., 1994. Wetlands of Old and New Worlds : Ecology and Management. I (Ed. Mitsch, WJ), Elsevier, Amsterdam.

Mittermeier, R.A., Mittermeier, C.G. and Cyril F. Mormos, 2002. Setting Priorities for saving life on earth : megadiversity countries, hotspots and wilderness areas. In *A Better Future for the Planet Earth,* Vol. II, 2002, 33-37. Asahi Glass Foundation, Tokyo.

Mobius, K., 1977. Die Auster und die Austernwirtschaft, Berlin.

Mooney and Billings, 1960. The annual carbohydrate cycle of alpine plants as related to growth, *Amer. Jour. Bot.,* 47 : 594-598.

Mulchi, C.L., Lee, E.H., Tuthill, K. and Olinick, E.V., 1988. Influence of ozone stress on growth processes, yields and grain quality characteristics among soybean cultivars. In : Dempster, J.P. and Manning, W.J. (eds.), Spl. Vol. Environmental Pollution, 53 : 151-170.

Mueller-Dombois, D. and H. Ellenberg, 1974. *Aims and Methods of Vegetation Ecology,* Wiley, NY.

Muller, C.H., 1966. The role of chemical inhibition (allelopathy) *in vegetational composition. Bull Torrey Bot. Club,* 93 : 332-351.

Musil, C.F. and S.J.E. Wand, 1999. Impact of UV-radiation on South African Mediterranean ecosystems. In (Ed. J. Rozema) Stratospheric ozone depletion : the effects of enhanced UV-B radiation on terrestrial ecosystems, 265-291. Backhuys Pub. Leiden.

Nelson, R.W., R.S. Ambasht, C. Amoros, G.W. Begg, A.A. Bonnetto, E. Dister, C.M. Finlayson, J.K. Handoo, K.M. Mavuti, N.K. Panddit, D. Parish, I.R. Waris and Wenger, 1991. River Floodplains and Delta Wetlands Management Team; A Project of the World Wetlands Partnership. Proc. Internat. Conference on Wetlands and River Corridor Management, USA, 75-82 (Ed. J. Kusler).

Newbould, P.J., 1967. *Methods of estimating the primary productivity of forest.* I.B.P. Handbook No. 2, Blackwell Scientific Pub, Oxford.

Odum, E.P., 1961. The strategy of ecosystem development *Science,* 164 : 264-270.

Odum, E.P., 1963. *Ecology* : Holt Reinhart and Winston Inc.

Odum, E.P., 1971. *Fundamentals of Ecology*, 3rd Edn. W.B. Saunders & Co., Philadelphia, 574.

Odum, E.P., 1983. *Basic Ecology*. Saunders College Pub., NY, 613.

Odum, H.T., 1971. Trophic structure and productivity of Silver Springs, Florida, *Ecol. Monogr.*, 27 : 55-112.

Odum, H.T., J.E. Cantlon and L.S. Kornicker, 1960. An organisational hierarchy postulate for the imterpretation of species individuals distribution, species entropy and ecosystem evolution and the meaning of species variety index. *Ecology*, 41 : 395-399.

Oertel, N., 1994. Biomonitoring in water quality control with particular reference to biomonitoring techniques used in the river Danube for detection of heavy metals. Acta. Biol. Debr. Oecol. Hung., 5, 81-90.

Oertel, N., 1996. Use of zebra mussel (*Dreissena polymorpha*) to assess heavy metal pollution in the river Danube (Hungary), 31, Konferenz der IAD, Baja-Ungarn, Wissen Schaftliche Referate, 405-410.

Oertel, N. and Salanki, J., 2003. Biomonitoring and bioindicators in aquatic ecosystems. In : Ambasht, R.S. and Ambasht, N.K. (eds.), *Modern Trends in Applied Aquatic Ecology*, Kluwer Academic/ Plenum Publishers, New York.

Ovington, J.D., 1962. Quantitative ecology and the woodland ecosystem concept. In : J.B. Cragg (Ed.) *Advances in Ecological Research*, Vol. I, 103-192; Acad. Press, London.

Ozborne, P.L., 2000. *Tropical Ecosystems and Ecological Concepts.* Cambridge University Press, 464.

Paine, R.T., 1966. Food web complexity and species diversity. *Am. Nat.*, 100 : 65-75.

Pal, D., 1974. Effects of fluoride pollution on plants and cattle, *Ph.D. thesis*, Banaras Hindu University.

Pandey, A.N. and J.S. Singh, 1984-85. Mechanism of ecosystem recovery : A case study from Kumaun Himalaya. *Reclamation and Revegetation Research*, 3 : 271-292.

Pandey, T.N. and R.S. Ambasht, 1979. Efficiency of *Aristida cyanantha* grassland in relation to ecological energetics, *Ind. Jour. Ecol.*, 6 : 22-26.

Pandeya, S.C., 1988. *Status of the Indian Rangelands*. Pres. address III, International Rangeland Congress, 213.

Pandeya, S.C., B.R. Pandit and S.C. Sharma, 1971. Biomass and Production correlation of Teak (*Tectona grandis*) in natural forest in River Narmada upper catchment area in Central India and a comparison thereof with the plantations, Abs. *Inter Symp. Tropical Ecology Emphasizing Organic Production* (New Delhi).

Pell, E.J., Pearson, N.S. and Johansen-Vinten, C., 1988. Qualitative and quantitative effects of ozone and sulphur dioxide on field grown potato plants. In : Dempster, J.P. and Manning, W.J. (eds.), Spl. Vol. *Environmental Pollution*, 53 : 171-186.

Perrings, C., Maler, K.G., Folke, C., Holling, C.S. and Jansson, B.O., 1995. *Biodiversity Loss,* Cambridge Univ. Press, Cambridge, 332.

Phillipson, J., 1966. *Ecological energetics*. Edward Arnold : UK.

Pielou, E.C., 1966. Species diversity and pattern diversity in the study of ecological succession : *J. Theoret Biol.*, 10 : 370-383.

Piper, C.S., 1944. *Soil and Plant analysis*, NY, Inter Science.

Polunin, N., 1960. *Introduction to Plant Geography*. McGraw Hill, NY, 640.

Puri, G.S., 1960. The distribution of conifers in the Kulu Himalayas with special relation to geology. *Ind. For.* 70 : 144-153.

Puri, G.S., 1951. The amount of foliar ash in Sal (*Shorea robusta*) trees of different quality classes in India. *Jour. Ind. Bot. Soc.*, 31: 82-85.

Puri, G.S., 1960. *Indian Forest Ecology* : 2 volumes, Oxford Book and Stationery Co., New Delhi.

Puri, G.S.V.M., Meher Homji, R.K. Gupta and S. Puri, 1983. *Forest Ecology, phytogeography and Forest Conservation,* Vol. I, 571. Oxford and I.B.H. Co., New Delhi.

Ramakrishnan, P.S., 1959. Contributions to the ecological flora of Varanasi district. *Ph.D. Thesis,* Banaras Hindu University.

Ramam, S.S., 1966. Organisation of grass communities on Western and Vindhyan uplands of Varanasi. *Jour. Ind. Bot. Soc.*, 45 : 266-276.

Rao, C.C., 1969. Phytosociological studies on a piece of Watershed land in the vicinity of rivers Karamanasa and Chandraprabha (Varanasi district), *Ph.D. thesis,* Banaras Hindu University.

Rao, D.N., 1971. A study of Air Pollution Problem due to coal unloading in Varanasi, India, Proc. II Internat. Clean Air Congress, 270-273.

Rao, D.N. and Leblanc, F., 1966. Effect of sulphur dioxide of the Lichen alga with special reference to chlorophyll. *The Bryologist*, 69 : 69-75.

Raunkiaer, C., 1934. *The life forms of plants and statistical plant geography* being the collected papers of C. Raunkiaer. Clarendon Press, Oxford, 639.

Rawat, V.S. and R.S. Ambasht, 1958. Root relation of *Orobanche* and its hosts *Curr. Sci.*, 445-446.

Read, D., 1994. Plant microbe mutualism and community structure. In Schulze E.D., and Mooney, H.A. (eds.), Biodiversity and Ecosystem functions. Ecological Studies, Vol. 99, Springer-Berlin.

Reid, 1992. Conserving life's diversity. *Env. Sci. Technol.*, 26(6) : 1090-1095.

Reid, W.V., 1992a. How many species will there be? In : Whitemore, T.C. and Sayer, J.A. (eds.), *Tropical Deforestation and Species Extinction*, 55-74, Chapman and Hall, London.

Revelle, R. et al., 1965. Atmospheric carbon dioxide, 111-133. Restoring quality of our Environment, Report on Environmental Pollution Panel, President's Science Advisory Committee, Washington, USA, 317.

Riley, G.A., 1957. Phytoplankton of the North Central Sargasso sea. *Limnol. Oceanogr.*, 2 : 253-270.

Rodin, L.E. and N.I. Basilevic, 1968. World distribution of plant biomass. In Proceedings Copenhagen Symposium. Functioning of terrestrial ecosystem at the primary production level (Ed. R.E. Eckhardt Natural Resources Research, 5 : 45-52, Paris, UNESCO.

Sahai, R., G. Kaur, P.S. Roy, 1977. Effect of some regulators and chemicals on seed germination and early seedling growth of *Lathyrus aphaca* Linn. *Ind. Jour. Ecol.*, 4(2) : 55-59.

Salisbury, E., 1942. *The Reproductive capacity of plants*, G. Bell and Sons Ltd., London.

Sandermann, H. Jr., 1994. Higher plant metabolism of Xenobiotics : The "green liver" concept. *Pharmacognetics*, 4 : 225-241.

Sandhu, J., M. Sinha and R.S. Ambasht (1990). Nitrogen release in *Leucaena leucocephala* decomposition. *Soil Biology and Biochemistry*, UK, 22 : 859-863.

Savill, P.S. and Fox, J.E.D., 1970. *Trees in Sierra Leone.* Crown Agents, London.

Schulze, E.D., E. Beck and K. Muler-Hohenstein, 2004. *Plant Ecology.* Springer Heidelberg, 702.

Schulze, E.D. and Mooney, H.A. (eds.), 1994. *Biodiversity and Ecosystem Functions.* Springer Verlag, NY.

Seiler, W., 1974. The cycle of atmospheric CO_2. Tellus, 26 : 116-135.

Shannon, C.E. and W. Weaver, 1949. *The mathematical theory of communication.* Urbana, Univ. Illinois Press, 117.

Shardendu and R.S. Ambasht (1988a). Limnological studies of a rural and an urban tropical aquatic ecosystem : Oxygen forms and ionic strength. *Tropical Ecology*, 29(2), 98-109.

Shardendu and R.S. Ambasht (1988b). Impact of Urban waste waters on primary production and nutrient status of *Hydrilla verticillata* (L.F.), Royle. Proceedings of Conference on aquatic Biology. In *Perspectives in Aquatic Biology* Papyrus Publication, New Delhi, 187-196.

Shardendu and R.S. Ambasht, 1991. Impact of urban surroundings on some limnological characteristics of a Tropical pond. *Acta Hydrochimica et Hydrobiologica* (Germany). 10(4) : 399-410.

Shardendu and R.S. Ambasht, 1989. Limnological study of a Tropical Pond I. Phytoplanktonic productivity. Proc. Phycotalk inter. Symposium (Ed. H.D. Kumar), 85-90.

Shardendu and R.S. Ambasht, 1990. Relationship of nutrients in water with biomass and nutrient accumulation of submerged macrophytes of a tropical wetland. *New Phytologist*, UK, 117(3) : 493-500.

Sharma, A.K., 1981. Impact of the development of science and technology on environment. Presidential Address *Ind. Sci. Congr.*, 1-43.

Sharma, B.K. and G.S. Lavania, 1978. Effect of some hormones on seed germination and early seedling growth of *Vicia* spp *Geobios*, 5(4) : 176-178.

Sharma, E. and R.S. Ambasht, 1984. Seasonal variation in Nitrogen fixation by different ages of root nodules of *Alnus nepalensis* plantation in Eastern Himalayas. *Jour. Applied Ecology*, UK, 21 : 265-270.

Sharma, E. and R.S. Ambasht, 1986. Root nodule age class transition, production and decomposition in an age sequence of *Alnus nepalensis* plantation stands in the Eastern Himalayas, *Jour. Applied Ecology* (UK), 23 : 689-701.

Sharma, E. and R.S. Ambasht, 1986. Symbiotic nitrogen fixation by actinorhizal plants : In Ambasht R.S. (ed.), Recent Advances in Environmental Biology, ERL, BHU, 34-45.

Sharma, E. and R.S. Ambasht, 1987. Litterfall, decomposition and nutrient release in an age sequence of *Alnus nepalensis* plantation stands in the Eastern Himalayas. *Journal of Ecology*, UK, 75 : 997-1010.

Sharma, E. and R.S. Ambasht, 1988. Nitrogen accretion and its energetics in the Himalayan alder. *Functional Ecology*, UK, 2 : 229-235.

Sharma, E., 1988. Altitudinal variation in nitrogenase activity of the Himalayan alder naturally regenerating on landslide affected sites. *New Phytologist*, 108, 411-416.

Sharma, E. and R.S. Ambasht, 1991. Biomass productivity and energetics in the Himalayan Alder Plantations. *Annals of Botany*. (UK), 67 : 285-293.

Shelford, V.E., 1943. The abundance of collared lemming in Churchill area, 1929-40, *Ecology*, 24 : 472-484.

Silvertown, J.W., 1987. *Introduction to plant population*. Ecology II, Ed. ELBS, London, 229.

Simmons, I.C., 1974. The ecology of natural resources. Edward Arnold (Publ.) Ltd., London, 424.

Simpson, E.J., 1949. Measurement of diversity. *Nature*, 163-188.

Singal, S.P., 2000. Noise pollution in Indian cities. In Environmental Hazards—Plants and People (Eds. Iqbal, Srivastava and Siddiqui. CBS Publishers & Distributors, New Delhi, pp. 38-51.

Singh, A.K., R.S. Amnasht and K.N. Misra, 1979. Weight loss and energy release of decomposing grass litter in a Savanna ecosystem, *Jap. Jour. Ecol.*, 29 : 369-374.

Singh, A.K., R.S. Ambasht and K.N. Misra, 1979. Photosynthetic structure in relation to organic matter production of a grassland community, Varanasi. *Proc. Nat. Acad. Sci.,* India. 49(B) : 144-150.

Singh, A.K. and R.S. Ambasht, 1980. Production and decomposition rate of litter in teak (*Tectona grandis*) plantation of Varanasi (India), *Rev. Ecol. Biol. Sol.* (Paris), 17 : 13-22.

Singh, A.K., K.N. Misra and R.S. Ambasht, 1980. Soil conservation efficiency of three dominant grasses of Vindhyan Hills, India. *Jour. Japanese forestry,* 62(7) : 273-275.

Singh, J.S., 1967. Seasonal Variation in composition, plant biomass and net community production in the grassland at Varanasi, *Ph.D. Thesis,* Banaras Hindu University.

Singh, J.S., Y. Hanxi and P.E. Sajise, 1985. Structural and functional aspects of Indian and S.E. Asian savanna ecosystems, 34-51. In J.C. Tothill and J.J. Mott, 1985. *Ecology and Management of World's Savannas.* Australian Acad. Sci. Canberra, 384.

Singh, J.S., 1973. A compartment model of herbage dynamics of Indian tropical grasslands. *Oikos,* 24 : 367-372.

Singh, J.S., 1992 (Ed.). *Restoration of Degraded Land : Concept and Strategies.* Rastogi Publ., Meerut, 321.

Singh, J.S., 1976. Structure and function of tropical grassland vegetation of India. *Polish Ecological Studies,* 2(2) : 17-34.

Singh, J.S. and M.C. Joshi, 1979. Primary production : In R.T. Coupland (Ed.), *Grassland Ecosystems of the World.* IBP Vol. 18, 197-218, Cambridge Univ. Press, London.

Singh, J.S., K.P. Singh and P.S. Yadava, 1979. Ecosystem synthesis. In R.T. Coupland (Ed.). Grassland Ecosystems of the World. IBP Vol. 28, 231-239. Cambridge Univ. Press, London.

Singh J.S. and P.S. Yadava, 1974. Seasonal variation in composition, plant biomass and net primary productivity of a tropical grassland at Kurukshetra, India. *Ecological Monograph,* 44 : 351-375.

Singh, J.S. and S.P. Singh, 1987. Forest vegetation of the Himalaya. *Botanical Rev.,* 53 : 80-192.

Singh, K.P., 1967. Production, chemical composition and rate of decomposition of litter, and soil properties in deciduous forest communities at Varanasi. *Ph.D. Thesis*, Banaras Hindu University.

Singh, K.P., 1972. Ecological variation of seed weight in populations of *Anagallis arvensis* L. *Curr. Sci.*, 41 : 425-426.

Singh, K.P., 1972. Effect of different photoperiods on growth and flowering in *Portulaca oleracea* L. *Curr. Sci.*, 41 : 573-574.

Singh, K.P., 1972(a). Effect of temperature and light on seed germination of two ecotypes of *Portulaca oleracea. New Phytol.*, 72(2): 289-295.

Singh, K.P. and G. Misra, 1985. Nitrogen use of a temperate and two tropical grasses as affected by clipping of varying moisture levels, 145-152. In : K.C. Misra (Ed.), *Ecology and resource management in tropics*.

Singh, K.P. and O.N. Pandey, 1981. Cycling of nitrogen in a tropical deciduous forest, 123-130. In : R. Wetsellar, J.R. Simpson and T. Rosswall (Eds.). *Nitrogen cycling in S.E. Asian wet Monsoonal Ecosystems*. The Australian Acad. Sci. Canberra.

Singh, K.P. and S.K. Srivastava, 1985. Seasonal variations in the spatial distribution of root tips in teak (*Tectona grandis* Linn.). Plantations in the Varanasi forest division, India, *Plant and Soil*, 84 : 93-104.

Singh, K.P. and S.K. Srivastava, 1986. Recent trends in fine root research in tropical forests, 22-23. In : R.S. Ambasht (Ed.). Recent advances in Environmental Biology, Banaras Hindu University.

Singh, K.P. and S.K. Srivastava, 1986. Seasonal variation in the total non-structural carbohydrate of fine root in teak (*Tectona grandis* Linn f.), *Tree Physiology*, 1 : 21-36.

Singh, M.P., R.S. Ambasht and E. Sharma, 1984. Ecodevelopment of a riparian ecosystem. Proc. Seminar Environmental management, Dept. of Ecology, UP Govt., Lucknow.

Singh, M.P. and R.S. Ambasht, 1986. Analysis of primary productivity pattern in a riparian agroecosystem. *Acta Botanica Indica*, 14(20) : 195-202.

Singh, M.P. and R.S. Ambasht, 1987. Edaphic dynamic aspects of a riparian ecosystem. *DEI Jour. Sci & Engineering Research*, 5(1): 1-12.

Singh, U.N., 1972. Phytosociology and productivity of grazed and ungrazed grasslands of Chakia, Varanasi, *Ph.D. Thesis,* Banaras Hindu University.

Singh, U.N. and R.S. Ambasht, 1975. Relationship among diversity, dominance, stability and net production of Indian grassland. *Ind. Jour., Ecology,* 2(2) : 110-114.

Singh, U.N. and R.S. Ambasht, 1975. Biotic stress and variability in structure and organic (net primary) production of grassland communities at Varanasi, India, *Trop. Ecol.,* 16 : 86-95.

Singh, U.N. and Ambasht, 1975. Energy conserving efficiency of a forest grassland at Varanasi, *Acta Botanica,* 3 : 132-135.

Singh, U.N. and R.S. Ambasht, 1980. Floristic composition and phytosociological analysis of three grass stands in Naugarh forest of Varanasi Division. *Ind. J. Forestry,* 3(2) : 143-147.

Smith, S.S.E., 1980. Mycorrizae in autotrophic plants, *Bio. Rev.,* 55: 475-510.

Smith, W.H. Pollutant uptake by plants. In : M. Treshwo (Ed.), *Air Pollution and Plant Life.* John Wiley & Sons, NY, 417-450.

Solbrig, O.T., Emden van, H.M. and Oordt van, P.G.W.J., 1994. *Biodiversity and Global Change,* CAB Internat., Oxford, UK, 227.

Sorensen, T., 1948. A method of establishing groups of equal amplitude in plant society based on similarity of species content K. Danske, Vidensk. Selsk, 5 : 1-34.

Srivastava, A.K. and R.S. Ambasht (1989). Energy dynamics of *Ceratophyllum demersum* in river Ganga at Varanasi, *Bio-Journal* (1), 51-55.

Srivastava, Alok K. and R.S. Ambasht, 1994a. Soil moisture control of nitrogen fixation activity in dry tropical *Casuarina* Plantation forest. *Journal of Environmental Management,* 42 : 49-54, Academic Press, London.

Srivastava, A.K. and R.S. Ambasht, 1994(b). Nitrogen deposition in *Casuarina equisetifolia* plantation stands in dry tropics of Sonbhadra, India. *Forest Ecology and Management.*

Srivastava, A.K. and R.S. Ambasht, 1995. Biomass, production, decomposition and nitrogen release from root nodules in *Casuarina equisetifolia* plantation in Sonbhadra, India. *Jour. Appl. Ecology,* 32 : 121-127.

Srivastava, N.K. and R.S. Ambasht, 1990. Impact of industrial effluents on productivity and nutrient and heavy metal accumulation in wetland marginal plant. Final Tech. Report D.O. En. Project P. 0765.

Srivastava, N.K., R.S. Ambasht and R. Kumar, 1991. Water quality, phytoplankton diversity and production in G.B. Pant Sagar at Rihand Dam, Pipri. *Acta Hydrochimica et Hydrobiologica*, 19(5): 529-539 (Germany).

Srivastava, N.K., R.S. Ambasht, R. Kumar and Shardendu, 1993. Effect of thermal power effluents on the community structure and primary production of phytoplankton. *Environment International*, USA, 19 : 79-90.

Srivastava, P.B.L., Ahmad Dhanarajah and Hamraz, 1979. Symposium on mangrove and esturine vegetation in South east Asia : Univ. Pertanian, Malaysia, 227.

Stearn, S.S., 1976. Life history tactics, a review of ideas. *Quartz. Rev. Biol.* 51 : 3-47.

Sukachev, V.N., 1944. (On principles of genetic classification in biocenology) in Russian Zur. Obsenei. Biol. 5 : 213-227.

Summerhayes, V.S., 1968. *Orchids of Britain.* II Ed. Collins, London.

Sytnik, K.M. (Ed.), 1985. Living in the *Environment.* Translated from Russian UNESCO, Naukova, Dumka Pub.

Tabazadeh, A. Santee, M.L., Purphrey, P.A., Newman, P.J., Hamill and J.L. Margenthaler 2002. *Science*, 288 : 1407-1411.

Tansley, A.G., 1935. The use and abuse of vegetational concepts and terms. *Ecology,* 42 : 237-245.

Teal, J.M., 1957. Community metabolism in a temperate Root Spring, *Ecol. Monogr.,* 27 : 283-302.

Teramura, 1983. Effect of Ultraviolet-B radiation on the growth and yield of crop plants. *Physiol. Plant.* 58 : 415-427.

Tewari, D.N., 1993. Conservation of biodiversity. In : *Biodiversity Conservation* (Ed. Bob Frame and Joe Victor). British High Commission, New Delhi, 153.

Tilman, D., 1985. The resource ratio hypothesis of plant succession. *American Naturalist,* 125 : 827-852.

Thornton, I.W.B., 1984. Krakatau, the development and repair of a tropical ecosystem, *Ambio.* 13 : 216-225.

Tolba, M.K. and O.A. Elkholy, 1992. The World Environment, 1972-1992. Chapman & Hall, 884.

Tripathi, R.S. and J.L. Harper, 1973. The comparative biology of *Agropyron repens* L. (Beav) and A. *canium* (Beav) Growth of mixed population established from tillers and seeds. *J. Ecol.* 61 : 353-368.

Tripathi, R.S., 1965. Tuber sprouting in *Cyperus rotundus* L.; in relation to temperature. Ann. Report PL/480 Ecol. Project, Banaras Hindu University, 55-59.

Trollope, W.S.W., 1984. Fire behaviour. In : P. de V. Booysen and N.M. Tainton (eds.). *Ecological Effects of Fire in South African Ecosystems,* Springer Verlag, NY, 199-217.

Turk, A., Turk, J. and Wittes, J.T., 1972. *Ecology, Pollution, Environment.* W.B. Saunders Co., Philadelphia, 217.

Turesson, G., 1922. The genotypic response of the plant species to habitat *Heriditas*, 3 : 211-350.

Turesson, G., 1930. The selective effect of climate upon the plant species. *Heriditas*, 14 : 99-152.

UNEP, 1995. *Global Biodiversity Assessment*, W.H. Heywood (Ed.), Cambridge Univ. Press, 1140.

UNESCO, 1994. Biodiversity science, conservation and sustainable use, Paris.

Varshney, C.K., 1968. Plant succession on walls : *Proc. Symp. Recent Adv. Tropical Ecology*, 2 : 471-481.

Varshney, C.K. and K.P. Singh, 1973. A Survey of Aquatic Weeds Problems in India, Abs. Proc. Regional Seminar on Noxious Aquatic Vegetation. D.S.T. Govt. of India, New Delhi (full paper published 1976 : W. Junk and Co.).

Vickery, M.L., 1984. *Ecology of Tropical Plants*. John Wiley & Sons, NY, 170.

Vavilov, N.I., 1926. Studies of the origin of cultivated plants. *Bull. Biol. Breed*, 16 : 1-245.

Vollenweider, R., 1968. Method of assessment of primary productivity in Freshwater, Blackwell Scientific Pub., UK.

Volkens, G., 1887. Die Flora der agyptisch arabischen Wilste auf Greendlage anatomisch - physiologischer Forschung. Gebruder Borntragar, Berlin, 151.

Wangeri, E. and W. Sanford, 1985. Tropical grasslands (Advance personal communication).

Watson, R., 1988. *Executive Summary, Ozone Trends Panel Report*, NASA H.Q., Washington, D.C.

Whittaker, R.H., 1954. Plant populations and the basis of plant indication. Festschrift Aichinger, Vol. I.

Whittaker, R.H., 1965. Dominance and diversity in land plant communities, *Science*, 147 : 250-260.

Whittaker, R.H., 1970. *Communities and Ecosystem*, Macmillan & Co. NY, 2nd Ed., 1975.

Wiegert, R.G. and D.F. Owen, 1971. Trophic structure, available resources and population density in terrestrial US aquatic ecosystems. *J. Theoret Biol.*, 30 : 69-81.

Wilson, E.O. and Peter, F.M. (eds.), 1988. *Biodiversity,* NASc Press, Washington, D.C.

WRI, IUCN and UNEP, 1992. Global Biodiversity Strategy, 244.

Wulff, J.C., 1922. *Age and Area*—A study in geographical distribution and origin of species, Cambridge, England.

Wulff, J.C., 1943. *An introduction of historical plant geography*, Waltham-Mass.

Yeaton, R.I., 1978. *A cyclic relationship between Larrea tridentata and Opuntia lepticaulis* in the northern Chihuahuan Desert. of Texas, USA, *Journal of Ecology* : 66 : 651-656.

Younes, T. and F. di Castri, 1996. Biodiversity, the emergence of a new scientific field, its properties and constraints. In : P.S. Ramakrishnan, A.K. Das and K.G. Saxena (eds.), *Conserving Biodiversity for Sustainable Development*, INSA, New Delhi, 246.

Index